AF411406

Polymorphism in Crystals

Polymorphism in Crystals

Editors

Jingxiang Yang
Xin Huang

Basel • Beijing • Wuhan • Barcelona • Belgrade • Novi Sad • Cluj • Manchester

Editors
Jingxiang Yang
State Key Laboratory of
Elemento-Organic Chemistry,
College of Chemistry,
Nankai University,
Tianjin, China

Xin Huang
National Engineering
Research Center of Industrial
Crystallization Technology,
School of Chemical
Engineering and Technology,
Tianjin University,
Tianjin, China

Editorial Office
MDPI
St. Alban-Anlage 66
4052 Basel, Switzerland

This is a reprint of articles from the Special Issue published online in the open access journal *Crystals* (ISSN 2073-4352) (available at: https://www.mdpi.com/journal/crystals/special_issues/ Polymorph).

For citation purposes, cite each article independently as indicated on the article page online and as indicated below:

Lastname, A.A.; Lastname, B.B. Article Title. *Journal Name* **Year**, *Volume Number*, Page Range.

ISBN 978-3-0365-8556-7 (Hbk)
ISBN 978-3-0365-8557-4 (PDF)
doi.org/10.3390/books978-3-0365-8557-4

Contents

Preface

Polymorphism, the property of a compound to crystallize in more than one distinct crystal form, is indispensable in researching and developing pharmaceuticals, agrochemicals, materials, and food. Polymorphs exhibit different properties, such as crystal habit, solubility, dissolution rate, melting point, stability, mechanical properties, and even bioavailability, which may influence product quality. Therefore, the study of polymorphs' behavior can provide a theoretical basis for selecting optimal solid forms and serve as a primary method for the polymorphic control and optimization of products. Recently, significant progress has been made in the experimental discovery and theoretical prediction of crystal polymorphs. Many molecules have been discovered to have polymorphs mainly attributed to the molecule's conformational flexibility and various functionalities in the molecule that could act as hydrogen bond donors/acceptors. In addition to conventional solution crystallization, more polymorphisms have been found in the melt, in confinement, and in the presence of ultrasound/lasers. Moreover, computational predictions usually yield far more possible polymorphs than are known. The ultimate limitations of experimental reachable polymorphs and thermodynamic and structure–activity relationships of the polymorphs remain an open question. In this book, some of the latest research on "Polymorphism in Crystals" is collected, and the goal of this book is to enhance our understanding of polymorphism in crystals and their remarkable potential applications in the pharmaceuticals industry, fine chemicals industry, photoelectric industry, etc.

Jingxiang Yang and Xin Huang
Editors

crystals

Article

Enhancement of Crystallization Process of the Organic Pharmaceutical Molecules through High Pressure

Yaoguang Feng [1], Hongxun Hao [1,2], Yiqing Chen [1], Na Wang [1,2], Ting Wang [1,2] and Xin Huang [1,2,*]

[1] National Engineering Research Center of Industrial Crystallization Technology, School of Chemical Engineering and Technology, Tianjin University, Tianjin 300072, China; 2019207028@tju.edu.cn (Y.F.); hongxunhao@tju.edu.cn (H.H.); 2019207051@tju.edu.cn (Y.C.); wangna224@tju.edu.cn (N.W.); wang_ting@tju.edu.cn (T.W.)
[2] School of Chemical Engineering and Technology, Hainan University, Haikou 570228, China
* Correspondence: x_huang@tju.edu.cn

Abstract: The enhancement of the crystallization process through high pressures was studied by using ribavirin (RVB) as a model compound. The effects of high pressure on crystallization thermodynamics, nucleation kinetics, and process yield were evaluated and discussed. The solubility of ribavirin in three pure solvents was measured at different pressures from 283.15 to 323.15 K. The results indicate that the solubility data of ribavirin decreased slightly when pressure was increased. The induction time of the cooling crystallization of ribavirin under different pressures was measured. The results show that high pressure could significantly reduce the nucleation induction period. Furthermore, the nucleation kinetic parameters under different pressures were calculated according to the classical nucleation theory. The effect of high pressure on the anti-solvent crystallization of ribavirin was also studied.

Keywords: high pressure; ribavirin; solubility; induction time; nucleation

Citation: Feng, Y.; Hao, H.; Chen, Y.; Wang, N.; Wang, T.; Huang, X. Enhancement of Crystallization Process of the Organic Pharmaceutical Molecules through High Pressure. *Crystals* **2022**, *12*, 432. https://doi.org/10.3390/cryst12030432

Academic Editor: Etsuo Yonemochi

Received: 22 February 2022
Accepted: 16 March 2022
Published: 20 March 2022

Publisher's Note: MDPI stays neutral with regard to jurisdictional claims in published maps and institutional affiliations.

1. Introduction

Crystallization is a very important unit operation in the pharmaceutical industry because more than 90% of active pharmaceutical ingredients (API) are crystalline products [1]. Recently, researchers have studied the crystallization process of pharmaceutical molecules under physical fields, including the electric field [2], magnetic field [3], laser [4], ultrasonic [5], gravity [6], pressure [7], etc. As a basic thermodynamic variable, pressure is a powerful tool for exploring new materials [8]. As early as 2003, Fabiani et al. obtained one methanol solvate by compressing the methanol solution of paracetamol at 0.62 GPa [9]. In the following 20 years, researchers have conducted a lot of work in the high-pressure crystallization of API [7]. Wierschem et al. found that high pressure can significantly reduce the induction time and improve the crystal growth rate under 200–450 MPa [10]. Elena Boldyreva have systematically studied the solid-phase transformation of a variety of amino acids under several GPA, which provides an effective method for the understanding of the mechanism of phase transitions between polymorphs of small organic molecules [11–13]. Andrzej Katrusiak obtained different polymorphisms and solvates of organic molecules such as xylazine hydrochloride and triiodoimidazole under several GPA and provided guidance for predicting the formation and structure of solvates under high pressure [14,15]. However, from the literature review, the high-pressure crystallization of API is mainly carried out by diamond anvil cell (DAC) under the pressure of hundreds of MPa to several GPa. There are few studies on the solution crystallization process of API under several MPa, although several MPa is easier to achieve on an industrial scale.

In this paper, the enhancement crystallization process of API at several MPa was studied, and whether this enhancement will bring the risk of polymorphic transformation was also considered, which can enrich the control methods of the crystallization process that are easily realized in industry. Ribavirin (RBV, $C_8H_{12}N_4O_5$, CAS No. 36791-04-5) is an antiviral drug widely used in hepatitis C virus infection treatment [16]. The chemical structure of ribavirin is presented in Figure 1. Three crystal forms of RBV have been previously reported, conventionally referred to as Form I, Form II, and DMSO solvate [17,18]. However, its crystallization behaviors and its polymorphic forms under high pressure have not been studied. In this work, the solubility of RBV in three solvents (water, dimethylformamide, and dimethylacetamide) was firstly measured from 283.15 to 323.15 K at 12.0 MPa and 0.1 MPa since the solubility is basic data of crystallization thermodynamics. Then, the induction time of cooling crystallization of RBV in water was measured under different temperatures, pressures, and supersaturations. Furthermore, nucleation parameters under different pressures were calculated based on the induction time data. The yield and crystal form of anti-solvent crystallization under different pressures were characterized and compared.

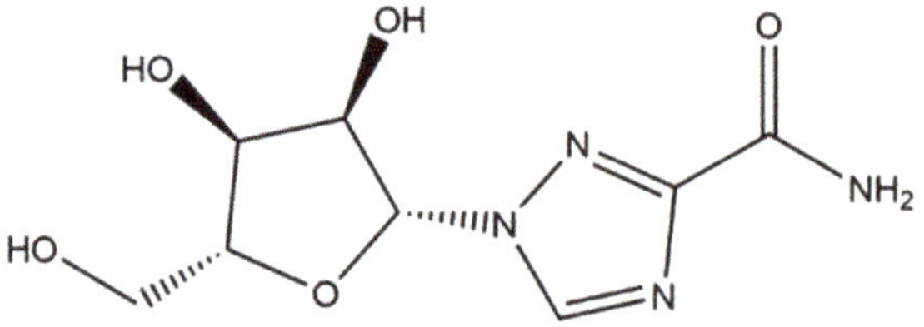

Figure 1. Chemical structure of ribavirin.

2. Experimental

2.1. Materials

Ribavirin (Form II, mass purity $\geq$ 99.0%) was purchased from Shanghai Xianding Biotechnology Co., Ltd. (Shanghai, China). Ultrapure water with a resistivity of 18.2 MΩ·cm was prepared in our laboratory. Other solvents were obtained from Lianlong Bohua Chemical Co., Ltd. and were directly used without further purification.

Ribavirin (Form I) was prepared by anti-solvent crystallization. The raw material (100 g) was dissolved in dimethylacetamide (300 mL) at 313.15 K and crystallized by gradually adding n-butanol (2000 mL, 5 mL/min). The product was washed with ethanol and dried in a vacuum-drying oven.

2.2. High-Pressure Device

The device used in the experiment was a high-pressure stainless-steel reactor (Xi'an Taikang Biotechnology Co., Ltd., Xi'an, China) with a volume of 100 mL and a maximum pressure of 12.5 MPa. As shown in Figure 2, the reactor was equipped with an inflation valve, liquid-taking valve, magnetic–mechanical coupled stirring, and a sapphire window for observing the crystallization process. The high pressure of the reactor was provided by a nitrogen steel cylinder (Tianjin Liufang Industrial Gas Distribution Co., Ltd., Tianjin, China) with the nitrogen purity $\geq$ 99.999% and the initial pressure $\geq$ 14.5 MPa. The pressure of the reactor was controlled by the pressure-reducing valve, and the control accuracy was 0.1 MPa. All experiments at 0.1 MPa in this paper were carried out at atmospheric pressure. A thermostat (CF41, Julabo Technology Co., Ltd., Seelbach, Germany) and jacket were used to control the temperature of the reactor with a temperature accuracy of 0.1 K.

Figure 2. High-pressure experimental device.

2.3. Solubility Measurement

The solubility of RBV in three solvents (water, dimethylformamide, and dimethylacetamide) at 12.0 MPa was determined by the gravimetric method in the above reactor [19]. Firstly, pure solvent (about 50 mL), excess RBV, and a magneton were added to the reactor. After the device was assembled, the pressure-reducing valve was opened to pressurize the reactor to 12.0 MPa. Then, the solid-liquid system in the reactor was continuously stirred with a magnetic stirrer at 300 rpm for 8 h to ensure that the system reached equilibrium, which had been proven by preliminary experiments. Then, the magnetic stirrer was stopped, and the solid–liquid system was kept still at a constant temperature for 4 h to obtain a clear, saturated solution. Finally, the liquid-taking valve was opened, and the supernatant liquid (about 5 mL) was pressed into the pre-weighed beaker through a polytetrafluoroethylene (PTFE) membrane filter (0.45 μm, Tianjin Jinteng Experimental Equipment Co., Ltd., Tianjin, China).

For comparison, the solubility of RBV in four solvents at 0.1 MPa was also determined by the gravimetric method. A jacketed crystallizer was used for measurements, and the steps were similar to the above description. The difference was that the supernatant was extracted through a syringe, and then was quickly filtered into a pre-weighed beaker through the above PTFE membrane filter.

The beaker containing the supernatant was weighted quickly and dried in a vacuum oven at 333.15 K for 72 h to ensure that the solvent was completely volatilized. The above operations were performed three times, and the average value was taken to calculate the solubility data. The solubility of ribavirin expressed in the molar fraction was calculated according to the following equation [20]:

$$x_1 = \frac{m_1/M_1}{m_1/M_1 + m_2/M_2} \tag{1}$$

where m_1 and M_1 are the mass and molar mass of RBV, respectively, and m_2 and M_2 are the mass and molar mass of the pure solvent, respectively.

2.4. Induction Time Measurement

The induction time (t_{ind}) is defined as the time from the generation of constant supersaturation to the appearance of crystals [21,22]. In this paper, the induction time of the cooling crystallization of RBV in water was measured by naked eye observation [23]. All measurements were carried out in the same high-pressure reactor, with pressures of 0.1 MPa, 5.0 MPa and 10.0 MPa and temperatures of 283.15 K, 293.15 K and 303.15 K, respectively. Firstly, RBV (Form II) aqueous solution with different supersaturation (S) was prepared. The preparation temperatures of supersaturated solutions at 283.15 K, 293.15 K and 303.15 K were 308.15 K, 313.15 K and 318.15 K, respectively, to ensure that the solute was completely dissolved. Then, 50 mL of supersaturated solution was quickly transferred into the pre-cooled reactor. The high-pressure nitrogen was added to the reactor when the solution was cooled down to the preset temperature. Then, the mechanical stirring was started at speed of 200 rpm and the timing was started. After the crystals appeared, the timing was stopped, and the crystals were separated immediately and characterized by microscope and PXRD. The above operations were carried out three times, and the average value was taken to calculate the induction time.

The supersaturation was calculated by Equation (2).

$$S = \frac{c}{c_0} \tag{2}$$

where c is the actual concentration of the solute, and c_0 is the equilibrium concentration of the solute.

2.5. Anti-Solvent Crystallization

The yield and crystal form of anti-solvent crystallization at 10.0 MPa and 0.1 MPa were compared. Firstly, RBV (5 g, Form II) was dissolved in solvent (15 g, water, dimethyl sulfoxide, and dimethylformamide). Then, anti-solvent (15 mL, alcohol solvents such as methanol, ethyl acetate, and butyl acetate) was added to the solution. The solution was stood for 24 h at 10.0 MPa and atmospheric pressure at 20 °C to allow crystal nucleation and growth. Finally, the solution was filtered, and the crystals were washed with ethanol and dried at room temperature. The crystals were weighed and characterized by PXRD. The above operations were carried out three times and the average value was taken for further calculation.

3. Results and Discussion

3.1. Effect of High Pressure on Solubility

The solubility data of RBV in four solvents at 12.0 MPa and 0.1 MPa are shown in Tables S1 and S2 and Figures 3 and 4. The experimental solubility data were correlated by the modified Apelblat model Equation (3), and the fitting curves are also given in Figures 3 and 4.

$$\ln x_1 = A + \frac{B}{T} + C \ln T \tag{3}$$

where x_1 represents the mole fraction solubility of RBV. T is the temperature (K). A, B, and C are model parameters [24].

The results indicate that all the solubilities of the two crystal forms of RBV at 12.0 MPa decrease slightly when compared with those at 0.1 MPa. The solubility data of RBV increase significantly with the increase in temperature whether at 12.0 MPa or 0.1 MPa. The effect of pressure on solubility values is weaker than the effect of temperature. In addition, the solubility of Form I in water is smaller than that of Form II, which is consistent with Form I being a thermodynamically stable form.

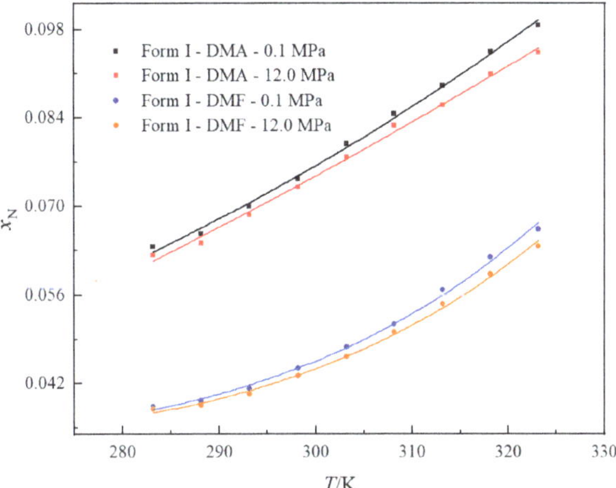

Figure 3. Mole fraction solubility of RBV (Form I) in dimethylacetamide (DMA) and dimethylformamide (DMF) at 0.1 MPa and 12.0 MPa: symbol—experimental values, curve—calculated values by the modified Apelblat model.

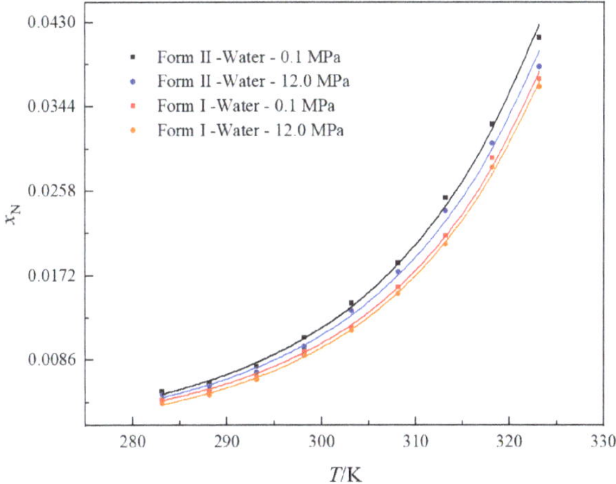

Figure 4. Mole fraction solubility of RBV (Form I and Form II) in water at 0.1 MPa and 12.0 MPa: symbol—experimental values, curve—calculated values by the modified Apelblat model.

The PXRD patterns and DSC plots of the RBV (Form I and Form II) used in the solubility experiment are shown in Figures S1 and S2. The PXRD patterns of undissolved wet solid taken out after the solubility experiments are shown in Figure S3. It can be seen that the PXRD patterns of the samples before and after the experiments are consistent, and they are all pure crystal forms. Therefore, no polymorphic transformation or solvate formation happened in the solubility experiments at two pressures. Furthermore, the decrease in solubility at high pressure is not caused by crystal transformation.

3.2. Effect of High Pressure on the Induction Time

The influence of high pressure on the induction time of the cooling crystallization of RBV in water is shown in Figure 5. The specific induction time data are listed in Table S3. For the convenience of calculation and comparison, all the calculations of supersaturation under high pressure in this section were based on the solubility data of RBV (Form II) at atmospheric pressure. At the same time, it is also because Form II is the polymorph always obtained in the induction time experiment. It can be seen that the induction time decreases significantly with the increasing concentration of RBV, whether at atmospheric pressure or high pressure. More importantly, the induction time of RBV cooling crystallization decreases significantly at high pressure. The PXRD characterization shows that all the crystals obtained in the induction time experiment are RBV (Form II). Therefore, the reduction in the induction time is independent of the crystal form of RBV. The high pressure may enhance the collision probability of solute molecules in the solution, thus promoting the nucleation process.

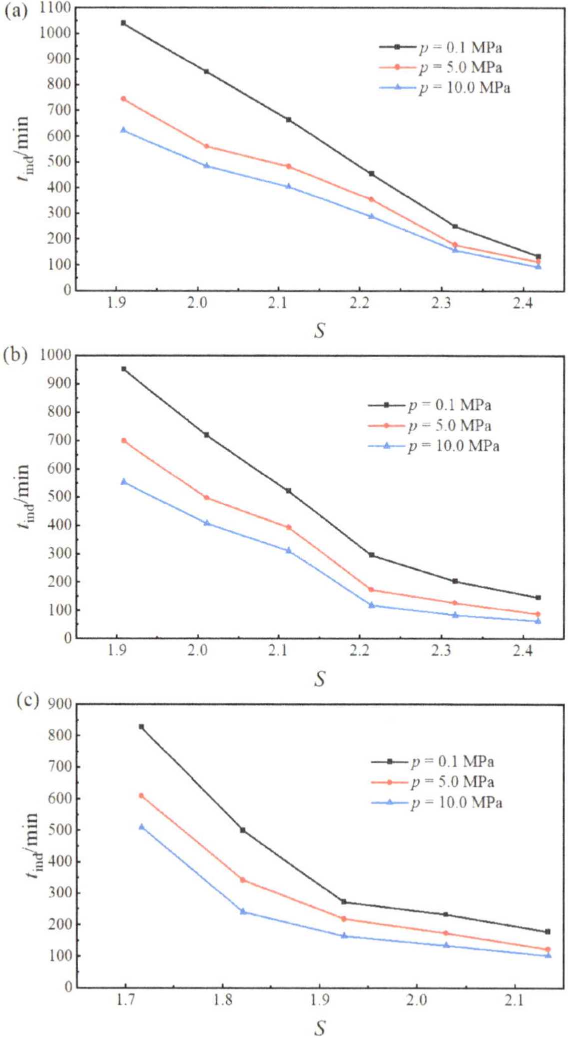

Figure 5. The induction time of cooling crystallization of RBV (Form II) in water at different supersaturation and pressure: (**a**) 283.15 K, (**b**) 293.15 K, (**c**) 303.15 K.

3.3. Effect of High Pressure on the Nucleation Kinetics

To quantitatively explain the effect of pressure on nucleation rate, the nucleation kinetic parameters under different pressures were calculated according to the classical nucleation theory (CNT) [25–27]. The crystal nucleation rate equation in the classical nucleation theory is given by the following formula [28,29]:

$$J = A \exp\left(-\frac{16\pi\gamma^3 V_m^2}{3\, k^3 T^3 \ln^2 S}\right) \tag{4}$$

where J represents the nucleation rate ($m^{-3}\,s^{-1}$), A is the pre-exponential factor, T is the nucleation temperature (K), κ refers to the Boltzmann constant ($J\,K^{-1}$), S is the supersaturation ratio, γ is the crystal-solution interfacial tension ($J\,K^{-1}$), and V_m is the molecular volume of the solute molecule (m^3) [30]. In this paper, the molecular volume of RBV is 1.943×10^{-28} m^3, which was calculated according to the Avogadro constant and the molar volume obtained from CAS SciFinder^n Database.

There is an inverse relationship between nucleation rate and induction time, which can be expressed follows:

$$J \propto \frac{1}{t_{ind}} \tag{5}$$

Therefore, the relationship between supersaturation and induction time can be written as follows:

$$\ln(t_{ind}) \propto \frac{16\pi\gamma^3 V_m^2}{3\, k^3 T^3} \ln^{-2} S = \alpha \ln^{-2} S \tag{6}$$

A linear relationship exists between $\ln t_{ind}$ and $\ln^{-2} S$. By linear fitting, the interfacial tension (γ) can be obtained from the slope (α):

$$\gamma = kT \left(3\alpha / 16\pi V_m^2\right)^{1/3} \tag{7}$$

Generally, the lower the interfacial energy, the easier the nucleation will be [31]. Based on the interfacial tension and supersaturation, the other four nucleation parameters, including the change of Gibbs free energy per unit volume (ΔG_V), the critical nucleation radius (r^*), the change of critical Gibbs free energy (ΔG^*), and the number of molecules constituting the critical nucleus (i^*) can be calculated by the following equations [32,33]:

$$\Delta G_V = \frac{-kT \ln S}{V_m} \tag{8}$$

$$r^* = \frac{-2\gamma}{\Delta G_V} \tag{9}$$

$$\Delta G^* = \frac{4}{3}\pi (r^*)^2 \gamma \tag{10}$$

$$i^* = \frac{4\pi (r^*)^3}{3 V_m} \tag{11}$$

Among them, ΔG_V can represent the driving force of nucleation, where ΔG^* represents the critical energy barrier that must be overcome in the nucleation process [32].

The linear fitting between $\ln(t_{ind})$ and $\ln^{-2} S$ is shown in Figure 6. It can be found that there is more than one linear relationship between $\ln(t_{ind})$ and $\ln^{-2} S$ at 283.15 K, 293.15 K, and the experimental supersaturation. According to the classical nucleation theory, homogeneous nucleation is dominant in the nucleation process at high supersaturation, whereas heterogeneous nucleation will be dominant at low supersaturation [34,35]. Therefore, there may be two different linear relationships. It can be found from Figure 6 that there are two linear lines between $\ln(t_{ind})$ and $\ln^{-2} S$ at 283.15 K, 293.15 K. The high supersaturation range with a large slope indicates the homogeneous nucleation, and the low supersaturation range with a small slope indicates the heterogeneous nucleation. At 303.15 K, there is only one linear relationship between $\ln(t_{ind})$ and $\ln^{-2}S$ and the slope is small, indicating the heterogeneous nucleation. In the homogeneous nucleation region, the slope (α) can be used to calculate the interfacial tension (γ) according to Equation (6) above [30].

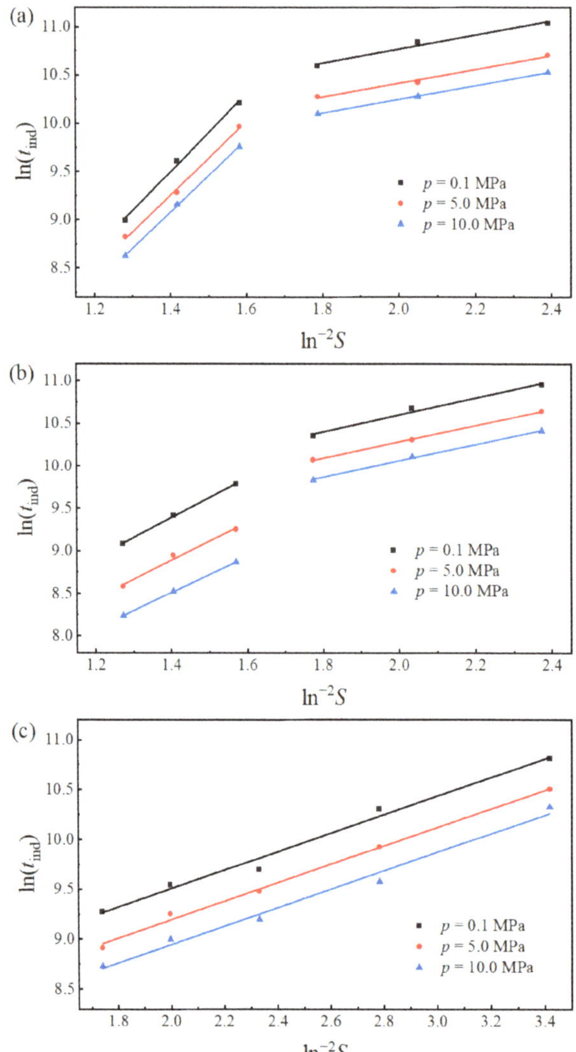

Figure 6. The plot of $\ln(t_{ind})$ versus $\ln^{-2}S$ for RBV (Form II) in water at different supersaturations and pressures: (**a**) 283.15 K, (**b**) 293.15 K, (**c**) 303.15 K.

The results of linear fitting slope and interfacial tension calculated according to the slope are shown in Table 1. The results show that the value of γ decreases with the increase in pressure, which further proves that high pressure can promote the nucleation of RBV.

Table 1. The fitting parameters of $\ln(t_{\text{ind}})$ and $\ln^{-2}S$ and calculated interfacial tension.

Pressure/MPa	α	R^2	Nucleation	$\gamma/\text{mJ m}^{-2}$
283.15 K				
0.1	4.053	0.9956	Homogeneous	7.007
	0.7348	0.9816	Heterogeneous	/
5.0	3.829	0.9970	Homogeneous	6.876
	0.7207	0.9903	Heterogeneous	/
10.0	3.776	0.9996	Homogeneous	6.844
	0.7137	0.9997	Heterogeneous	/
293.15 K				
0.1	2.345	0.9996	Homogeneous	6.053
	0.9966	0.9865	Heterogeneous	/
5.0	2.235	0.9888	Homogeneous	5.957
	0.9665	0.9995	Heterogeneous	/
10.0	2.133	0.9999	Homogeneous	5.865
	0.9569	0.9981	Heterogeneous	/
303.15 K				
0.1	0.9307	0.9870	Heterogeneous	/
5.0	0.9281	0.9961	Heterogeneous	/
10.0	0.9272	0.9871	Heterogeneous	/

Based on the interfacial tension and supersaturation, the calculation results of the other four nucleation parameters are listed in Table 2. Firstly, these four nucleation parameters decrease with the increase in supersaturation under both atmospheric pressure and high pressure, which proves that high supersaturation is conducive to nucleation. In addition, at the same supersaturation, the nucleation parameters of r^*, ΔG^*, and i^* decrease slightly under high pressure compared with atmospheric pressure, which is consistent with the decrease in induction time under high pressure.

Table 2. The calculation results of nucleation parameters.

S	$\Delta G_V \times 10^{-7}/\text{J m}^{-3}$	r^*/nm	$\Delta G^* \times 10^{-16}/\text{J}$	i^*
$T = 283.15\text{ K}, p = 0.1\text{ MPa}$				
2.215	-1.544	0.9080	4.926	16.13
2.316	-1.631	0.8594	4.412	13.67
2.418	-1.715	0.8174	3.992	11.77
$T = 283.15\text{ K}, p = 5.0\text{ MPa}$				
2.215	-1.544	0.8910	4.833	15.24
2.316	-1.631	0.8432	4.329	12.92
2.418	-1.715	0.8021	3.917	11.12
$T = 283.15\text{ K}, p = 10.0\text{ MPa}$				
2.215	-1.544	0.8868	4.811	15.03
2.316	-1.631	0.8393	4.309	12.74
2.418	-1.715	0.7983	3.899	10.96
$T = 293.15\text{ K}, p = 0.1\text{ MPa}$				
2.221	-1.606	0.7537	3.929	9.226
2.324	-1.697	0.7133	3.519	7.819
2.427	-1.785	0.6784	3.183	6.727
$T = 293.15\text{ K}, p = 5.0\text{ MPa}$				
2.221	-1.606	0.7417	3.867	8.793
2.324	-1.697	0.7019	3.463	7.452
2.427	-1.785	0.6676	3.132	6.411
$T = 293.15\text{ K}, p = 10.0\text{ MPa}$				
2.221	-1.606	0.7303	3.807	8.392
2.324	-1.697	0.6911	3.409	7.112
2.427	-1.785	0.6573	3.084	6.119

3.4. Effect of High Pressure on Anti-Solvent Crystallization

The yield and crystal form of anti-solvent crystallization at 10.0 MPa and 0.1 MPa are shown in Tables 3–5. It can be found that the yield of anti-solvent crystallization under high pressure is higher than that under atmospheric pressure, and it is independent of the choice of solvent. Furthermore, the increase in the pressure of several MPa does not affect the final crystal polymorph.

Table 3. The results of anti-solvent crystallization at 10.0 MPa and 0.1 MPa (DMSO as solvent).

Anti-Solvent	p = 0.1 MPa		p = 10.0 MPa	
	Yield/%	Crystal Form	Yield/%	Crystal Form
Methanol	18.8	DMSO solvate	20.1	DMSO solvate
Ethanol	48.9	DMSO solvate	50.1	DMSO solvate
N-propanol	49.2	DMSO solvate	50.6	DMSO solvate
Isopropanol	51.4	DMSO solvate	52.2	DMSO solvate
N-butanol	48.4	DMSO solvate	49.3	DMSO solvate
Isobutanol	76.1	DMSO solvate	85.8	DMSO solvate
Ethyl acetate	51.4	DMSO solvate	61.9	DMSO solvate
Butyl acetate	35.7	DMSO solvate	38.8	DMSO solvate

Table 4. The results of anti-solvent crystallization at 10.0 MPa and 0.1 MPa (DMF as solvent).

Anti-Solvent	p = 0.1 MPa		p = 10.0 MPa	
	Yield/%	Crystal Form	Yield/%	Crystal Form
Methanol	8.83	Form I	12.1	Form I
Ethanol	9.28	Form I	14.0	Form I
N-propanol	26.1	Form I	43.5	Form I
Isopropanol	10.3	Form I	13.5	Form I
N-butanol	33.3	Form I	42.7	Form I
Isobutanol	28.1	Form I	38.1	Form I
Ethyl acetate	14.8	Form I	27.1	Form I
Butyl acetate	15.6	Form I	31.2	Form I

Table 5. The results of anti-solvent crystallization at 10.0 MPa and 0.1 MPa (water as solvent).

Anti-Solvent	p = 0.1 MPa		p = 10.0 MPa	
	Yield/%	Crystal Form	Yield/%	Crystal Form
Methanol	25.9	Form II	33.0	Form II
Ethanol	23.9	Form II	39.2	Form II
N-propanol	11.8	Form II	19.4	Form II
Isopropanol	26.5	Form II	30.9	Form II

4. Conclusions

The crystallization process of RBV at high pressure was systematically studied. It was found that the solubility data of ribavirin in three pure solvents decrease under high pressure. In terms of crystallization kinetics, high pressure can reduce the nucleation induction period, interfacial tension, and nucleation energy barrier. Therefore, high pressure is conducive to nucleation. In addition, the results of anti-solvent crystallization under high pressure show that high pressure can improve the crystallization yield without changing the crystal form. In short, whether it is the cooling crystallization of a single solvent system or the anti-solvent crystallization of a variety of solvents system, the yield and the nucleation of the crystallization process of RBV could be enhanced.

Supplementary Materials: The following supporting information can be downloaded at: https://www.mdpi.com/article/10.3390/cryst12030432/s1, Table S1: Mole fraction solubility of RBV (Form I) at 0.1 MPa and 12.0 MPa; Table S2: Mole fraction solubility of RBV (Form I and Form II) in water; Figure S1: The DSC plots of the RBV raw material (Form I and Form II); Figure S2: The PXRD patterns of the RBV raw material (Form I and Form II); Figure S3: The PXRD patterns of residual solid in solubility experiment; Table S3: The induction time of cooling crystallization of RBV in water at different pressures.

Author Contributions: Conceptualization, Y.F. and H.H.; Methodology, Y.F.; Software, N.W.; Validation, Y.F., H.H. and X.H.; Formal Analysis, T.W.; Investigation, Y.C.; Resources, H.H.; Data Curation, Y.F.; Writing—Original Draft Preparation, Y.F.; Writing—Review and Editing, H.H.; Visualization, N.W.; Supervision, X.H.; Project Administration, X.H.; Funding Acquisition, H.H. All authors have read and agreed to the published version of the manuscript.

Funding: This research was funded by National Natural Science Foundation of China, grant number 21978201 and 21908159.

Institutional Review Board Statement: Not applicable.

Informed Consent Statement: Not applicable.

Data Availability Statement: Not applicable.

Acknowledgments: The author thanks the National Engineering Research Center of Industrial Crystallization Technology of Tianjin University for equipment support.

Conflicts of Interest: The authors declare no competing financial interest.

References

1. Orehek, J.; Teslić, D.; Likozar, B. Continuous Crystallization Processes in Pharmaceutical Manufacturing: A Review. *Org. Process Res. Dev.* **2021**, *25*, 16–42. [CrossRef]
2. Parks, C.; Koswara, A.; Tung, H.-H.; Nere, N.; Bordawekar, S.; Nagy, Z.K.; Ramkrishna, D. Molecular Dynamics Electric Field Crystallization Simulations of Paracetamol Produce a New Polymorph. *Cryst. Growth Des.* **2017**, *17*, 3751–3765. [CrossRef]
3. Potticary, J.; Hall, C.L.; Guo, R.; Price, S.L.; Hall, S.R. On the Application of Strong Magnetic Fields during Organic Crystal Growth. *Cryst. Growth Des.* **2021**, *21*, 6254–6265. [CrossRef]
4. Yu, J.; Yan, J.; Jiang, L. Crystallization of Polymorphic Sulfathiazole Controlled by Femtosecond Laser-Induced Cavitation Bubbles. *Cryst. Growth Des.* **2021**, *21*, 3202–3210. [CrossRef]
5. Los Santos Castillo-Peinado, L.; Luque de Castro, M. The role of ultrasound in pharmaceutical production: Sonocrystalization. *J. Pharm. Pharmacol.* **2016**, *68*, 1249. [CrossRef] [PubMed]
6. Wu, K.; Wu, H.; Dai, T.; Liu, X.; Chen, J.; Le, Y. Controlling Nucleation and Fabricating Nanoparticulate Formulation of Sorafenib Using a High-Gravity Rotating Packed Bed. *Ind. Eng. Chem. Res.* **2018**, *57*, 1903. [CrossRef]
7. Guerain, M. A Review on High Pressure Experiments for Study of Crystallographic Behavior and Polymorphism of Pharmaceutical Materials. *J. Pharm. Sci.* **2020**, *109*, 2640. [CrossRef]
8. Zhang, L.; Wang, Y.; Lv, J.; Ma, Y. Materials discovery at high pressures. *Nat. Rev. Mater.* **2017**, *2*, 17005. [CrossRef]
9. Fabbiani, F.; Allan, D.; Dawson, A.; David, W.; McGregor, P.; Oswald, I.; Parsons, S.; Pulham, C. Pressure-induced formation of a solvate of paracetamol. *Chem. Commun.* **2003**, *3*, 3004. [CrossRef]
10. Ferstl, P.; Eder, C.; Ruß, W.; Wierschem, A. Pressure-induced crystallization of triacylglycerides. *High Press. Res.* **2011**, *31*, 339–349. [CrossRef]
11. Zakharov, B.A.; Boldyreva, E.V. A high-pressure single-crystal to single-crystal phase transition inDL-alaninium semi-oxalate monohydrate with switching-over hydrogen bonds. *Acta Crystallogr. Sect. B Struct. Sci. Cryst. Eng. Mater.* **2013**, *69*, 271–280. [CrossRef] [PubMed]
12. Minkov, V.S.; Goryainov, S.V.; Boldyreva, E.V.; Görbitz, C.H. Raman study of pressure-induced phase transitions in crystals of orthorhombic and monoclinic polymorphs of L-cysteine: Dynamics of the side chain. *J. Raman Spectrosc.* **2010**, *41*, 1748–1758. [CrossRef]
13. Fisch, M.; Lanza, A.; Boldyreva, E.; Macchi, P.; Casati, N. Kinetic Control of High-Pressure Solid-State Phase Transitions: A Case Study on l-Serine. *J. Phys. Chem. C* **2015**, *119*, 18611–18617. [CrossRef]
14. Olejniczak, A.; Krūkle-Bērziṇa, K.; Katrusiak, A. Pressure-Stabilized Solvates of Xylazine Hydrochloride. *Cryst. Growth Des.* **2016**, *16*, 3756–3762. [CrossRef]
15. Marciniak, J.; Kaźmierczak, M.; Rajewski, K.W.; Katrusiak, A. Volume and Pressure Effects for Solvation: The Case Study on Polymorphs of Neat Triiodoimidazole Replaced by Its Solvate. *Cryst. Growth Des.* **2016**, *16*, 3917–3923. [CrossRef]
16. Clercq, E.D.; Li, G. Approved Antiviral Drugs over the Past 50 Years. *Clin. Microbiol. Rev.* **2016**, *29*, 695. [CrossRef] [PubMed]

17. Vasa, D.M.; Wildfong, P.L. Solid-state transformations of ribavirin as a result of high-shear mechanical processing. *Int. J. Pharm.* **2017**, *524*, 339–350. [CrossRef] [PubMed]

18. Xing, C.; Song, J.; Zhang, L.; Yang, S.; Shi, Y.; Du, G.; Lu, Y. Polymorphism and Pharmacokinetic Research of Ribavirin. *Chin. Pharm. J.* **2013**, *48*, 621.

19. Li, X.; Huang, X.; Luan, Y.; Li, J.; Wang, N.; Zhang, X.; Ferguson, S.; Meng, X.; Hao, H. Solubility and thermodynamic properties of 5-nitrofurazone form γ in mono-solvents and binary solvent mixtures. *J. Mol. Liq.* **2019**, *275*, 815–828. [CrossRef]

20. Ji, X.; Wang, J.; Yang, J.; Wang, N.; Li, X.; Tian, B.; Huang, X.; Hao, H. Solubility and isoelectric point of cefradine in different solvent systems. *J. Mol. Liq.* **2020**, *300*, 112312. [CrossRef]

21. Li, D.; Wang, Y.; Zong, S.; Wang, N.; Li, X.; Dong, Y.; Wang, T.; Huang, X.; Hao, H. Unveiling the self-association and desolvation in crystal nucleation. *IUCrJ* **2021**, *8*, 468–479. [CrossRef] [PubMed]

22. de Castro, M.L.D.; Priego-Capote, F. Ultrasound-assisted crystallization (sonocrystallization). *Ultrason. Sonochemistry* **2007**, *14*, 717–724. [CrossRef] [PubMed]

23. Su, M.; Han, J.; Li, Y.; Chen, J.; Zhao, Y.; Chadwick, K. Ultrasonic Crystallization of Calcium Carbonate in Presence of Seawater Ions. *Desalination* **2015**, *369*, 85–90. [CrossRef]

24. Baluja, S.; Hirapara, A. Solubility and solution thermodynamics of novel pyrazolo chalcone derivatives in various solvents from 298.15 K to 328.15 K. *J. Mol. Liq.* **2019**, *277*, 692–704. [CrossRef]

25. Teychené, S.; Biscans, B. Nucleation Kinetics of Polymorphs: Induction Period and Interfacial Energy Measurements. *Cryst. Growth Des.* **2008**, *8*, 1133–1139. [CrossRef]

26. Zhao, Y.; Hou, B.; Liu, C.; Ji, X.; Huang, Y.; Sui, J.; Liu, D.; Wang, N.; Hao, H. Mechanistic Study on the Effect of Magnetic Field on the Crystallization of Organic Small Molecules. *Ind. Eng. Chem. Res.* **2021**, *60*, 15741–15751. [CrossRef]

27. Liu, X.; Wang, Z.; Duan, A.; Zhang, G.; Wang, X.; Sun, Z.; Zhu, L.; Yu, G.; Sun, G.; Xu, D. Measurement of l-arginine trifluoroacetate crystal nucleation kinetics. *J. Cryst. Growth* **2008**, *310*, 2590–2592. [CrossRef]

28. Hao, H.; Wang, J.; Wang, Y. Determination of induction period and crystal growth mechanism of dexamethasone sodium phosphate in methanol–acetone system. *J. Cryst. Growth* **2005**, *274*, 545–549. [CrossRef]

29. Zhou, L.; Wang, Z.; Zhang, M.; Guo, M.; Xu, S.; Yin, Q. Determination of metastable zone and induction time of analgin for cooling crystallization. *Chin. J. Chem. Eng.* **2017**, *25*, 313–318. [CrossRef]

30. Lenka, M.; Sarkar, D. Determination of metastable zone width, induction period and primary nucleation kinetics for cooling crystallization of l-asparaginenohydrate. *J. Cryst. Growth* **2014**, *408*, 85. [CrossRef]

31. Karthika, S.; Radhakrishnan, T.; Kalaichelvi, P. Measurement of nucleation rate of ibuprofen in ionic liquid using induction time method. *J. Cryst. Growth* **2019**, *521*, 55–59. [CrossRef]

32. Devi, K.R.; Raja, A.; Srinivasan, K. Ultrasound assisted nucleation and growth characteristics of glycine polymorphs—A combined experimental and analytical approach. *Ultrason. Sonochemistry* **2015**, *24*, 107–113. [CrossRef] [PubMed]

33. Shang, Z.; Li, M.; Hou, B.; Zhang, J.; Wang, K.; Hu, W.; Deng, T.; Gong, J.; Wu, S. Ultrasound assisted crystallization of cephalexin monohydrate: Nucleation mechanism and crystal habit control. *Chin. J. Chem. Eng.* **2022**, *41*, 430–440. [CrossRef]

34. Li, X.; Yin, Q.; Zhang, M.; Hou, B.; Bao, Y.; Gong, J.; Hao, H.; Wang, Y.; Wang, J.; Wang, Z. Antisolvent Crystallization of Erythromycin Ethylsuccinate in the Presence of Liquid–Liquid Phase Separation. *Ind. Eng. Chem. Res.* **2016**, *55*, 766. [CrossRef]

35. Cui, P.; Zhang, X.; Yin, Q.; Gong, J. Evidence of Hydrogen-Bond Formation during Crystallization of Cefodizime Sodium from Induction-Time Measurements and In Situ Raman Spectroscopy. *Ind. Eng. Chem. Res.* **2012**, *51*, 13663–13669. [CrossRef]

Article

Baloxavir Marboxil Polymorphs: Investigating the Influence of Molecule Packing on the Dissolution Behavior

Xinbo Zhou [1], Kaxi Yu [2], Jiyong Liu [2], Zhiping Jin [1] and Xiurong Hu [2,*]

[1] Zhejiang Jingxin Pharmaceutical Co., Ltd., Shaoxing 312500, China; xinbo.zhou@jingxinpharm.com (X.Z.); hechengsuo@jingxinpharm.com (Z.J.)

[2] Department of Chemistry, Zhejiang University, Hangzhou 310027, China; yukaxi@zju.edu.cn (K.Y.); liujy@zju.edu.cn (J.L.)

* Correspondence: huxiurong@zju.edu.cn; Tel.: +86-0571-8827-3491

Abstract: Baloxavir marboxil (BXM) is a new blockbuster FDA-approved anti-influenza virus agent. However, its poor solubility has limited its oral bioavailability. In this study, BXM was crystallized from several organic solvents, obtaining three polymorphs, and their dissolution behaviors were studied. Detailed crystallographic examination revealed that Form I is monoclinic, space group $P2_1$, with unit cell parameters a = 7.1159 (3) Å, b = 20.1967 (8) Å, c = 9.4878 (4) Å, β = 109.033 (1)°, V = 1289.02 (9) Å^3, and Z = 2, and Form II is monoclinic, space group $P2_1$, with unit cell parameters a = 7.1002 (14) Å, b = 39.310 (7) Å, c = 9.7808 (18) Å, β = 110.966 (5)°, V = 2549.2 (8) Å^3, and Z = 4. Form I has a rectangular three-dimensional energy frameworks net, while Form II has a two-dimensional net. On the other hand, Form II has a much larger percentage of its surface area of exposed hydrogen bond acceptors than Form I. These crystallographic features offered increased solubility and dissolution rate to Form II. The results of stability and solubility experiments suggest that Form II may be preferred in the solid form used for the industrial preparation of BXM medicinal products.

Keywords: baloxavir marboxil; crystallization; polymorphism; crystal structure; energy framework; solubility

Citation: Zhou, X.; Yu, K.; Liu, J.; Jin, Z.; Hu, X. Baloxavir Marboxil Polymorphs: Investigating the Influence of Molecule Packing on the Dissolution Behavior. *Crystals* **2022**, *12*, 550. https://doi.org/10.3390/cryst12040550

Academic Editors: Jingxiang Yang, Xin Huang and Duane Choquesillo-Lazarte

Received: 22 March 2022
Accepted: 13 April 2022
Published: 15 April 2022

Publisher's Note: MDPI stays neutral with regard to jurisdictional claims in published maps and institutional affiliations.

1. Introduction

Epidemic and pandemic influenza is a contagious respiratory illness caused by influenza viruses and has become a major public health concern [1,2]. Treatment of influenza has relied heavily on neuraminidase (NA) inhibitors, which target the viral neuraminidase activity of the NA protein [3].

Fortunately, in 2018, baloxavir marboxil (abbreviated as BXM, Scheme 1) was developed as a first in class, orally active, cap-dependent endonuclease inhibitor, which has a unique mechanism of action when compared with the currently existing neuraminidase inhibitor drug class used to treat influenza infections [4,5]. BXM is a prodrug that is metabolized into the active baloxavir acid (BXA, Scheme 1) and directly inhibits the cap-dependent endonuclease activity of the polymerase acidic protein of influenza A and B viruses [6]. The drug was approved in Japan and other countries, including the U.S., and marked under the brand name Xofluza [7–9]. This is the first new antiviral flu treatment with a novel mechanism of action approved by the FDA in nearly 20 years [7]. Recent studies by experimental and computational techniques indicate that BXM is a valuable candidate treatment for human patients suffering from the highly pathogenic H7N9 virus, H3N2 virus, and COVID-19 virus infection [10–13]. However, BXM is insoluble in an aqueous medium in its crystalline form, and its poor solubility has limited its oral bioavailability [14]. The average oral bioavailability of BXM was only approximately 14.7% [14–16].

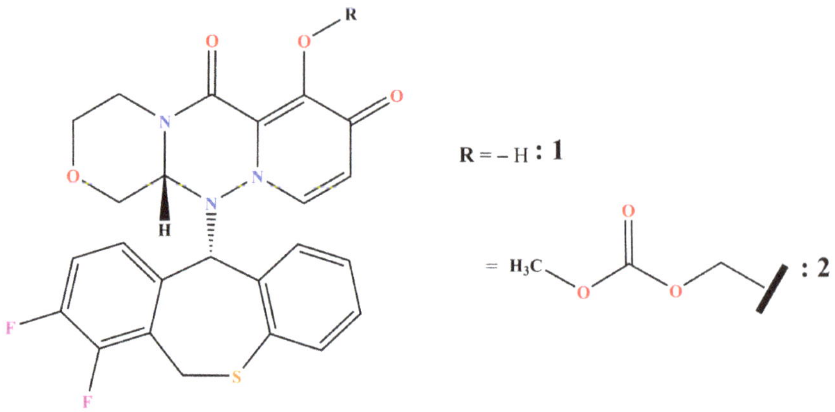

Scheme 1. Chemical structure of baloxavir acid (BXA, 1) and baloxavir marboxil (BXM, 2).

Polymorphism means the potential for a drug to form one or more crystalline solids that differ by the molecular arrangement of drug molecules in the crystal lattice [17,18]. Polymorphic forms of drugs that give a difference in thermodynamic and physicochemical properties, such as melting point, density, stability, and in particular, solubility, can offer an improvement/reduction on the original form [19–26]. The most notorious example of the impact of polymorphs on solubility and dissolution rate is the protease inhibitor Ritonavir [27]. Thus, the evaluation and application of polymorphs that are able to improve the solubility and dissolution rate of BXM are of paramount importance [26,28]. However, to the best of our knowledge, studies about its crystalline structures, morphology, thermodynamic stability, and dissolution properties have not yet been reported. Structure–property relationship is helpful in understanding the critical aspects during formulation development. Herein the study considers in depth the polymorphic forms of BXM with particular emphasis on their molecule packing, solubility, and the relationship between crystal packing and dissolution behavior.

2. Materials and Methods

2.1. Materials

Baloxavir marboxil (purity > 99.5%, Form I) was supplied by Zhejiang Jingxin Pharmaceutical Co., Ltd. (Shaoxing, China) and was used without further purification. All other solvents and chemicals were of analytical grade or chromatographic grade and were purchased from Shanghai Aladdin bio-chem technology company Ltd. (Shanghai, China), Sinopharm chemical reagent Co., Ltd. (Shanghai, China), Shanghai lingfeng chemical reagent Co., Ltd. (Shanghai, China) and used as received.

2.2. Methods

2.2.1. BXM Polymorphs Preparation and Single Crystal Growth

Form I was prepared by recrystallization of BXM in methanol. In particular, 500 mg of BXM was dissolved in 35 mL of methanol, the solution was filtered, and slow cooling of the solution yielded Form I. For crystal structure determination, good quality crystals of Form I were produced by dissolving 20 mg of BXM in 10 mL methanol. Slow evaporation at 25 °C produced block crystals after approximately 1 week.

Form II was prepared from acetonitrile solution of BXM by adding n-heptane as an antisolvent. In particular, 500 mg of BXM was dissolved in 10 mL of acetonitrile, and the resultant clear solution cooled to 0 °C; 15 mL of n-heptane was dropped over 10 min. The prepared mixture became cloudy and then filtered. Crystals of Form II for crystal structure determination were produced by dissolving 20 mg of BXM in a mixture of 3 mL acetonitrile and 7 mL n-heptane. Slow evaporation at 25 °C produced rod-like crystals after about 10 days.

Form III was prepared by recrystallization of BXM in ethyl acetate. In particular, 500 mg of BXM was dissolved in 50 mL of ethyl acetate. The prepared solution was then stirred and quickly cooled to 0 °C. The obtained cloudy solution was filtered. Unfortunately, slow crystallization in ethyl acetate does not result in Form III but in Form I.

2.2.2. Crystal Habit Observation

Microscopic examination and photomicroscopy were performed using a Bresser microscope equipped with a CMOS camera. The sample was put on the objective glass and was examined directly, without cover glass.

2.2.3. Powder X-ray Diffraction (PXRD)

The PXRD patterns were obtained on a Rigaku D/Max-2550PC diffractometer (Rigaku Co., Tokyo, Japan), using a CuKα X-ray radiation source (λ = 1.5418 Å) and generator operated at 40 kV and 250 mA. The scans were run from 3.0 to 40.0° (2θ), with an increasing step size of 0.02° and a count time of 1 s.

2.2.4. Differential Scanning Calorimetry (DSC)

The DSC analysis was performed on a TA DSC Q100 differential scanning calorimeter. Approximately 4–7 mg powder samples were placed in an aluminum pan, and the heating was carried out at a rate of 10 °C/min under a nitrogen flow of 50 mL/min. A temperature range of 25–250 °C was scanned. The data were managed using TAQ Series Advantage software 4.7 (Universal analysis 2000).

2.2.5. Thermogravimetric Analysis (TGA)

TGA was performed on an SDT Q600 instrument from 25 °C to 450 °C, at a heating rate of 10 °C/min, and under nitrogen purge at a flow rate of 50 mL/min. The data were managed using TAQ Series Advantage software 4.7 (Universal analysis 2000).

2.2.6. Single-Crystal X-ray Diffraction (SCXRD)

Single crystal X-ray diffraction data collection was performed on a Bruker Apex II CCD diffractometer (Karlsruhe, Germany) with Mo-Kα radiation (λ = 0.71073 Å). Integration and scaling of intensity data were accomplished using the SAINT V8.38A program [29]. The crystal structure was solved by direct methods using SHELXT [30] and refined by a full-matrix least-squares method with anisotropic thermal parameters for all non-hydrogen atoms on F2 using SHELX-L [31] in Olex 2 [32]. Hydrogen atoms were placed in the position of metrically calculation or difference Fourier map and were refined isotropically using a riding model. Diamond and Olex 2 [32] were used to draw figures. The simulated PXRD pattern was calculated using Mercury [33]. The crystallographic data are listed in Table 1. Crystal structures are deposited as part of the supporting information and may be accessed at www.ccdc.cam.ac.uk/data_request/cif (accessed on 22 March 2022, CCDC 2088906–2088907).

Table 1. Crystallographic data of Form I and II.

	Form I	Form II
Empirical formula	$C_{27}H_{23}F_2N_3O_7S$	$C_{27}H_{23}F_2N_3O_7S$
Formula weight	571.54	571.54
Temperature (K)	296	170
Crystal size (mm)	$0.50 \times 0.46 \times 0.16$	$0.35 \times 0.08 \times 0.06$
Crystal system	Monoclinic	Monoclinic
Space group	$P2_1$	$P2_1$
a (Å)	7.1159 (3)	7.1002 (14)
b (Å)	20.1967 (8)	39.310 (7)
c (Å)	9.4878 (4)	9.7808 (18)
β (°)	109.033 (1)	110.966 (5)
Volume (Å^3)	1289.02 (9)	2549.2 (8)
Z	2	4
ρ_{calc} g/cm^3	1.473	1.489
μ (mm^{-1})	0.193	0.20
F (000)	592.0	1184
Reflections collected	19,783	26,198
Independent reflections (Rint)	5239 (0.023)	10,025 (0.052)
Data/restraints/parameters	5239/121/400	10,025/1/723
R_1, wR_2 [I > 2_(I)]	0.0285, 0.0732	0.06650.1608
R_1, wR_2 [all data]	0.0293,0.0739	0.0717, 0.1637
Goodness-of-fit on F^2	1.05	1.19
Largest diff. peak/hole/e Å^{-3}	0.17, −0.18	0.40, −0.34
Flack parameter	0.039 (18)	0.04 (3)
CCDC No.	2088906	2088907

2.2.7. Powder Dissolution Studies and Stability Tests

The powder dissolution experiments were carried out in pH 1.2 simulated gastric fluids (water was added to 2.0 g of sodium chloride in 7.0 mL of hydrochloride acid to reach 1000 mL). A series of known BMX concentrations in pH 1.2 simulated gastric fluids were prepared to generate a calibration curve. The powder dissolution experiments were carried out in a Tianda Tianfa Technology RC806D dissolution tester with a paddle rotation speed of 100 rpm at 37 °C. Precisely weighted samples (1.0 g) were added into the dissolution vessels with 900 mL pH 1.2 simulated gastric fluids. Prior to powder dissolution experiments, all solid-phase samples were sieved through a 300-mesh screen to eliminate the effect of size on the results. Five milliliters of the aliquot were collected at specific time intervals and filtered via an organic membrane (0.22 μm), and the concentration of the aliquots was determined with appropriate dilutions from the predetermined standard curves of the respective compounds. Moreover, the undissolved samples were collected after solubility experiments for measurement via PXRD. Three crystal forms were stored at 25 ± 2 °C and 60% RH ± 5% RH and then analyzed after 12 months of storage.

3. Results

3.1. Crystallization

According to the Q3C guideline for residual solvents [34], a systematic polymorph screening of BXM was performed by recrystallization from several organic solvents at variable conditions. The identity and purity of the prepared solid forms were verified by PXRD (Table S1 in the Supplementary Materials). Through the screening process, BXM was found to crystallize in three crystal forms, named 'Form I', 'Form II', and 'Form III'. Forms I and II, prepared by the solvent evaporation method, were crystallized and suitable for single-crystal X-ray diffraction analysis (Figure 1a,b). Despite all our efforts, the structure solution of Form III was not successful thus far because of the small size and poor quality of the crystals. Crystal size and quality were unable to be improved upon, as it was obtained only by stir precipitation (Figure 1c).

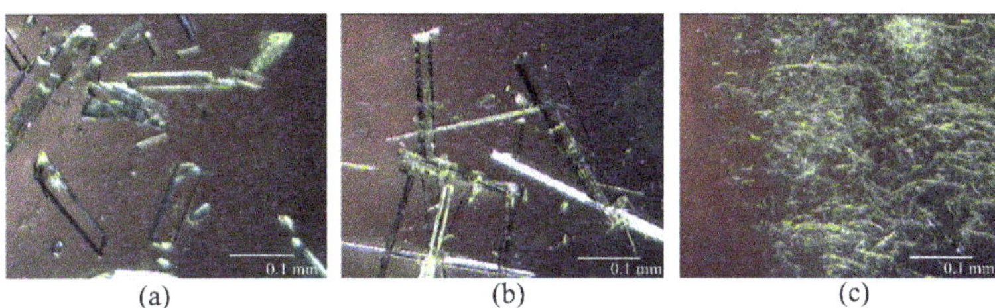

(a)   (b)   (c)

Figure 1. Photographs of (**a**) Form I, (**b**) Form II, and (**c**) Form III.

3.1.1. Crystal Structure of Form I and II

X-ray crystal structures were determined for Forms I and II, and their crystallographic parameters and structure refinement details are listed in Table 1. Forms I and II crystallized in a monoclinic crystal system and $P2_1$ space group. Additionally, the asymmetric unit of I consisted of one BMX molecule (Figure 2a), while Form II contained two molecules (Figure 2b,c). Homochirality of the BMX molecules in both polymorphic forms had been determined from polarimetric data, all molecules having the S-configuration at C10 and the R-configuration at C15. The methymethyl carbonate chain displays positional disorder in the single crystal structure of Form I; these atoms were split into two positions, with site-occupation factors of 0.606(5):0.394(5). Overlaying the molecular conformations found in Form II with that of I shows that the only difference is the rotation of the methymethyl carbonate chain, while the other bonds overlay nicely (Figure 2d). This result indicates that the formation of different solid forms is not due to changes in molecular conformation.

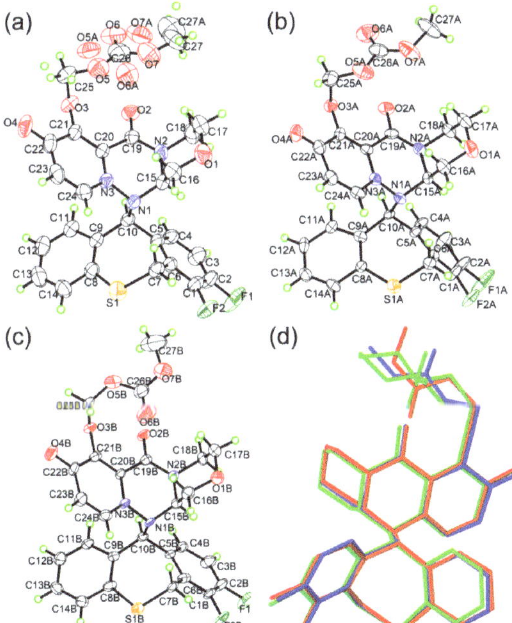

Figure 2. The ORTEP figures of (**a**) Form I, (**b**) molecule A of Form II, and (**c**) molecule B of Form II drawn at the 50% probability level. H atoms are shown as small spheres at a 0.15 Å radius. (**d**) Superposition of the BXM in I (green), molecule A (red), and molecule B (blue) in II. Hydrogen atoms were omitted for clarity.

Since the absence of a typical O-H or N-H donor but with 10 hydrogen bond acceptors in the BXM molecule, the formation of crystal packing arrangement for BXM is governed by a network of nonclassic hydrogen bonds. In the crystal structure of Form I, BXM molecules are linked via C(sp^3)-H ... O and C(sp^2)-H ... F interactions, which develop into a one-dimensional chain and run parallel to the *c*-axis, as shown in Figure 3a and Table S2. Neighboring chains are then packed head-to-tail via short contacts in the *bc* plane (Figure 3c). The major intermolecular interaction between molecules in Form II is a C(sp^2)-H ... F hydrogen bond, as shown in Figure 3b and Table S3. The major difference from Form I is the relative orientation of adjacent chains stacking. In Form II, the neighboring chains are packed head-to-head via C(sp)-H ... π interactions between aromatic rings, forming a hydrogen-bonded bilayered structure (Figure 3d). Additionally, auxiliary C(sp^3)-H ... O interactions are also contributing to the stability of the crystal structure in both polymorphs. In order to visualize and quantify the similarities/dissimilarities of intermolecular contacts in two polymorphs, Hirshfeld surface analysis [35,36] was calculated with the aid of CrystalExplorer 17.5 [37]. The results were visualized in the two-dimensional fingerprint plots, and there is no significant difference between the % distributions in different interactions (Figure S1 in the Supplementary Materials).

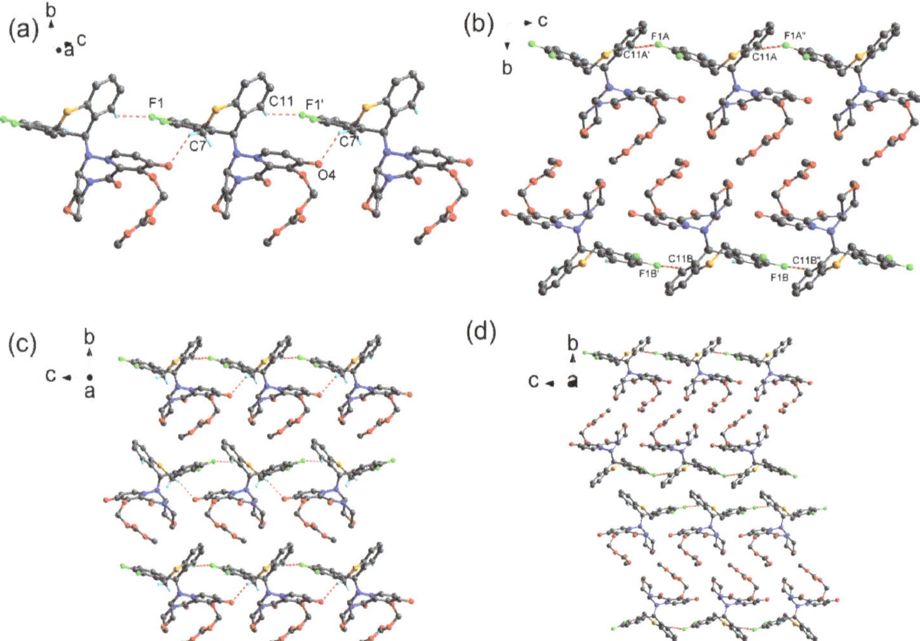

Figure 3. (**a**) 1D chain constructed by C11-H11 ... F1 and C7-H7B ... O4 interactions run parallel to the c-axis in Form I; (**b**) 1D chain constructed by C11-H11 ... F1 interactions run parallel to the c-axis in Form II; (**c**) view of the three-dimensional packing of Form I (viewed along the a-axis direction); (**d**) view of the three-dimensional packing of Form II (viewed along the a-axis direction).

3.1.2. PXRD

The excellent accordance between the experimental and crystal structure simulated PXRD patterns of Form I and II corroborated the identity of the precipitated powder (Figure 4). The major characteristic PXRD peaks of Form I appear at 2θ values of 8.7°, 10.8°, 14.3°, 17.5°, 21.6°, 24.1°, 26.4°, and 29.7°. Form II showed the peaks at 2θ values of 4.4°, 8.9°, 11.7°, 14.0°, 17.8°, 22.3°, 24.3°, and the characteristic peaks of Form III are at 2θ values of 4.3°, 8.7°, 10.6°, 13.4°, 17.1°, and 22.0°. Specifically, Form II has distinctive peaks at 11.7°, 14.0°, 22.3°, and 24.3° that are absent in Form III (Figure S2).

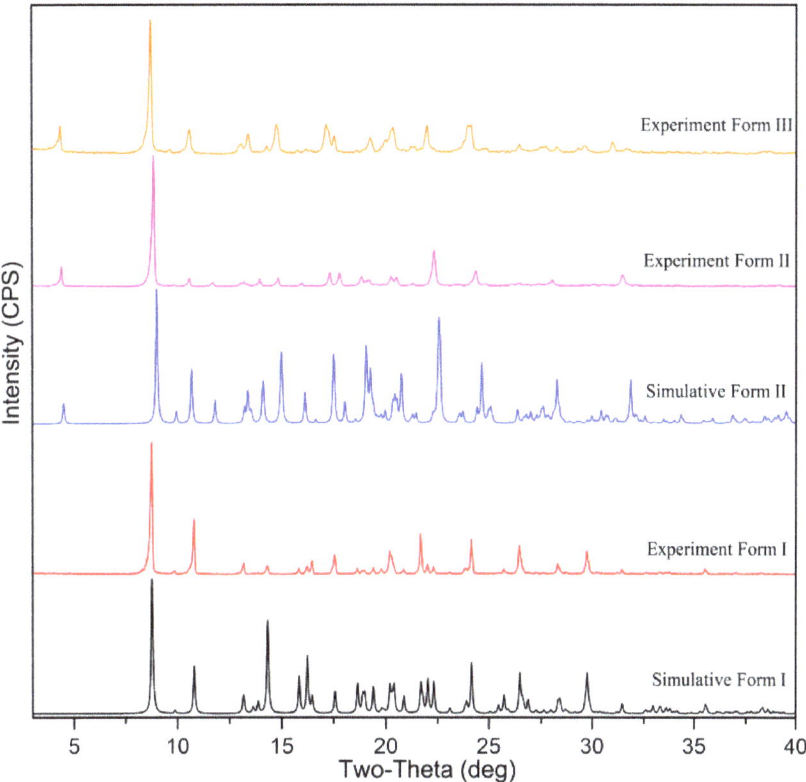

Figure 4. PXRD patterns of all here described solid forms of BXM.

3.1.3. TGA-DSC

To evaluate the thermal properties of the prepared solids, we performed DSC and TGA measurements. As shown in Figure 5, the DSC curve of Form I exhibited two thermal events: a small endothermic event at 230.4 °C followed by a sharply defined melting event at 238.7 °C. The first endothermic peak was associated with a phase change since there was no mass loss in this temperature range. The second one resulted from the melting of the obtained solid in the phase transition process. DSC curve of Form II and III shows stability up to approximately 235.5 °C and 238.7 °C, respectively. The DSC thermal analysis shows that three polymorphs exhibit an explosive type of degradation at melting, which is accompanied by a large release of heat. The melting was irreversible and resulted in a dark-brown tar. The TGA curve exhibits no weight loss until decomposition, which proves all three polymorphs are in anhydrate form (Figure S3).

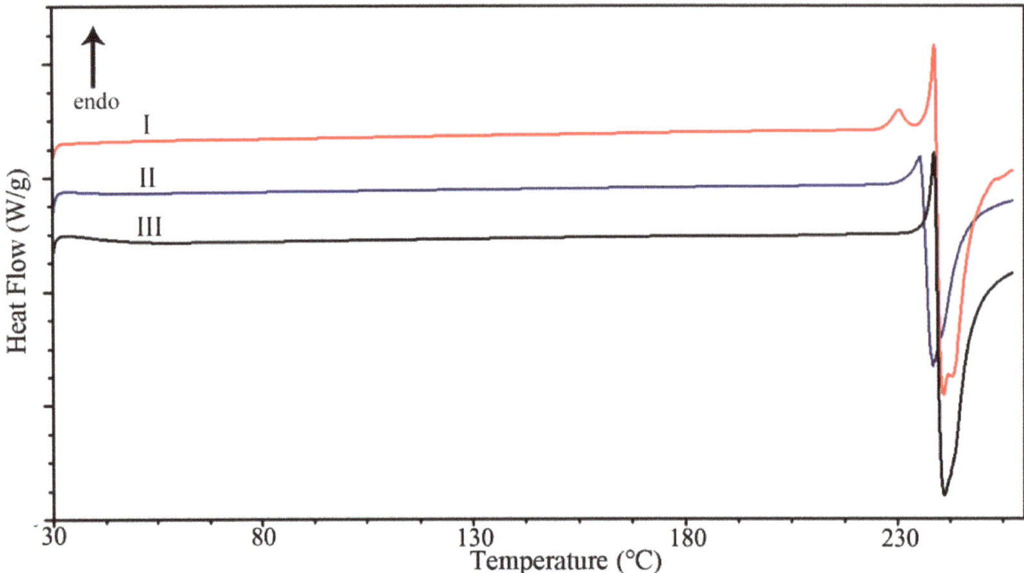

Figure 5. DSC curves of BXM in Form I, Form II, and Form III at a 10 °C/min heating rate.

3.2. Dissolution Profile of Polymorph and Stability

The powder dissolution behaviors of the BXM polymorphs determined in pH 1.2 simulated gastric fluids at 37 °C are demonstrated in Figure 6. Form III was dissolved faster than Form I and II at the first stage. After 10 min, the solubility of Form III displays a "spring and parachute" effect [38]. With increasing time, BXM concentration in solution decreases, and after 40 min, the concentration is close to that of Form I, indicating that the precipitation process (Form I) is happening, and Form III converts into less soluble Form I under suspension condition. In addition, Form II presents an extreme increase in solubility when compared with Form I. The S_{max} values of II and III are 1.4 and 1.3 times as high as that of Form I. After the dissolution study, the remaining solid phase was characterized by PXRD analysis (Figure S10). It was confirmed that Form III almost converted to Form I as expected, while Form I and II retained their respective crystal forms, and there is no change even if suspended in water for seven days. The equilibrium solubility of Form I and II was 21.2 ± 0.2 and 29.4 ± 0.4 µg/mL, respectively. All three polymorphs have been shown to be as stable on the shelf for 12 months under the conditions of 25 ± 2 °C and 60% RH ± 5% RH (relative humidity) (Figure S11). The particle size is an important factor in the dissolution behavior of the API [39]. Therefore, to exclude the impact of particle size on the dissolution, crystals of a similar size range were used in the experiment. Thus, the difference in dissolution behavior between BXM polymorphs, especially for Form I and II, is due to the differences in the molecule packing rather than the particle size of the material studied.

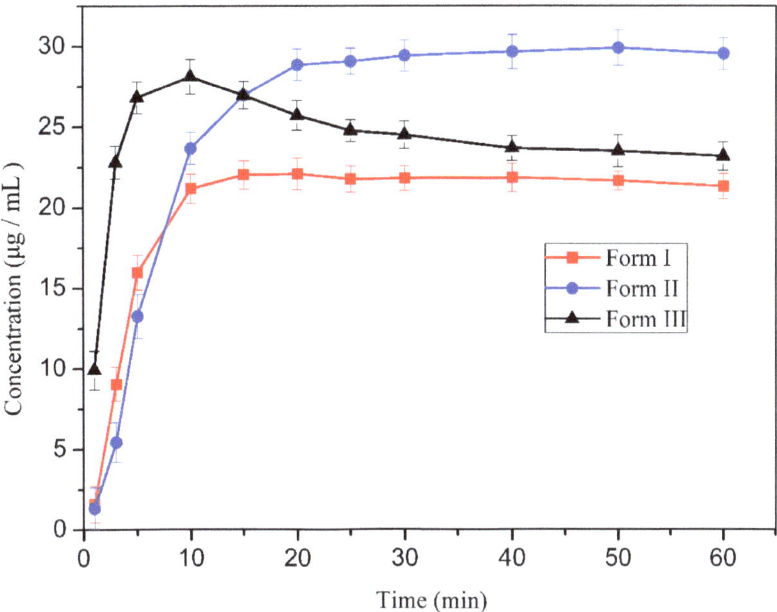

Figure 6. Powder dissolution profiles of BXM Form I, II, and III in pH 1.2 simulated gastric fluids at 37 °C. Data were shown as mean ± SD (*n* = 3).

4. Discussion

There has been an increasing interest in the correlations between the crystal structure, crystal morphology, and physicochemical properties of an investigational pharmaceutical compound [22,40]. Crystal morphology plays an important role in the dissolution behavior of drugs, depending on the functional groups exposed on the surface [41–43]. The theoretical crystal morphology of Form I and II was simulated by the Bravais-Friedel-Donnay-Harker (BFDH) model and growth morphology (GM) models with the aid of Materials Studio [38] (Figure 7, Table S4). By comparing Figures 7 and 1a,b, it can be seen that the simulated morphologies are similar to the crystal habit of the experimentally obtained crystals, but there are still some differences because the simulated morphologies only predict the crystal habit from the perspective of the internal structure of the crystal itself and do not consider the influence of the external environment on the crystal growth. The crystal habit obtained in the actual crystallization process is simultaneously affected by the combined effects of thermodynamics and kinetics. Therefore, the actual crystal habit obtained under different crystallization conditions may be different from the theoretical crystal habit calculated by the model. Vacuum morphology of Form I and II was generated by the BFDH model. The benefit of this method was to identify important faces in the growth process [44]. Results show that (0–20), (001), (0–11), and (100) are among the major surfaces, with (0–20) being the most dominant face. The calculated aspect ratio by BFDH morphology for Form I and II was 1.778 and 3.254, respectively. Surface structures of all important facets of Form I and II given by the MG model were studied. For Form I, (0–20) occupies 35% of the surface area and exhibits an attachment energy (E_{att}) value of −47.906 kcal/mol, while for Form II, (0–20) occupies 62% of the surface area and exhibits an E_{att} value of −31.244 kcal/mol. Because of the structural dissimilarity as described earlier, the orientation of molecules on the exterior of each facet of a crystal was diverse. For Form I, the (0–20) facet has a bed of phenyl moieties, making it relatively more hydrophobic (Figures 8 and S4), while, in Form II, the exposure of methymethyl carbonate chain makes it more hydrophilic (Figures 8 and S5). Thus Form II has a much greater percentage of its surface area with exposed hydrogen bond acceptors than Form I. Since the solubility and dissolution rates are proportional to the polarity of the crystal surface [45,46], the bulk solubility in aqueous media is much greater for Form II than for Form I.

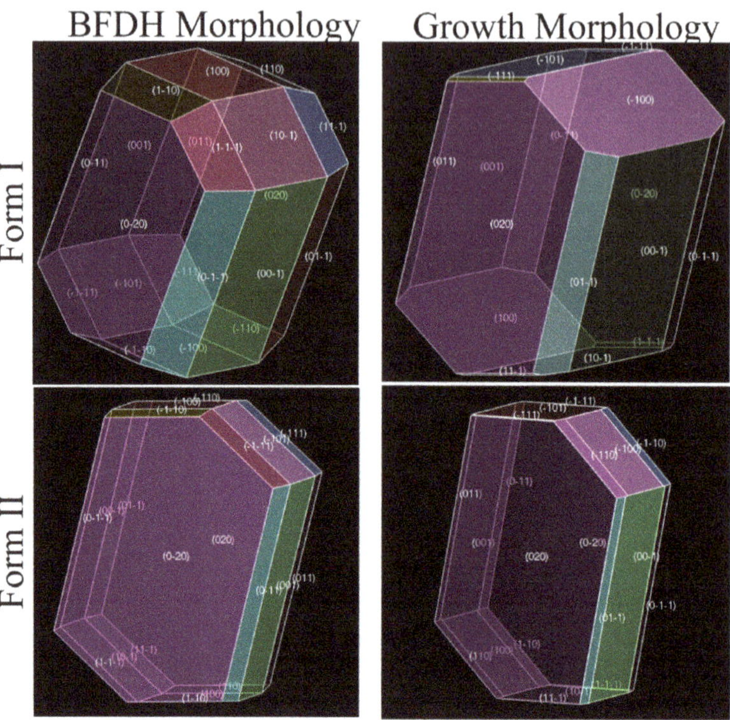

Figure 7. The BFDH and growth morphology of Form I and II. The morphologically important faces are evidenced together with their Miller indices.

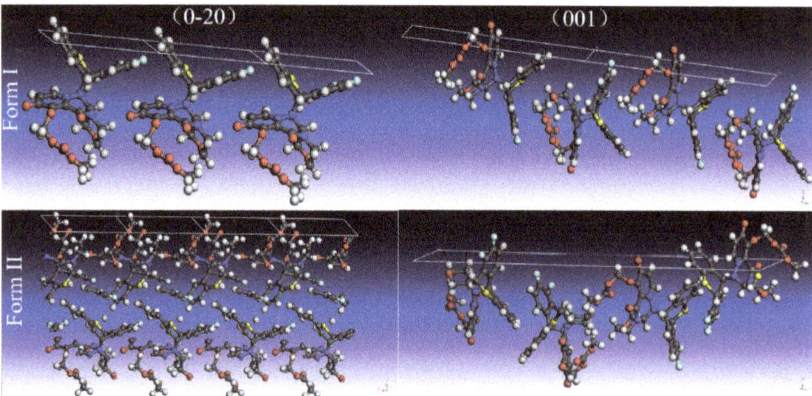

Figure 8. Crystal slices of different facets of BXM Form I and Form II expressing the presence of functional groups. The most dominant facet (0–20) is covered by 1 phenyl, and the second dominant facet (001) is covered by 1 carbonyl in Form I. In Form II, the most dominant facet (0–20) is covered by 1 methoxyl, and the second dominant facet (001) is covered by 2 fluorines, 2 carbonyl, and 1 methymethyl carbonate chain.

To further understand the relationship between crystal packing and dissolution behavior, energy frameworks were generated to visualize the differences in the supramolecular architecture of Form I and II in terms of energetics [47,48]. CrystalExplorer 3.1 [37] was used to construct energy frameworks at the B97D/6-31G** level of theory [47,49]. For Form I, energy bonds are evenly distributed over the whole structure such that they make a rectangular three-dimensional net (Figure S6, Table S5), while, in Form II, the energy bonds are ordered between aromatic rings such that they make rectangular two-dimensional layers (Figure S7, Table S6). The energy frameworks suggest relatively weaker interactions between the layers in Form II. The methymethyl carbonate segment of BXM oriented right and left the energy layers, rendering the surface of the layers hydrophilic. The progress of the dissolution requires easily accessible hydrophilic groups after the initially exposed groups have dissolved. For Form II, after one layer of BXM molecules is dissolved, the subsequent layers are more easily split along the b-axis due to no energy bonds between layers. In addition, the stripped bilayer surface covered hydrophilic groups. This result may also account for the higher solubility of Form II.

5. Conclusions

In this contribution, we prepared and characterized the solid-state behavior of the polymorphs of BXM. Three polymorphs have chemically the same building blocks, but their dissolution behaviors differ significantly. The only difference which can be related to the observed change in the dissolution behaviors is their significantly different packing features. The crystal structure of Form I and II were identified by SXRD. Using the crystal structure data, we confirmed that Form I contains strong energy bonds between the molecular layers, while in Form II, there are no energy bonds observed between the layers. Moreover, the theoretical crystal habits of Form I and II show that Form II has a much greater percentage of its surface area with exposed hydrogen bond acceptors than Form I. The results of stability and solubility experiments suggest that Form II may be a preferred solid form.

Supplementary Materials: The following are available online at https://www.mdpi.com/article/10.3390/cryst12040550/s1, Table S1: screening of BXM crystal form (Part of the experimental process), Table S2: Hydrogen bonds geometry (Å, °) for Form I, Table S3: Hydrogen-bond geometry (Å, °) for Form II, Figure S1: 2D fingerprint plots of the Hirshfeld surfaces for components of BXM Form I and II, Figure S2: showing the comparison of observed PXRD patterns of Form II and III. The variations in peak positions across the polymorphs are pointed by the arrows, Figure S3: TG curves of BXM in I, II, and III at a 10°C/min heating rate, Table S4: morphology predictions for Form I and II by means of BFDH and GM calculations, Figure S4: crystal slices of different facets of BXM Form I expressing the presence of functional groups. The most dominant facet (020) is covered by 1 phenyl, and the second dominant facet (001) is covered by 1 carbonyl, Figure S5: crystal slices of different facets of BXM Form II expressing the presence of functional groups. The most dominant facet (020) is covered by 1 methoxyl, and the second dominant facet (001) is covered by 2 fluorines, 2 carbonyl, and 1 methymethyl carbonate chain, Figure S6: energy frameworks corresponding to the different energy components and the total interaction energy in BXM Form I, Figure S7: energy frameworks corresponding to the different energy components and the total interaction energy in BXM Form II, Figure S8: color coding for the neighboring molecules around a molecule of BXM Form I. The molecule is shown with atom type color, Figure S9: color coding for the neighboring molecules around molecule A (top) and molecule B (bottom) of BXM Form II. The molecules are shown with atom type color, Table S5: molecular pairs and the interaction energies (kJ/mole) obtained from energy framework calculation for BXM Form I, Table S6: molecular pairs and the interaction energies (kJ/mole) obtained from energy framework calculation for BXM Form II, Figure S10: the PXRD pattern of the source BXM, Figure S11: Stability of Form I, Form II and Form III after storage of 12 months at 25 ± 2 °C and 60% RH ± 5% R shown in the PXRD pattern.

Author Contributions: Conceptualization, X.Z. and X.H.; software, K.Y. and X.Z.; Single-crystal XRD measurement, structure solution, and refinement, J.L.; writing—original draft preparation, X.Z.; writing—review and editing, X.Z. and X.H.; supervision, X.H. and Z.J. All authors have read and agreed to the published version of the manuscript.

Funding: This research received no external funding.

Institutional Review Board Statement: Not applicable.

Informed Consent Statement: Not applicable.

Data Availability Statement: Not applicable.

Acknowledgments: The authors thank Linshen Chen of Zhejiang University for his help with the TG/DSC measurements and analyses.

Conflicts of Interest: The authors declare no conflict of interest.

References

1. Heo, Y.A. Baloxavir: First Global Approval. *Drugs* **2018**, *78*, 693–697. [CrossRef] [PubMed]
2. Baxter, D. Evaluating the case for trivalent or quadrivalent influenza vaccines. *Hum. Vaccines Immunother.* **2016**, *12*, 2712–2717. [CrossRef] [PubMed]
3. Kiso, M.; Yamayoshi, S.; Murakami, J.; Kawaoka, Y. Baloxavir Marboxil Treatment of Nude Mice Infected With Influenza A Virus. *J. Infect. Dis.* **2020**, *221*, 1699–1702. [CrossRef] [PubMed]
4. Noshi, T.; Kitano, M.; Taniguchi, K.; Yamamoto, A.; Omoto, S.; Baba, K.; Hashimoto, T.; Ishida, K.; Kushima, Y.; Hattori, K.; et al. In vitro characterization of baloxavir acid, a first-in-class cap-dependent endonuclease inhibitor of the influenza virus polymerase PA subunit. *Antivir. Res.* **2018**, *160*, 109–117. [CrossRef] [PubMed]
5. Dufrasne, F. Baloxavir Marboxil: An Original New Drug against Influenza. *Pharmaceuticals* **2022**, *15*, 28. [CrossRef]
6. Baloxavir Marboxil FDA Label. Available online: https://s3-us-west-2.amazonaws.com/drugbank/cite_this/attachments/files/000/002/079/original/Baloxavir_Marboxil_FDA_label.pdf?1543256133 (accessed on 16 May 2021).
7. FDA Approves New Drug to Treat Influenza. Available online: https://www.fda.gov/news-events/press-announcements/fda-approves-new-drug-treat-influenza (accessed on 16 May 2021).
8. Abraham, G.M.; Morton, J.B.; Saravolatz, L.D. Baloxavir: A Novel Antiviral Agent in the Treatment of Influenza. *Clin. Infect. Dis.* **2020**, *71*, 1790–1794. [CrossRef]
9. Fujita, J. Introducing the new anti-influenza drug, baloxavir marboxil. *Respir. Investig.* **2020**, *58*, 1–3. [CrossRef]
10. Taniguchi, K.; Ando, Y.; Nobori, H.; Toba, S.; Noshi, T.; Kobayashi, M.; Kawai, M.; Yoshida, R.; Sato, A.; Shishido, T.; et al. Inhibition of avian-origin influenza A(H7N9) virus by the novel cap-dependent endonuclease inhibitor baloxavir marboxil. *Sci. Rep.* **2019**, *9*, 3466. [CrossRef]
11. Takashita, E.; Ichikawa, M.; Morita, H.; Ogawa, R.; Fujisaki, S.; Shirakura, M.; Miura, H.; Nakamura, K.; Kishida, N.; Kuwahara, T.; et al. Human-to-Human Transmission of Influenza A(H3N2) Virus with Reduced Susceptibility to Baloxavir, Japan, February 2019. *Emerg. Infect. Dis.* **2019**, *25*, 2108–2111. [CrossRef]
12. Kiso, M.; Yamayoshi, S.; Furusawa, Y.; Imai, M.; Kawaoka, Y. Treatment of Highly Pathogenic H7N9 Virus-Infected Mice with Baloxavir Marboxil. *Viruses* **2019**, *11*, 1066. [CrossRef]
13. Parveen, S.; Alnoman, R.B.; Bayazeed, A.A.; Alqahtani, A.M. Computational Insights into the Drug Repurposing and Synergism of FDA-approved Influenza Drugs Binding with SARS-CoV-2 Protease against COVID-19. *Am. J. Microbiol. Res.* **2020**, *8*, 93–102.
14. Kawai, M.; Tomita, K.; Akiyama, T.; Okano, A.; Miyagawa, M. Pharmaceutical Compositions Containing Substituted Polycyclic Pyridone Derivatives and Prodrug Thereof. U.S. Patent 10,759,814, 1 September 2020.
15. Shibahara, S.; Fukui, N.; Maki, T. Method for Producing Substituted Polycyclic Pyridone Derivative and Crystal of Same. Patent WO2017221869, 28 December 2017.
16. Kawai, M. Substituted Polycyclic Pyridone Derivatives and Prodrugs Thereof. U.S. Patent 10,392,406, 27 August 2019.
17. Raw, A.S.; Furness, M.S.; Gill, D.S.; Adams, R.C.; Holcombe, F.O., Jr.; Yu, L.X. Regulatory considerations of pharmaceutical solid polymorphism in Abbreviated New Drug Applications (ANDAs). *Adv. Drug Deliv. Rev.* **2004**, *56*, 397–414. [CrossRef] [PubMed]
18. Bucar, D.K.; Lancaster, R.W.; Bernstein, J. Disappearing polymorphs revisited. *Angew. Chem. Int. Ed. Engl.* **2015**, *54*, 6972–6993. [CrossRef] [PubMed]
19. Bannigan, P.; Zeglinski, J.; Lusi, M.; O'Brien, J.; Hudson, S.P. Investigation into the Solid and Solution Properties of Known and Novel Polymorphs of the Antimicrobial Molecule Clofazimine. *Cryst. Growth Des.* **2016**, *16*, 7240–7250. [CrossRef]
20. Censi, R.; Di Martino, P. Polymorph Impact on the Bioavailability and Stability of Poorly Soluble Drugs. *Molecules* **2015**, *20*, 18759–18776. [CrossRef]
21. Li, L.; Yin, X.H.; Diao, K.S. Improving the Solubility and Bioavailability of Pemafibrate via a New Polymorph Form II. *ACS Omega* **2020**, *5*, 26245–26252. [CrossRef] [PubMed]
22. Prado, L.D.; Rocha, H.V.A.; Resende, J.A.L.C.; Ferreira, G.B.; Figueiredo, T. An insight into carvedilol solid forms: Effect of supramolecular interactions on the dissolution profiles. *CrystEngComm* **2014**, *16*, 3168–3179. [CrossRef]
23. Fandaruff, C.; Rauber, G.S.; Araya-Sibaja, A.M.; Pereira, R.N.; de Campos, C.E.M.; Rocha, H.V.A.; Monti, G.A.; Malaspina, T.; Silva, M.A.S.; Cuffini, S.L. Polymorphism of Anti-HIV Drug Efavirenz: Investigations on Thermodynamic and Dissolution Properties. *Cryst. Growth Des.* **2014**, *14*, 4968–4975. [CrossRef]
24. Pudipeddi, M.; Serajuddin, A.T. Trends in solubility of polymorphs. *J. Pharm. Sci.* **2005**, *94*, 929–939. [CrossRef]

25. Sathisaran, I.; Dalvi, S.V. Engineering Cocrystals of PoorlyWater-Soluble Drugs to Enhance Dissolution in Aqueous Medium. *Pharmaceutics* **2018**, *10*, 108. [CrossRef]

26. Blagden, N.; Matas, M.; Gavan, P.T.; York, P. Crystal engineering of active pharmaceutical ingredients to improve solubility and dissolution rates. *Adv. Drug Deliv. Rev.* **2007**, *59*, 617–630. [CrossRef] [PubMed]

27. Bauer, J.; Spanton, S.; Henry, R.; Quick, J.; Dziki, W.; Porter, W.; Morris, J. Ritonavir: An Extraordinary Example of Conformational Polymorphism. *Pharm. Res.* **2001**, *18*, 859–866. [CrossRef] [PubMed]

28. Singhal, D.; Curatolo, W. Drug polymorphism and dosage form design: A practical perspective. *Adv. Drug Deliv. Rev.* **2004**, *56*, 335–347. [CrossRef] [PubMed]

29. Bruker AXS Inc. Bruker AXS announces novel APEX(TM) DUO, the most versatile system for small molecule X-ray crystallography. *Anti-Corros. Methods Mater.* **2007**, *54*, 375. [CrossRef]

30. Sheldrick, G.M. SHELXT-integrated space-group and crystal-structure determination. *Acta Crystallogr. A Found Adv.* **2015**, *71*, 3–8. [CrossRef]

31. Sheldrick, G.M. Crystal structure refinement with SHELXL. *Acta Crystallogr. C Struct. Chem.* **2015**, *71*, 3–8. [CrossRef]

32. Dolomanov, O.V.; Bourhis, L.J.; Gildea, R.J.; Howard, J.A.K.; Puschmann, H. OLEX2: A complete structure solution, refinement and analysis program. *J. Appl. Crystallogr.* **2009**, *42*, 339–341. [CrossRef]

33. Macrae, C.F.; Bruno, I.J.; Chisholm, J.A.; Edgington, P.R.; McCabe, P.; Pidcock, E.; Rodriguez-Monge, L.; Taylor, R.; van de Streek, J.; Wood, P.A. Mercury CSD 2.0–new features for the visualization and investigation of crystal structures. *J. Appl. Crystallogr.* **2008**, *41*, 466–470. [CrossRef]

34. Impurities: Guideline For Residual Solvents Q3C(R5). Available online: http://www.pmda.go.jp/files/000156308.pdf (accessed on 17 July 1997).

35. Bojarska, J.; Fruziński, A.; Sieroń, L.; Maniukiewicz, W. The first insight into the supramolecular structures of popular drug repaglinide: Focus on intermolecular interactions in antidiabetic agents. *J. Mol. Struct.* **2019**, *1179*, 411–420. [CrossRef]

36. Qi, M.H.; Zhang, Q.D.; Liu, Y.; Ren, F.Z.; Ren, G.B. Four solid forms of filgotinib hydrochloride: Insight into the crystal structures, properties, stability, and solid-state transitions. *J. Mol. Struct.* **2019**, *1178*, 242–250. [CrossRef]

37. Spackman, P.R.; Turner, M.J.; McKinnon, J.J.; Wolff, S.K.; Grimwood, D.J.; Jayatilaka, D.; Spackman, M.A. CrystalExplorer: A program for Hirshfeld surface analysis, visualization and quantitative analysis of molecular crystals. *J. Appl. Cryst.* **2021**, *54*, 575–587. [CrossRef] [PubMed]

38. Bavishi, D.D.; Borkhataria, C.H. Spring and parachute: How cocrystals enhance solubility. *Prog. Cryst. Growth Charact. Mater.* **2016**, *62*, 1–8. [CrossRef]

39. Chu, K.R.; Lee, E.; Jeong, S.H.; Park, E.S. Effect of particle size on the dissolution behaviors of poorly water-soluble drugs. *Arch. Pharmacal Res.* **2012**, *35*, 1187–1195. [CrossRef]

40. Destri, L.G.; Marrazzo, A.; Rescifina, A.; Punzo, F. How molecular interactions affect crystal morphology: The case of haloperidol. *J. Pharm. Sci.* **2011**, *100*, 4896–4906. [CrossRef] [PubMed]

41. Mishnev, A.; Stepanovs, D. Crystal Structure Explains Crystal Habit for the Antiviral Drug Rimantadine Hydrochloride. *Z. Nat. B* **2014**, *69*, 823–828. [CrossRef]

42. Xu, T.; Jiang, Z.; He, M.; Gao, X.; He, Y. Effect of arrangement of functional groups on stability and gas adsorption properties in two regioisomeric copper bent diisophthalate frameworks. *CrystEngComm* **2019**, *21*, 4820–4827. [CrossRef]

43. Modi, S.R.; Dantuluri, A.K.R.; Perumalla, S.R.; Sun, C.C.; Bansal, A.K. Effect of Crystal Habit on Intrinsic Dissolution Behavior of Celecoxib Due to Differential Wettability. *Cryst. Growth Des.* **2014**, *14*, 5283–5292. [CrossRef]

44. Kumar, D.; Thipparaboina, R.; Shastri, N.R. Can vacuum morphologies predict solubility and intrinsic dissolution rate? A case study with felodipine polymorph form IV. *J. Comput. Sci.* **2015**, *10*, 178–185. [CrossRef]

45. Chen, J.; Trout, B.L. Computer-Aided Solvent Selection for Improving the Morphology of Needle-like Crystals: A Case Study of 2,6-Dihydroxybenzoic Acid. *Cryst. Growth Des.* **2010**, *10*, 4379–4388. [CrossRef]

46. Moreno-Calvo, E.; Calvet, T.; Cuevas-Diarte, M.A.; Aquilano, D. Relationship between the Crystal Structure and Morphology of Carboxylic Acid Polymorphs. Predicted and Experimental Morphologies. *Cryst. Growth Des.* **2010**, *10*, 4262–4271. [CrossRef]

47. Jha, K.K.; Dutta, S.; Kumar, V.; Munshi, P. Isostructural polymorphs: Qualitative insights from energy frameworks. *CrystEngComm* **2016**, *18*, 8497–8505. [CrossRef]

48. Mackenzie, C.F.; Spackman, P.R.; Jayatilaka, D.; Spackman, M.A. CrystalExplorer model energies and energy frameworks: Extension to metal coordination compounds, organic salts, solvates and open-shell systems. *IUCrJ* **2017**, *4*, 575–587. [CrossRef] [PubMed]

49. Turner, M.J.; Thomas, S.P.; Shi, M.W.; Jayatilaka, D.; Spackman, M.A. Energy frameworks: Insights into interaction anisotropy and the mechanical properties of molecular crystals. *Chem. Commun.* **2015**, *51*, 3735–3738. [CrossRef] [PubMed]

crystals

Article

Crystal Structure Determination and Hirshfeld Analysis of a New Alternariol Packing Polymorph

Kelly L. Rue [1], Guodong Niu [2], Jun Li [2,3,]* and Raphael G. Raptis [1,3,]*

[1] Department of Chemistry and Biochemistry, College of Arts, Science & Education, Modesto A. Maidique Campus, Florida International University, Miami, FL 33199, USA; krue001@fiu.edu

[2] Department of Biological Sciences, College of Arts, Science & Education, Modesto A. Maidique Campus, Florida International University, Miami, FL 33199, USA; gniu@fiu.edu

[3] Biomolecular Sciences Institute, College of Arts, Science & Education, Modesto A. Maidique Campus, Florida International University, Miami, FL 33199, USA

* Correspondence: lij@fiu.edu (J.L.); rraptis@fiu.edu (R.G.R.)

Abstract: A new polymorph of the mycotoxin alternariol is reported and characterized by single crystal X-ray diffraction. Structural data, Hirshfeld surface analysis, and 2D fingerprint plots are used to compare differences in the intermolecular interactions of the orthorhombic $Pca2_1$ Form I (previously reported) and the monoclinic $P2_1/c$ Form II (herein reported). The polymorphs have small differences in planarity—7.55° and 2.19° between the terminal rings for Form I and Form II, respectively—that brings about significant differences in the crystal packing and O-H ... H interactions.

Keywords: polymorph; Hirshfeld analysis; alternariol

Citation: Rue, K.L.; Niu, G.; Li, J.; Raptis, R.G. Crystal Structure Determination and Hirshfeld Analysis of a New Alternariol Packing Polymorph. *Crystals* **2022**, *12*, 579. https://doi.org/10.3390/cryst12050579

Academic Editors: Jingxiang Yang and Xin Huang

Received: 31 March 2022
Accepted: 19 April 2022
Published: 21 April 2022

Publisher's Note: MDPI stays neutral with regard to jurisdictional claims in published maps and institutional affiliations.

1. Introduction

Alternariol (AOH; systematic name: 3,7,9-trihydroxy-1-methyl-6H-benzo[c]chromen-6-one; Scheme 1), a mycotoxin produced by various species of *Alternaria* molds, is an important contaminant in fruit, vegetable, and cereal products [1–3]. It possesses cytotoxic, genotoxic, and mutagenic properties in vitro [3–6]; however, these properties are still being studied in vivo. The underlying mechanism of toxicity is not yet fully established. AOH has been reported to induce the growth of the Alternaria species, *Alternaria alternata*, on various fruits [2,3].

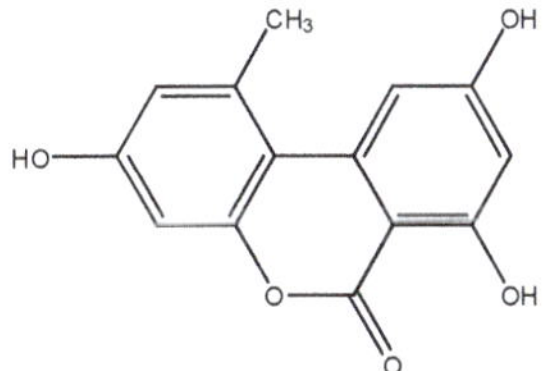

Scheme 1. Line drawing of alternariol structure.

Crystal polymorphs can exhibit different physical, chemical, and mechanical properties [7]. These are especially important for biologically active compounds, such as pharmaceuticals or mycotoxins [8–10]. The difference between polymorphic forms is in either the conformation or the packing arrangement of the molecules determining the intermolecular interactions [11–14]. An alternariol analogue, alternariol monomethyl ether (AME), with three reported structures in the Cambridge Crystallographic Data Centre (CCDC), exemplifies both polymorphism and solvomorphism (i.e., it is a compound which crystallizes in multiple space groups due to the presence of interstitial solvent molecules): two true polymorphs have crystallized in the *P*-1 and *Fdd*2 space groups, while the solvomorph has co-crystallized with dimethylsulfoxide in the *C*2/*m* space group [15–17].

In 2010, the crystal structure of AOH was published by Siegel et al. (CCDC refcode: TUPJOE), void of interstitial solvents; no polymorphs have been reported since [18]. Herein, we report the first polymorph of alternariol and discuss the differences in crystal structures and packing interactions. The previously reported structure will be identified as Form I and the one reported herein will be identified as Form II.

2. Materials and Methods

2.1. Materials

AOH was produced by the fungus *Purpureocillium lilacinum*, which was cultured on cereal. Secondary metabolites were extracted by ethyl acetate, and AOH was isolated by chromatography using the pulixin isolation procedure, as described previously [19]. All solvents were purchased from commercial sources and used without further purification.

2.2. Crystal Synthesis

Approximately 20 mg HPLC-pure AOH was dissolved in 2 mL methanol, and the solvent was allowed to evaporate at room temperature through two needle holes on the cover of the glass bottle. The colorless crystal used in the X-ray diffraction experiment was obtained on day three.

2.3. X-ray Crystallography and Data Collection

The slow evaporation of methanol under ambient conditions afforded colorless crystals of Form II. A suitable crystal was selected and mounted on a Bruker D8 Quest diffractometer equipped with a PHOTON 100 detector operating at T = 298 K (Bruker AXS, Madison, WI, USA). Data were collected with the shutterless ω-scan technique using graphite monochromated Mo-Kα radiation (λ = 0.71073 Å). The APEX3 [20] suite was used for collection, multiscan absorption corrections were applied, and structure solution was obtained using intrinsic phasing with SHELXT [21] (Bruker AXS, Madison, WI, USA). Data were then refined, using the Olex2 interface, by the least-squares method in SHELXL [22]. All hydrogen atoms were located in the difference map. Crystal data and structure refinement parameters are listed in Table 1. CCDC 2163068 contains the supplementary crystallographic data for this paper and can be obtained free of charge from The Cambridge Crystallographic Data Center via www.ccdc.cam.ac.uk/structures/. Hirshfeld surfaces were examined using *CrystalExplorer17* [23]. Interplanar geometric parameters were calculated using *Mercury 2020.3.0* [24].

2.4. Hirshfeld Surface Analysis

To illustrate differences in the intermolecular contacts of Form I and Form II, Hirshfeld surfaces were examined. Each surface has unique and well-defined points (d_i, d_e) where d_i represents a distance from a point on the Hirshfeld surface to the nearest nucleus internal to the surface and d_e represents a distance from a point on the surface to the nearest nucleus external to the surface. These points, along with the van der Waals radii, are normalized (d_{norm}) and mapped onto the three-dimensional (3D) Hirshfeld surface where red regions represent close contacts (shorter than the sum of van der Waals radii) and negative d_{norm} values, blue regions represent long contact (longer than the sum of van der Waals radii) and positive d_{norm} values, and white regions represent a d_{norm} value of 0 (i.e., the contact distance is equal to the sum of van der Waals radii) [25–28]. These points can also be incorporated into two-dimensional (2D) fingerprint plots in which data are binned into pairs (d_i, d_e). Each bin is colored from blue (few points) to green (moderate points) to red (many points), and each point on the plot represents a bin with a width of 0.1 Å.

Table 1. Crystal data and structure refinement parameters for alternariol—Form II.

Formula	$C_{14}H_{10}O_5$
$D_{calc.}/g\ cm^{-3}$	1.590
μ/mm^{-1}	0.122
Formula Weight	258.22
Color	Colorless
Shape	Plate
T/K	298
Crystal System	Monoclinic
Space group	$P2_1/c$
a/Å	7.2836(3)
b/Å	14.3875(5)
c/Å	10.5110(3)
$\beta/°$	101.621(1)
$V/Å^3$	1078.90(7)
Z	4
Wavelength/Å	0.71073
Radiation Type	Mo-$K\alpha$
$2\theta_{min}/°$	4.8
$2\theta_{max}/°$	52.8
Measured Refl.	23,218
Independent Refl.	2213
Reflections Used, $I_o > 2s(I_o)$	1435
R_{int}	0.083
Parameters	213
[a] GooF	1.021
[b] wR_2	0.1177
[c] R_1	0.0505

[a] $GooF = \left[\sum \left[w \left(F_o^2 - F_c^2 \right)^2 \right] / (N_o - N_v) \right]^{1/2}$; [b] $wR_2 = \sum \left| |F_o| - |F_c| \right| / \sum |F_o|$; [c] $R_1 = \left[\left(\sum w \left(F_o^2 - F_c^2 \right)^2 / \sum |F_o|^2 \right) \right]^{1/2}$.

3. Results and Discussion

3.1. Structure Description of Form II

The new AOH polymorph (Form II) crystallized in the $P2_1/c$ space group. The molecule consists of three fused, six-membered rings (Ring 1: C2, C3, C4, C5, C6, C7; Ring 2: O1, C1, C2, C7, C8, C9; and Ring 3: C8, C9, C10, C11, C12, C13); the best-fit planes defined by these rings will be referred to as Plane 1, Plane 2, and Plane 3, respectively (Figure 1). The molecule is nearly planar with a plane twist angle of 2.19(8)° between Plane 1 and Plane 3 and surrounded by six approximately coplanar molecules. A strong intramolecular H-bond (O3-H . . . O2) is present in the molecular structure. A packing diagram (Figure 2) shows there are four additional intermolecular hydrogen H-bonds (O-H . . . O) per molecule (Table 2) and one non-classical hydrogen bonding interaction with a distance of 2.986(3) Å (O3-H . . . O4). Thus, each molecule has classical H-bonding interactions with four of the surrounding alternariol molecules and non-classical hydrogen bonding interactions with the remaining two molecules (Figure 2). All H-bonding interactions are between approximately coplanar molecules. There are two pairs of closely π-stacked layers with interlayer distances of 3.391 Å and 3.322 Å between each pair, while alternating pairs form a dihedral angle of 7.10° (Figure 3).

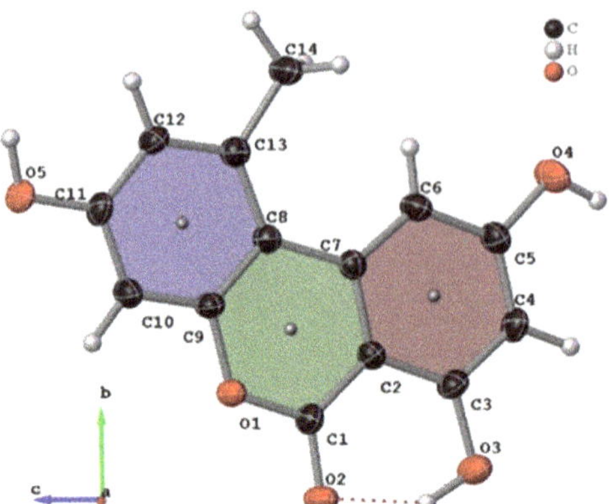

Figure 1. Crystal structure and labeling scheme of alternariol Form II with thermal ellipsoids at 50% probability. Plane 1 (red), Plane 2 (green) and Plane 3 (blue) are also shown.

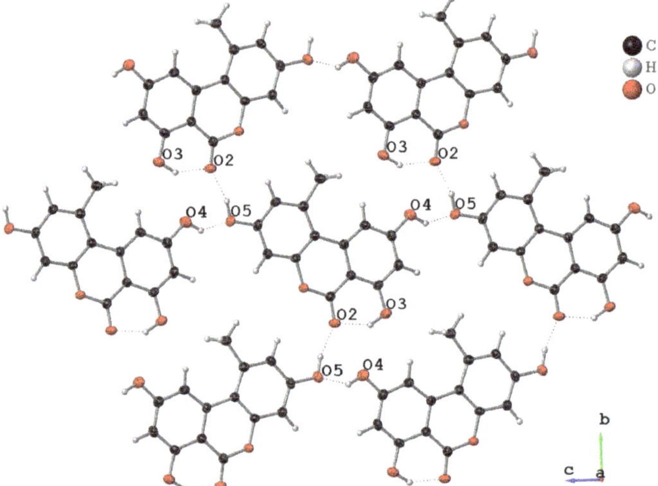

Figure 2. Hydrogen bonding network of alternariol—Form II. The middle molecule is surrounded by six other molecules and has classical H-bonding interactions with four of them.

Table 2. Hydrogen-bond geometry (Å, °) of alternariol—Form II.

D-H … A	D-H	H … A	D … A	D-H … A
O3-H3 … O2	0.94(4)	1.76(4)	2.590(2)	144(3)
O4-H4 … O5[i]	0.89(3)	1.99(3)	2.735(2)	140(3)
O5-H5 … O2[ii]	0.93(3)	1.74(3)	2.640(2)	163(3)
O3-H3 … O4[iii]	0.94(4)	2.34(3)	2.986(3)	125(3)

[i] $x - 1, y, z - 1$; [ii] $-x + 1, y + 1/2, -z + 3/2$; [iii] $-x, y - 1/2, -z + 1/2$.

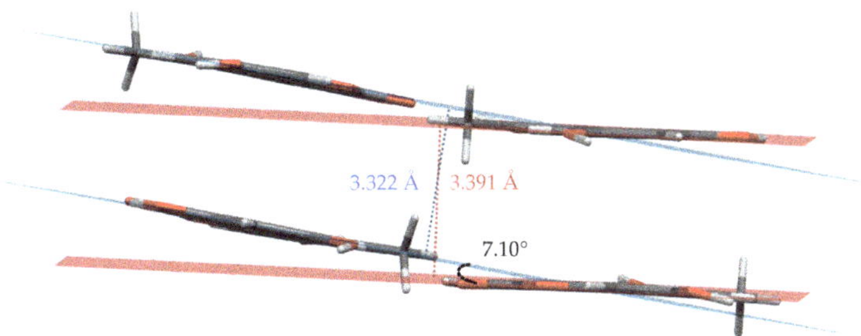

Figure 3. Perspective view depicting the interlayer distance and dihedral angle between best fit planes for two pairs of π-stacked molecules.

3.2. Comparison of Form I and Form II

Crystals of Form I were obtained by Siegel et al. via sublimation in an argon atmosphere and crystallized in the orthorhombic $Pca2_1$ space group, while crystals of Form II were obtained via slow evaporation of a methanolic solution, crystallizing in the monoclinic $P2_1/c$ space group. There are three rotatable hydroxyl groups in alternariol, giving rise to eight possible conformations (as optimized by Scharkoi et al.) [29]; however, Forms I and II crystallize in the same conformation, which is different than the calculated gas phase energy minimum [29,30].

The rings of Form I are not strictly coplanar—there is a plane twist angle of 7.6(1)° between Plane 1 and Plane 3. Initially, the lack of planarity was hypothesized to be due to the steric effects of the methyl group (C14) in relation to the hydrogen atom on C6 (H6A). A benzo[c]chromen-6-one analogue, 2-chloro-7-hydroxy-8-methyl-6H-benzo[c]chromen-6-one [31], which lacks a sterically incumbered methyl group, yet its rings are considered to be coplanar with a plane twist angle of 2.78(8)°, was used for justification. However, this analogue also lacks intermolecular hydrogen bonding that may exert forces onto the hydroxyl groups, causing their respective rings to twist slightly out of plane. Form II exhibits an even smaller plane twist angle of 2.19(8)° than the above reference analogue. Moreover, the distance between the methyl group and H6A are statistically the same—2.36(3) Å and 2.39(2) Å for Form I and II, respectively. Therefore, the lack of planarity in Form I must be due to factors other than the steric interaction between the methyl group and H6A. A previously mentioned alternariol analogue, AME [15], solidifies this notion: like alternariol, it has a methyl group in close proximity to a hydrogen atom and a plane twist angle of only 0.59(5)°, suggesting that rings 1 and 3 are coplanar. The difference is that AME has a more extended hydrogen bonding network (including C-H … O and O-H … O interactions) and stronger π-π stacking.

Superimposed images of Form I and Form II, matching atoms C11, C12, and C13 of each structure (Figure 4), allow for visualization of the maximal deviations between the polymorphs occurring at C4, C5, O4, and C6 with differences of 0.256(3) Å, 0.352(4) Å, 0.503(3) Å, and 0.306(4) Å, respectively. Clearly, forms I and II of alternariol do not show significant variance in conformation and can therefore be described as packing polymorphs. The different packing arrangements are attributed to a slight difference in planarity between the two forms (Figure 5). In Form I, parallel molecules from different layers are eclipsed (Figure 5a) while in Form II adjacent molecules are rotated 180° and offset by approximately 1.46 Å. (Figure 5d). Form I exhibits a zig-zag packing motif where interlayer H-bonding interactions are formed between adjacent layers (Figure 5b), while Form II forms parallel layers (Figure 5e) with no interlayer H-bonding interactions.

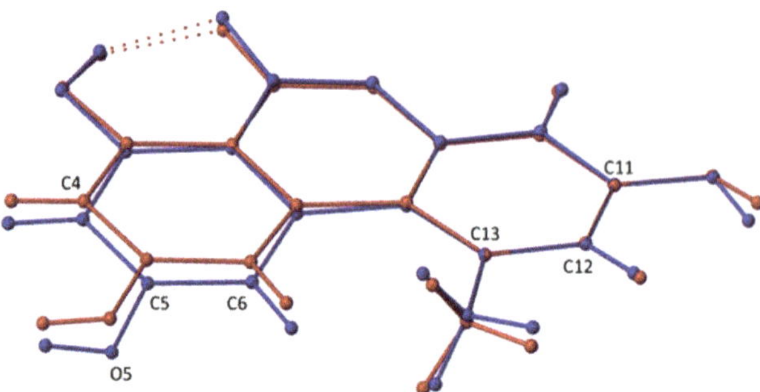

Figure 4. Overlayed wireframe images of alternariol Form I (blue) and Form II (red) matching C11, C12, and C13 of each structure.

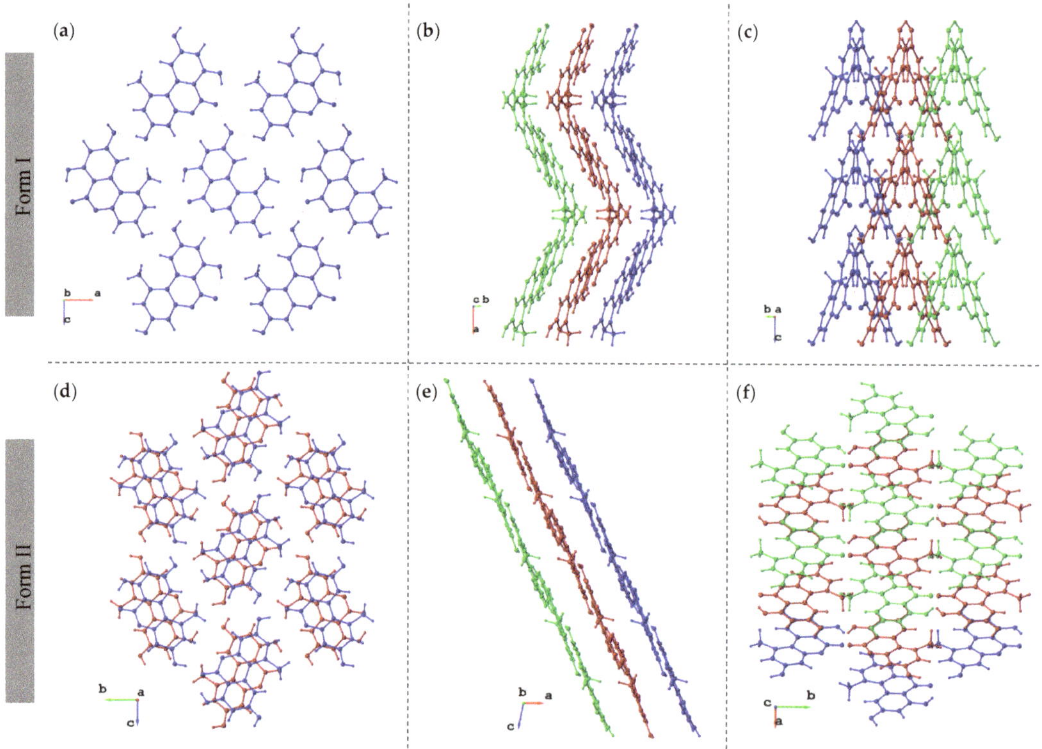

Figure 5. Crystal packing diagrams of Form I (upper) and Form II (lower) at different orientations. Left column (**a,d**): views perpendicular to molecular planes; Middle column (**b,e**): side views of the molecule; Right column (**c,f**): the same layers of **5b** and **5e** rotated by 90°.

Hirshfeld Surface Analysis

Intermolecular interactions were investigated for Form I and Form II of alternariol via analysis of Hirshfeld surfaces. Figure 6 shows the Hirshfeld surface for Form I, mapped over d_{norm} (from −0.6764 to 1.0428) along with the neighboring molecules associated with the closest contacts. Figure 7 shows the Hirshfeld surface for Form II, mapped over d_{norm} (from −0.7193 to 1.2040) along with the neighboring molecules associated with the closest contacts. Figures 6a and 7a illustrate the contact points. In both figures, the red regions on the surface represent the closest interactions between molecules. Four intermolecular H-bonded interactions, O5-H ... O4^i (2.809(3) Å; green), the reciprocal O4 ... H-O5^{ii} (2.809(3) Å; red), O4-H ... O2^{iii} (2.685(3) Å; orange), and O2 ... H-O4^{iv} (2.685(3) Å; purple) dominate Form I (Figure 6). Similarly, four intermolecular H-bonded contacts are encountered in Form II: O5-H ... O2^i (2.640(2) Å; green); O5 ... H-O4^{iv} (2.735(2) Å; purple); O4-H ... O5^{ii} (2.735(2) Å; red); and O2 ... H-O5^{iii} (2.640(2) Å; orange) (Figure 7). In both Forms I and II, there is a pattern of two shorter, 2.685 Å (I) and 2.640 Å (II), and two longer, 2.809 Å (I) and 2.735 Å (II), intermolecular H-bonds, the ones of Form II being approximately 0.050 Å shorter than the corresponding ones in Form I. Counterintuitively, the tighter H-bonded pattern of Form II does not render it denser than I, with calculated densities of 1.594 g cm^{-3} vs. 1.590 g cm^{-3} for I and II, respectively. In Form I, three smaller, less intense red spots can be observed, denoting long-range interactions: the first one between the methyl group (C14) of external molecule i and O5 of the central molecule with a distance of 3.768(3) Å. The second long range interaction can be seen between the methyl group of the central molecule and the methyl group of an external one directly above it (not shown in Figure 5) with a distance of 3.7244 (6) Å. The third long range interaction at 3.547(3) Å can be seen between C4 of external molecule iv and O1 of the central one. This last interaction can also be seen between C4 ... O1^{iii}. A number of long-range interactions can also be seen in Form II. The first is a non-classical hydrogen bond between O4 of the central molecule with O3-H of an external one (shown in Figure 2) with a distance of 2.986(3) Å. The second interaction is between O1 of external molecule i and C12 of the central one with a distance of 3.821(3) Å. Two additional reciprocal interactions between C4 and the methyl group are shown at 4.126(3) Å.

From the d_{norm} surface, 2D fingerprint plots are assembled in Figure 8. Here, dark blue squares represent the fewest concentration of points, green—moderate, and red—dense concentration of points. The shapes of the full fingerprint plots for Forms I and II share some similarities. They each have two sharp spikes in the bottom left quadrant of the plot that correspond to the shortest interactions (O ... H); however, the spikes of Form II are much sharper. Both plots also show a cluster of green and red squares roughly around $d_i = d_e \approx 1.8$–2.0 which indicates C ... C interactions and π-π stacking interactions; however, Form II has a larger concentration of these interactions.

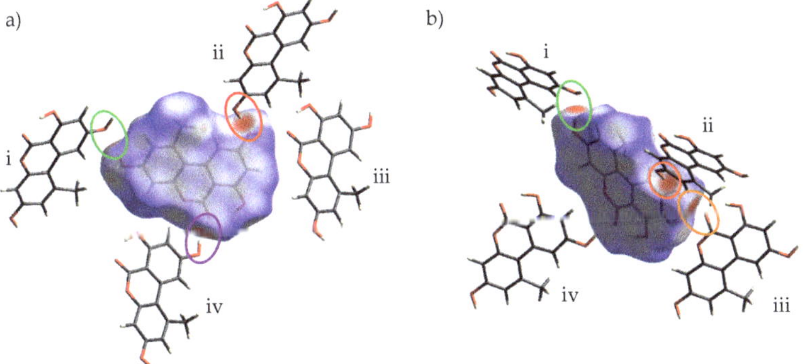

Figure 6. Hirshfeld surface of Form I mapped over d$_{norm}$ along with external molecules i–iv: a perpendicular view of the molecule and its surface (**a**) and an angled view (**b**). Red regions on the surface represent close contacts, blue regions represent long contacts, and white regions represent contacts in which the distance is equal to the sum of van der Waals radii.

The decomposition of the full fingerprint plot into the specific types of molecular interactions are shown in the Supplementary Materials. From the greatest to least contribution, the interactions are as follows (percent contributions are written in parentheses in the order of Form I, Form II): O ... H (38.6, 37.4); H ... H (28.9, 29.6); C ... C (16.0, 16.2); C ... H (9.1,

10.1); O . . . C (5.7, 4.7); O . . . O (1.7, 2.0). The order of contributions is the same for both Forms I and II; even the precent contributions of each interaction are similar for each Form.

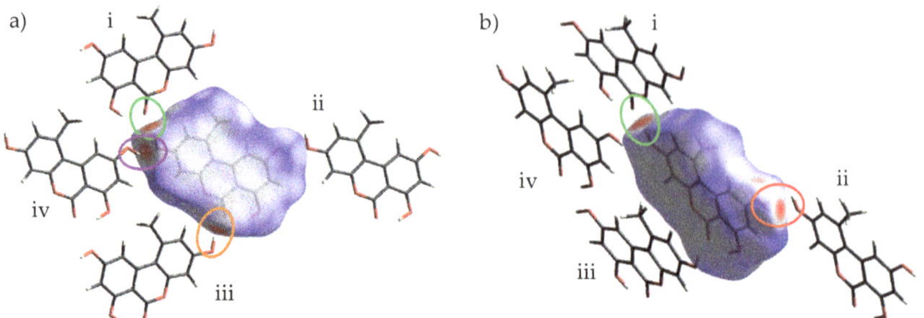

Figure 7. Hirshfeld surface of Form II mapped over d_{norm} along with external molecules i–iv: a perpendicular view of the molecule and its surface (**a**) and an angled view (**b**). Red regions on the surface represent close contacts, blue regions represent long contacts, and white regions represent contacts in which the distance is equal to the van der Waals radii.

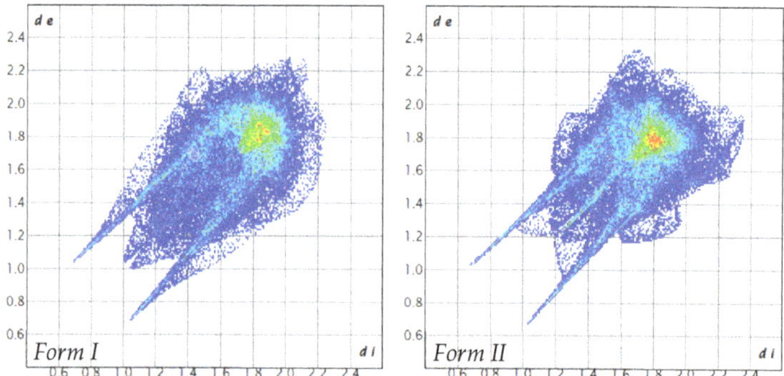

Figure 8. Full 2D fingerprint plot of Form I (**left**) and Form II (**right**) of alternariol.

4. Conclusions

Single crystals of two polymorphs of alternariol have been grown using two different methods—sublimation and recrystallization from a methanolic solution. As the bioavailability of polymorphs, and consequently their biological properties, vary among them, the possibility exists that recrystallization from solvents of a different polarity than methanol may result in additional polymorphic structures, which may exhibit differences in cytotoxicity.

Supplementary Materials: The following supporting information can be downloaded at: https://www.mdpi.com/article/10.3390/cryst12050579/s1, Figures S1 & S2: 2D fingerprint plots filtered by type of molecular interaction for Form I and Form II, respectively; Table S1: Selected bond lengths (Å) for alternariol—Form II.

Author Contributions: Formal analysis, K.L.R. and G.N.; investigation, K.L.R. and G.N.; writing—original draft preparation, K.L.R.; writing—review and editing, K.L.R., R.G.R. and J.L.; visualization, K.L.R.; supervision, R.G.R. and J.L.; funding acquisition, R.G.R. and J.L. All authors have read and agreed to the published version of the manuscript.

Funding: This work was supported by NIH NIAID R01AI125657.

Institutional Review Board Statement: Not applicable.

Informed Consent Statement: Not applicable.

Data Availability Statement: Not applicable.

Acknowledgments: K.L.R. was supported by U.S. Nuclear Regulatory Commission (NRC) fellowship grant No. 31310018M0012 awarded to FIU.

Conflicts of Interest: The authors declare no conflict of interest.

References

1. Scott, P.M. Analysis of Agricultural Commodities and Foods for Alternaria Mycotoxins. *J. AOAC Int.* **2001**, *84*, 1809–1817. [CrossRef]
2. Escrivá, L.; Oueslati, S.; Font, G.; Manyes, L. Alternaria Mycotoxins in Food and Feed: An Overview. *J. Food Qual.* **2017**, *2017*, e1569748. [CrossRef]
3. Solhaug, A.; Eriksen, G.S.; Holme, J.A. Mechanisms of Action and Toxicity of the Mycotoxin Alternariol: A Review. *Basic Clin. Pharmacol. Toxicol.* **2016**, *119*, 533–539. [CrossRef] [PubMed]
4. Brugger, E.-M.; Wagner, J.; Schumacher, D.M.; Koch, K.; Podlech, J.; Metzler, M.; Lehmann, L. Mutagenicity of the Mycotoxin Alternariol in Cultured Mammalian Cells. *Toxicol. Lett.* **2006**, *164*, 221–230. [CrossRef] [PubMed]
5. Fernández-Blanco, C.; Juan-García, A.; Juan, C.; Font, G.; Ruiz, M.-J. Alternariol Induce Toxicity via Cell Death and Mitochondrial Damage on Caco-2 Cells. *Food Chem. Toxicol.* **2016**, *88*, 32–39. [CrossRef]
6. Grover, S.; Lawrence, C.B. The Alternaria Alternata Mycotoxin Alternariol Suppresses Lipopolysaccharide-Induced Inflammation. *Int. J. Mol. Sci.* **2017**, *18*, 1577. [CrossRef]
7. Braga, D.; Grepioni, F.; Maini, L.; Polito, M. Crystal Polymorphism and Multiple Crystal Forms. In *Molecular Networks*; Hosseini, M.W., Ed.; Structure and Bonding; Springer: Berlin/Heidelberg, Germany, 2009; pp. 87–95.
8. Variankaval, N.; Cote, A.S.; Doherty, M.F. From Form to Function: Crystallization of Active Pharmaceutical Ingredients. *AIChE J.* **2008**, *54*, 1682–1688. [CrossRef]
9. Vippagunta, S.R.; Brittain, H.G.; Grant, D.J.W. Crystalline Solids. *Adv. Drug Delivery Rev.* **2001**, *48*, 3–26. [CrossRef]
10. Lee, A.Y.; Erdemir, D.; Myerson, A.S. Crystal Polymorphism in Chemical Process Development. *Annu. Rev. Chem. Biomol. Eng.* **2011**, *2*, 259–280. [CrossRef] [PubMed]
11. Bernstein, J. *Polymorphism in Molecular Crystals*; Oxford University Press: Oxford, UK, 2007.
12. Nogueira, B.A.; Castiglioni, C.; Fausto, R. Color Polymorphism in Organic Crystals. *Commun. Chem.* **2020**, *3*, 34. [CrossRef]
13. Brittain, H.G. *Polymorphism in Pharmaceutical Solids*; CRC Press: Boca Raton, FL, USA, 2018.
14. Lee, E.H. A Practical Guide to Pharmaceutical Polymorph Screening & Selection. *Asian J. Pharm. Sci.* **2014**, *9*, 163–175.
15. Dasari, S.; Bhadbhade, M.; Neilan, B.A. Alternariol 9-O-Methyl Ether. *Acta Crystallogr. Sect. E Struct. Rep. Online* **2012**, *68*, o1471. [CrossRef] [PubMed]
16. Light, M.; Sudlow, L.; Ganesan. CCDC 1475142: Experimental Crystal Structure Determination, CSD Communication. 2016. Available online: https://www.ccdc.cam.ac.uk/structures/Search?Ccdcid=1475142&DatabaseToSearch=Published (accessed on 18 April 2022). [CrossRef]
17. Dasari, S.; Miller, K.I.; Kalaitzis, J.A.; Bhadbhade, M.; Neilan, B.A. Alternariol 9-O-Methyl Ether Dimethyl Sulfoxide Monosolvate. *Acta Crystallogr. Sect. E Struct. Rep. Online* **2013**, *69*, o872–o873. [CrossRef] [PubMed]
18. Siegel, D.; Troyanov, S.; Noack, J.; Emmerling, F.; Nehls, I. Alternariol. *Acta Crystallogr. Sect. E Struct. Rep. Online* **2010**, *66*, o1366. [CrossRef] [PubMed]
19. Niu, G.; Wang, X.; Hao, Y.; Kandel, S.; Niu, G.; Raptis, R.G.; Li, J. A Novel Fungal Metabolite Inhibits Plasmodium Falciparum Transmission and Infection. *Parasit. Vectors* **2021**, *14*, 177. [CrossRef]
20. Bruker. *APEX3*; Bruker AXS LLC: Madison, WI, USA, 2020.
21. Sheldrick, G.M. SHELXT—Integrated Space-Group and Crystal-Structure Determination. *Acta Crystallogr. Sect. A Found. Crystallogr.* **2015**, *71*, 3–8. [CrossRef]
22. Sheldrick, G.M. Crystal Structure Refinement with SHELXL. *Acta Crystallogr. Sect. C Cryst. Struct. Commun.* **2015**, *71*, 3–8. [CrossRef]
23. Spackman, P.R.; Turner, M.J.; McKinnon, J.J.; Wolff, S.K.; Grimwood, D.J.; Jayatilaka, D.; Spackman, M.A. CrystalExplorer: A Program for Hirshfeld Surface Analysis, Visualization and Quantitative Analysis of Molecular Crystals. *J. Appl. Crystallogr.* **2021**, *54*, 1006–1011. [CrossRef]
24. Macrae, C.F.; Sovago, I.; Cottrell, S.J.; Galek, P.T.A.; McCabe, P.; Pidcock, E.; Platings, M.; Shields, G.P.; Stevens, J.S.; Towler, M.; et al. Mercury 4.0: From Visualization to Analysis, Design and Prediction. *J. Appl. Crystallogr.* **2020**, *53*, 226–235. [CrossRef]
25. Spackman, M.A.; Jayatilaka, D. Hirshfeld Surface Analysis. *CrystEngComm* **2009**, *11*, 19–32. [CrossRef]
26. Sundareswaran, S.; Karuppannan, S. Hirshfeld Surface Analysis of Stable and Metastable Polymorphs of Vanillin. *Cryst. Res. Technol.* **2020**, *55*, 2000083.
27. McKinnon, J.J.; Fabbiani, F.P.A.; Spackman, M.A. Comparison of Polymorphic Molecular Crystal Structures through Hirshfeld Surface Analysis. *Cryst. Growth Des.* **2007**, *7*, 755–769. [CrossRef]
28. Spackman, M.A.; McKinnon, J.J. Fingerprinting Intermolecular Interactions in Molecular Crystals. *CrystEngComm* **2002**, *4*, 378–392. [CrossRef]

29. Scharkoi, O.; Fackeldey, K.; Merkulow, I.; Andrae, K.; Weber, M.; Nehls, I.; Siegel, D. Conformational Analysis of Alternariol on the Quantum Level. *J. Mol. Model.* **2013**, *19*, 2567–2572. [CrossRef]
30. Tu, Y.-S.; Yufeng, J.T.; Appell, M. Quantum Chemical Investigation of the Detection Properties of Alternariol and Alternariol Monomethyl Ether. *Struct. Chem.* **2019**, *30*, 1749–1759. [CrossRef]
31. Appel, B.; Saleh, N.N.R.; Langer, P. Domino Reactions of 1,3-Bis-Silyl Enol Ethers with Benzopyrylium Triflates: Efficient Synthesis of Fluorescent 6H-Benzo[c]Chromen-6-Ones, Dibenzo[c,d]Chromen-6-Ones, and 2,3-Dihydro-1H-4,6-Dioxachrysen-5-Ones. *Chem. Eur. J.* **2006**, *12*, 1221–1236. [CrossRef] [PubMed]

Review

High-Pressure Polymorphism in Hydrogen-Bonded Crystals: A Concise Review

Tingting Yan [1,*], Dongyang Xi [2], Qiuxue Fang [1], Ye Zhang [3], Junhai Wang [1] and Xiaodan Wang [2]

[1] School of Science, Shenyang Jianzhu University, Shenyang 110168, China; fqiuxue@sjzu.edu.cn (Q.F.); jhwang@sjzu.edu.cn (J.W.)
[2] School of Materials Science and Engineering, Shenyang Jianzhu University, Shenyang 110168, China; xidy12@sjzu.edu.cn (D.X.); happywxd0402@sjzu.edu.cn (X.W.)
[3] School of Civil Engineering, Shenyang Jianzhu University, Shenyang 110168, China; zy02470@sjzu.edu.cn
* Correspondence: yantt@sjzu.edu.cn; Tel.: +86-136-2406-3066

Abstract: High-pressure polymorphism is a developing interdisciplinary field. Pressure up to 20 GPa is a powerful thermodynamic parameter for the study and fabrication of hydrogen-bonded polymorphic systems. This review describes how pressure can be used to explore polymorphism and surveys the reports on examples of compounds that our group has studied at high pressures. Such studies have provided insight into the nature of structure–property relationships, which will enable crystal engineering to design crystals with desired architectures through hydrogen-bonded networks. Experimental methods are also briefly surveyed, along with two methods that have proven to be very helpful in the analysis of high-pressure polymorphs, namely, the ab initio pseudopotential plane–wave density functional method and using Hirshfeld surfaces to construct a graphical overview of intermolecular interactions.

Keywords: high pressure; diamond anvil cell; polymorphism; hydrogen bonds

Citation: Yan, T.; Xi, D.; Fang, Q.; Zhang, Y.; Wang, J.; Wang, X. High-Pressure Polymorphism in Hydrogen-Bonded Crystals: A Concise Review. *Crystals* **2022**, *12*, 739. https://doi.org/10.3390/cryst12050739

Academic Editors: Jingxiang Yang, Xin Huang and Borislav Angelov

Received: 15 April 2022
Accepted: 18 May 2022
Published: 20 May 2022

Publisher's Note: MDPI stays neutral with regard to jurisdictional claims in published maps and institutional affiliations.

1. Introduction

Polymorphism (Greek: poly = many, morph = form) is a term applied in several areas to describe the diversity of nature.

Mitscherlich (1822, 1823) is generally regarded as the first person to apply polymorphism to crystallography when he recognized various crystal structures of the same compound in several arsenates and phosphate salts. Polymorphism, like many other words in chemistry, lacks a comprehensive definition. McCrone (1965) addresses this issue, and his working definition of polymorphism and the attached stipulations are still as influential as they were when he initially enunciated them.

A polymorph, according to McCrone, is "a solid crystalline phase of a particular compound that results from the possibility of at least two distinct configurations of that compound's molecules in the solid form".

Solid forms of the same compound exhibit different physicochemical properties, such as melting point, boiling point, hardness, and conductivity [1–3]. These features are crucial in a wide range of industrial and commercial applications [4–7]. Given the multiple ways in which a new polymorph is innovative, it is vital to characterize and regulate its polymorphic behavior during its development and marketing. Drug goods, particularly in the pharmaceutical business, undergo various manufacturing processes before they reach the market [8–10]. A pharmaceutically active molecule's chemical structure affects its activity and toxicity [11–14]. Thus, understanding the structural stability of polymorphs is crucial to physics, chemistry, and pharmaceutics.

High pressure has been shown to be a powerful approach for investigating polymorphism [15–21]. Multitudinous new high-pressure polymorphs related to simple molecules (e.g., alcohols and carboxylic acids), more complex systems (e.g., amino acids), and substantially larger systems (e.g., energetic materials, pharmaceuticals, metal organic frameworks, and transition metal complexes) have been produced [22–32].

Under identical high-pressure circumstances, the orthorhombic and monoclinic polymorphs of L-cysteine are affected differentially. Although the two L-cysteine polymorphs coexisting under environmental conditions have different density values, increasing the pressure does not change the orthorhombic polymorph with lower density into the monoclinic polymorph with higher density. On the contrary, each of the two polymorphs has undergone its own series of structural changes and has become a unique set of high-pressure phases. The irreversibility of the phase transition in the orthorhombic polymorphs of L-cysteine and the formation of new phases during decompression or the second compression cycle illustrate the kinetic control of pressure-induced phase transitions in crystalline amino acids [33]. Glycine is a classic case study in high-pressure polymorphism research. The α–form of glycine is stable to 23 GPa, the β–form undergoes phase transition to a new form at 0.76 GPa, and the γ–phase transforms into the ε–phase at 2 GPa [34–37]. Two paracetamol polymorphs are studied to determine the importance of hydrogen bonding and crystal structure. At 443 K, piracetam changes from form–III to form–I with a compression of 0.2 GPa [38,39]. Therefore, it is concluded that pressure has an unmatched function to modify crystal structures, especially during phase transitions or the formation of new polymorphs.

Due to its directionality, reversibility, and saturability, hydrogen bonding, as an important intermolecular interaction, plays a vital role in polymorphism [40–43]. Pressure has a long history of decreasing the intermolecular gap between materials and bringing atoms closer together. This feature illustrates that pressure can alter the strength of hydrogen bonding [44]. Increased pressure has been shown to strengthen weak and moderate $D-H\cdots A$ connections (where D and A denote a donor and acceptor, respectively), resulting in a lengthened $D-H$ distance. By contrast, pressure has little effect on strong hydrogen bonding [45,46]. Pressure-induced changes in hydrogen bonding can also result in changes in crystal symmetry, most frequently in conjunction with the effects of π-stacking, van der Waals, and electrostatic interactions. Thus, the cooperativity of various noncovalent interactions and changes in hydrogen bonds as a function of pressure are crucial for material structural stability [4–50].

Extensive research has been conducted on hydrogen-bonded polymorphs under high pressures, particularly for diamond anvil cell (DAC) methods. Compressive strength, hardness, and transparency to visible light are not the only properties of diamond that make it such a good choice for anvils. Diamond is also extraordinarily transparent to parts of the electromagnetic spectrum that are invisible. Many kinds of in situ characterizations based on the DAC, including Raman scattering, infrared, angle-dispersive X-ray diffraction (ADXRD), UV absorption, and fluorescence spectroscopy, have gathered significant information about pressure-induced changes in intermolecular interactions. In researching polymorphic systems, these high-pressure techniques provide considerable precision, accuracy, and sensitivity. These polymorphic materials have demonstrated a variety of critical functions and properties [51–56].

High-pressure science was dominated by hydraulically powered anvil and piston–cylinder devices from the early 1900s to the 1960s. These machines are monstrous in size and require the operation of specialized laboratories. As the gasketed DAC developed in the mid-1960s, high-pressure research technology became less demanding [57,58]. As shown in Figure 1, the DAC's basic concept is as uncomplicated as previously described [59]. Samples are placed in the metal gasket hole between the flat and parallel culets of two diamond anvils. The sample sustains high pressure when an external force pushes the two opposed anvils together. By filling the pressure chamber with a liquid pressure-transmitting substance, hydrostatic pressure conditions can be achieved [60,61]. A well-calibrated ruby fluorescence pressure scale is used to determine the pressure within the DAC sample chamber.

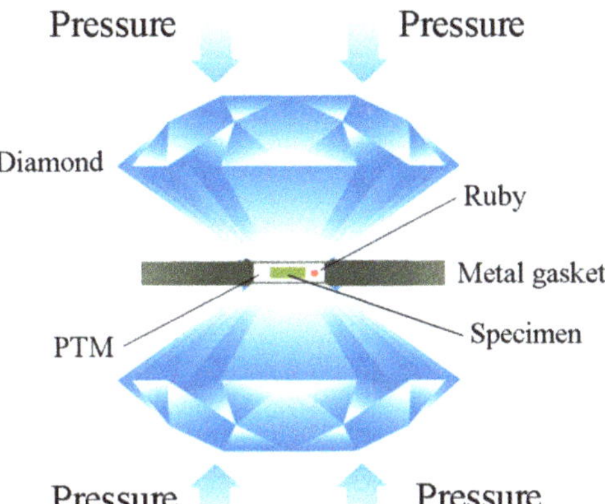

Figure 1. Schematic of DAC for high-pressure studies.

This review outlines recent advances in polymorphism in hydrogen-bonded crystals produced by the application of in situ high-pressure techniques and briefly discusses new development challenges. Additionally, potential obstacles and future directions are evaluated.

2. High-Pressure Polymorphism in Amides

Amides are particularly important because they are abundant in nature and are commonly employed in industry and technology as structural materials. In addition, the hydrogen bonds within the functional group of amides exhibit a similar directionality to those found in carboxylic acid synthons.

As a first-line antituberculosis medicine, pyrazinamide ($C_5H_5N_3O$, pyrazine-2-carboxamide, PZA) has been included on the World Health Organization's Model List of Essential Medicines. This medication is an infrequent instance of a conformationally stiff molecule with four polymorphs, namely, the α, β, γ, and δ forms [62–64]. Figure 2 illustrates the molecular packings in several polymorphs.

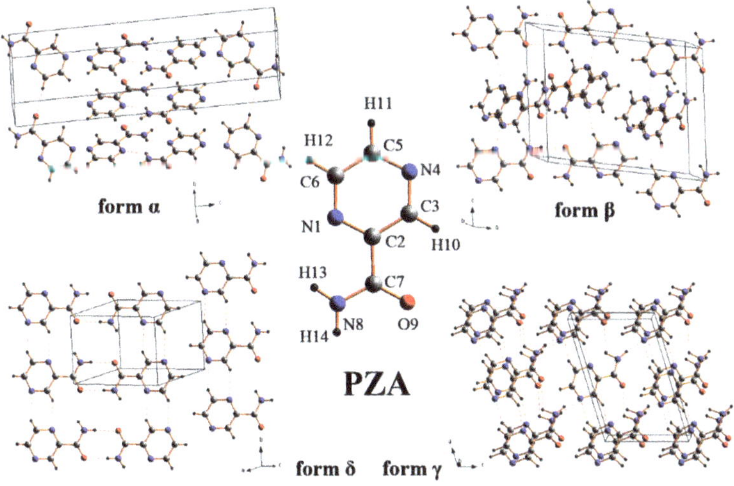

Figure 2. Crystal structures of PZA polymorphic forms: α, β, δ, and γ forms. Reprinted with permission from Ref. [65]. Copyright 2012 American Chemical Society.

The phase interactions between the four polymorphs have been widely researched at ambient pressure. Once the pure δ and γ forms undergo hand-grinding, they will be converted to the α–form in 45 min. The α, β, and δ forms will be converted to the γ-form upon being heated to 165 °C. After being cooled to room temperature, the γ-form can remain stable for up to 6 months before gradually converting to the α-form. In other words, a stable form at a high temperature can be retained at ambient temperature as a metastable form. Within the temperature range of −263 to −13 °C, the α-form changes into the δ-form.

We have adopted the in situ Raman spectroscopy and ADXRD techniques to examine the high-pressure responses of three forms (α, δ, and γ) of PZA [65]. At 4 GPa, the γ-phase transforms to the β-phase, while the other two forms preserve their original structures up to 14 GPa. The volume of the γ-phase is reduced by $\sim$3% at a pressure greater than 4 GPa. The bulk modulus and pressure derivatives for the γ-form are $B_0 = 6.3 \pm 0.4$ GPa and $B_0' = 10.0 \pm 0.3$. These parameters for the β-form are $B_0 = 7.9 \pm 0.3$ GPa and $B_0' = 5.8 \pm 0.2$. These equations imply that the β-form is more difficult to compress, which is consistent with the conclusion of lattice vibration analysis in the Raman spectra.

The β-form cannot be obtained entirely under ambient conditions. It always crystallizes alongside the γ-form. The α- and δ-forms' stability under compression is attributable to their unique dimer link. Previous studies have shown that the dimers are connected by relatively strong N–H $\cdots$ O hydrogen bonds in the α, β, and δ forms, whereas head-to-tail connection is only present in the γ-form through the N–H $\cdots$ N hydrogen bonds [66].

The pressure-induced γ-to-β phase change is exceptional in that it happens between two well-characterized polymorphs and is reversible. Additionally, the phase transition is detected at a wide range of pressures. Hysteresis behavior is very common in high-pressure investigations and is a characteristic of first-order phase transitions. The γ-form is not recovered until the pressure is close to the ambient level, which indicates an energy barrier to the transition [67]. An imbalance between hydrogen bonding and van der Waals interactions within γ-PZA could be responsible for the phase transition. The distances between PZA molecules inside the 3D structure decrease as pressure increases, strengthening hydrogen bonding and increasing overall Gibbs free energy. The free energy of the phase transition is reduced by rotating the PZA molecules and reassembling hydrogen bonds, resulting in a $\sim$4 GPa phase transition.

Malonamide ($C_3H_6N_2O_2$) is a model compound for investigating the hydrogen bonding interactions that occur when amides are compressed. This technique is used to manufacture pharmaceuticals and insecticides. This substance is analogous to a glycine residue with the peptide group inverted. Malonamide derivatives are of particular relevance because they are used in the manufacture of peptidomimetic compounds. Malonamide is a polymorphous compound that crystallizes into monoclinic, tetragonal, and orthorhombic crystal forms [68–70]. The unit cell parameters of the monoclinic forms are a = 15.78(0) Å, b = 4.63(7) Å, c = 9.33(1) Å, β = 133.84(0)°, V = 492.46(4) Å^3, and Z = 4. Tetragonal forms crystallize into a = 5.3140(3) Å, c = 15.5360(12) Å, V = 438.71(5) Å^3, and Z = 4. Crystallographic parameters of the orthorhombic forms are a = 5.3602(9) Å, b = 7.5178(8) Å, c = 11.791(2) Å, V = 475.14(12) Å^3, and Z = 4.

The monoclinic form Is chosen as the sample and compressed to a pressure of 10.4 GPa in a DAC at room temperature [71]. In situ Raman spectroscopy is utilized to detect structural changes caused by pressure. At 2.1 GPa, the Raman spectra show considerable changes, indicating the occurrence of a phase transition. At various pressures, the Raman spectrum is analyzed in terms of its fluctuations, which include the elimination of original modes, the development of new modes, and discontinuous changes in the Raman modes' pressure dependence.

Ab initio calculations are adopted to calculate the changes in molecular configurations and hydrogen-bonded networks. The computed results indicate that the molecules distort slightly in reaction to pressure. As a result, the original hydrogen bonds between N–H$\cdots$O are twisted. Additionally, the hydrogen-bonded patterns shift significantly. All amine groups participate in hydrogen bonding in the ambient structure. Eight hydrogen bonds are created between each molecule, including four donors and four acceptors. At a pressure of 10 GPa, however, one sort of asymmetric unit molecule establishes six hydrogen bonds with three donors and three acceptors. These findings corroborate the Raman spectra observed in the NH_2 stretching modes. The appearance of NH_2 asymmetric stretching modes indicates that the hydrogen bond networks have been reconstructed. Meanwhile, the modes that are related to N–H $\cdots$ O hydrogen bonds, including the NH_2 twisting, NH_2 rocking, NH_2 scissoring, NH_2 bending, CO bending, and CO stretching modes, all vary remarkably during phase transition. These variations indicate that the hydrogen bond donors and acceptors adopt new orientations.

We use Hirshfeld surfaces and fingerprint plots to compare changes in packing patterns and intermolecular interactions (Figure 3). This method simplifies the process of determining hydrogen bonding and van der Waals radius. On Hirshfeld surfaces, blue patches denote long interactions, and red regions denote short contacts. When the pressure is increased, the blue parts shrink, and the red regions expand. This development corresponds to the typical shortening of connections under conditions of high pressure. The two "spikes" reflect N–H$\cdots$O hydrogen bonding in both plots. The upper spike is the hydrogen bond donor (where $d_e > d_i$), whereas the lower spike is the hydrogen bond acceptor (where $d_i > d_e$). This conclusion is evident from the fingerprint plot: the N–H$\cdots$O spikes grow less prominent as the plot moves closer to the origin. The reduced maximum values of d_e between ambient pressure (2.399 Å) and 10 GPa (1.921 Å) are attributable to the overall shortening of the long contacts. The contribution of the H$\cdots$O connections remains stable at 23.9% at ambient pressure and 21.1% at 10 GPa, respectively. The H$\cdots$H contacts are compressed from 2.6 Å at ambient pressure to 2.2 Å at 10 GPa. Moreover, the contribution of the H$\cdots$H interaction changes from 32.3% to 30.2%. According to the computed results, fluctuations in the NH_2 stretching Raman vibrations, and the degree of freedom of the molecules, the phase transition of crystalline malonamide is likely induced by rearrangements of the hydrogen-bonded networks.

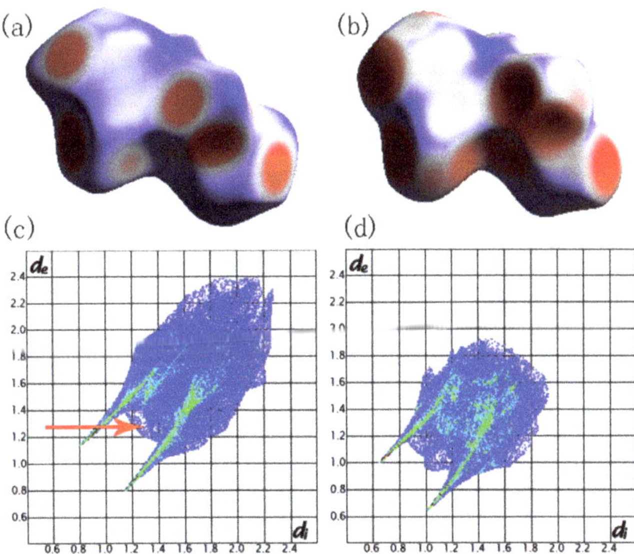

Figure 3. Hirshfeld surfaces mapped with d_{norm} and fingerprint plots for malonamide at ambient pressure(**a**,**c**) and at 10 GPa (**b**,**d**). Reprinted with permission from Ref. [71]. Copyright 2015 Royal Society of Chemistry.

Acetamide (H_3CCONH_2) is a hydrogen-bonded system with a basic structure that can be used as a model system for researching hydrogen-bonded systems under high pressures. Additionally, since acetamide contains just one peptide link, it can shed light on the structures of complicated peptides and proteins. Acetamide crystallizes into two crystal forms under ambient conditions, namely, the stable rhombohedral space group R3c and the metastable orthorhombic space group Pccn [72,73].

At pressures up to 16 GPa, in situ ADXRD and Raman scattering are utilized to examine the vibrational and structural properties of rhombohedral acetamide [74]. The ambient unit cell is seen in Figure 4a. Each acetamide molecule includes two acceptors and two donors, resulting in the formation of four N–H···O hydrogen bonds with other molecules. As a result, the equilibrium between hydrogen bonding and van der Waals interactions has an effect on the acetamide structure's stability under high pressure. At 0.9 GPa and 3.2 GPa, two structural phase transitions are observed, as indicated by significant changes in the Raman spectrum and discontinuities in peak positions vs. pressure. Significant alterations in the ADXRD patterns, as seen in Figure 4b, further verify the phase transition. Pawley refinement is used to determine the lattice parameters of the high-pressure phase–II using ab initio calculations. The crystal structure of phase–II can be indexed and refined as a monoclinic system with a potential C2/c space group; the lattice constants are as follows: a = 7.34(2) Å, b = 17.26(1) Å, and unit cell volume V = 690.7(1) Å^3. Phase–III has similar diffraction patterns to phase–II, implying that the two high-pressure phases may have similar structures.

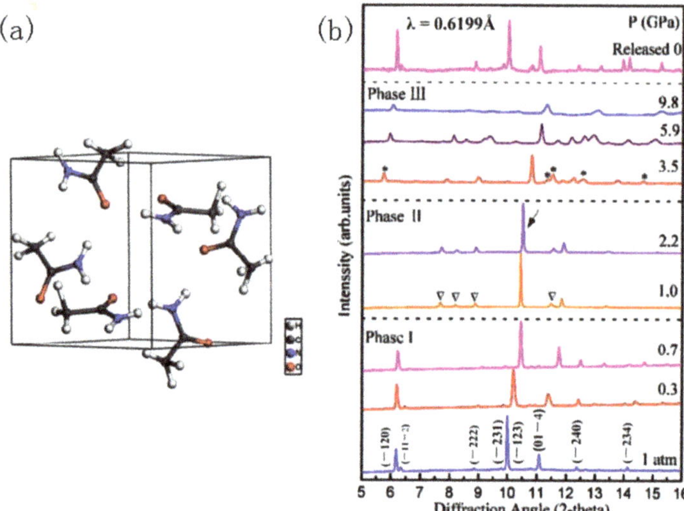

Figure 4. Ambient unit cell (a) and representative ADXRD patterns of acetamide crystal at different pressures (b). The peaks marked by asterisks and hollow triangles indicate the emergences of new phases. Reprinted with permission from Ref. [74] Copyright 2013 Royal Society of Chemistry.

The mechanism for pressure-induced phase changes has been postulated based on experimental and computational studies. Hydrogen bonding and van der Waals forces are the two key interactions that maintain acetamide's structural integrity at ambient temperature. The van der Waals forces between neighboring acetamide molecules increase as pressure increases because the distances between nearby molecules shrink. At the same time, the shorter hydrogen bonds make hydrogen bonding contact easier. The total Gibbs free energy available to the cosmos is enhanced by this method. With further compression, the acetamide crystal structures can no longer support the increasing Gibbs free energy. Thus, acetamide molecules slide and/or rotate, reorganizing hydrogen-bonded networks (through reconstruction and torsion) to lower the free energy. Thus, acetamide crystals undergo phase changes at various critical pressures (0.9, 3.2 GPa).

3. High-Pressure Polymorphism in Hydrazides

Maleic hydrazide ($C_4H_4N_2O_2$, MH), a well-known plant growth inhibitor in agriculture, comes in three polymorphic forms: triclinic MH1 and monoclinic MH2 and MH3 [75–77]. In their crystal structures, the three polymorphs exhibit identical patterns of hydrogen bonding, namely, an O−H···O motif that connects infinite chains of molecules. These polymorphs are connected into ribbons of double chains through N−H···O interactions.

The pressure-induced polymorphism of MH3 is investigated by in situ ADXRD and high-pressure Raman spectroscopy [78]. At 2 GPa, variations in the Raman spectrum suggest the presence of a pressure-induced phase transition. High-pressure ADXRD tests are carried out to further characterize this transition, as seen in Figure 5a. The monoclinic polymorphic form of MH2 with space group $P2_1/c$ is connected to the pressure-induced phase. The new MH2 phase remains when the pressure is increased even more. Further analysis indicates that this pressure-induced polymorphic transformation could be attributed to changes in the hydrogen-bonded ribbons. The ADXRD pattern indicates that this MH3 to MH2 polymorphism is partially reversible when the sample is returned to atmospheric pressure, with the majority of the sample reverting to the original MH3 structure. The quenched sample contains 68.3 wt% MH3 and 31.7 wt% MH2, according to the Rietveld quantitative phase analysis results in Figure 5b. The coexistence of these two polymorphic forms upon complete pressure release could be explained by their identical hydrogen-bonded aggregates and similar energies.

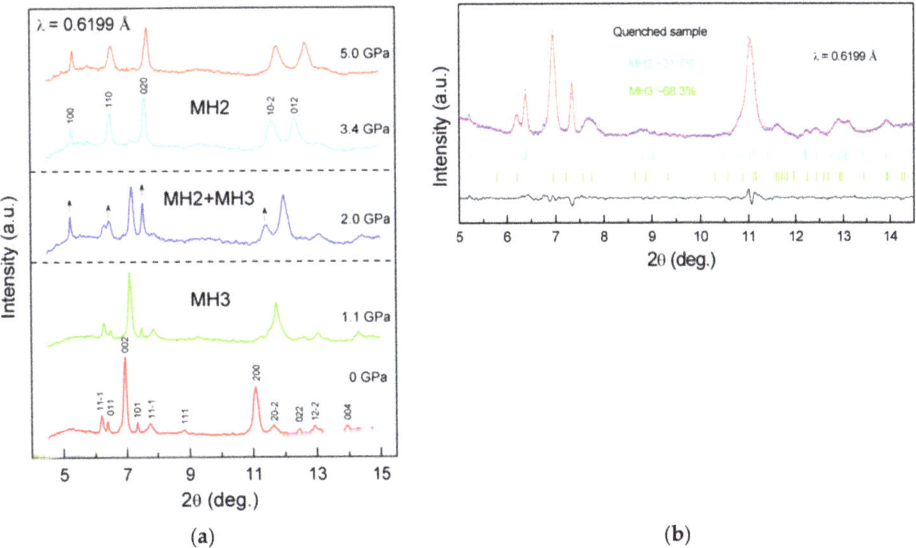

Figure 5. (**a**) Representative ADXRD patterns at elevated pressures. Arrows indicate new peaks. (**b**) Quantitative phase analysis based on the Rietveld fit of the diffraction patterns collected from the quenched sample: red line, experimental pattern; blue dotted line, simulated pattern; the black line is the fit residual. Reprinted with permission from Ref. [78]. Copyright 2014 American Chemical Society.

The conformational polymorphism oxalyl dihydrazide [$H_2N−NH−CO−CO−NH−NH_2$, ODH] has five known polymorphs. The ODH molecule is composed of six possible hydrogen-bond donors (N−H bonds of the NH and NH_2 group) and four possible hydrogen-bond acceptors (oxygen atoms in the C=O groups and nitrogen atoms in the

NH_2 groups). Numerous polymorphisms have been found, including α, β, γ, δ, and ε [79]. The crystal structures of the ODH polymorphic forms are depicted in Figure 6.

The molecule's N−N−C−C−N−N backbone is planar in all polymorphs and conforms to the trans-trans-trans conformation. The oxygen atoms of the O=C groups and the hydrogen atoms of the NH groups lie within this plane. The pyramidal structure of the NH_2 group indicates that the nitrogen atom has undergone sp3 hybridization, which enhances the nitrogen atom's capacity to act as a hydrogen-bond acceptor. Computational approaches are utilized to determine the energy sequence of the known polymorphs: α > ε > γ > δ > β [80,81].

At a pressure of 20 GPa, high-pressure Raman spectroscopy and ADXRD studies of the five types of ODH are carried out [82]. By applying pressure to the original counterparts, five novel forms are identified. Under high pressure, each polymorph yields a new form, resulting in a total of ten polymorphs, with the new forms being α′, β′, γ′, δ′, and ε′. The transition pressure points of the five ODH polymorphs are 12.0, 14.0, 11.3, 10.6, and 6.0 GPa for the α, β, γ, δ, and ε forms, respectively, as determined by Raman and ADXRD patterns. Under the influence of shear force, the β-form is changed into the α-form. As a result, high-pressure ADXRD experiments on the β-form cannot be carried out. The new forms might be indexed as $P2_1/c$ for α′, *Pmnb* for γ′, $P2_1/n$ for δ′, and *P-1* for ε′ via ab initio calculations. Several hydrogen-bond donors and acceptors, and low-energy conformational changes in NH and NH_2 groups, all play a role in the molecular conformation of ODH, resulting in a variety of crystal polymorphs. The transition pressures of the α, β, γ, and δ forms are all above 10 GPa, while the ε-form is changed into the ε′-form at about 6 GPa. This is attributable to the fact that the ε-form's structure is significantly different from that of the other polymorphs. The ε-form has ribbons that generally run perpendicularly to create a grid-like pattern when viewed in projection along the c-axis. The grid-like structure of the form is not as stable under pressure as the other polymorphs' sheet-like parallel ribbon structures. Therefore, the ε-form's phase transition takes place at 6 GPa, at least 4 GPa lower than those of the other four forms.

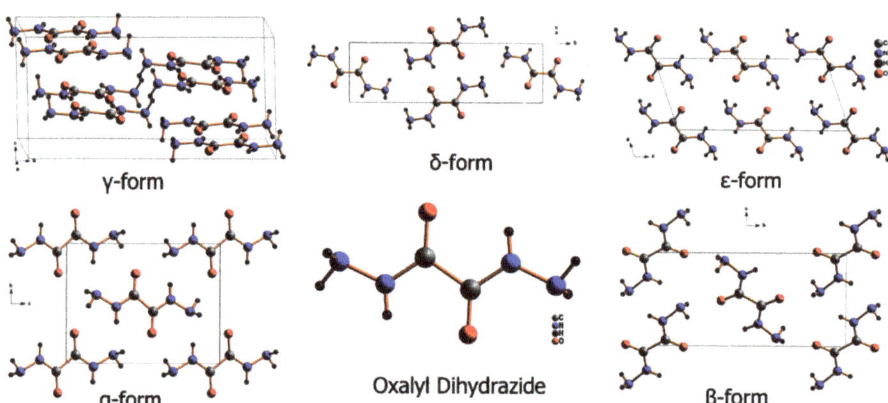

Figure 6. Crystal structures of ODH polymorphic forms: α, β, δ, γ, and ε forms. Reprinted with permission from Ref. [82]. Copyright 2015 American Chemical Society.

4. High-Pressure Polymorphism in Carboxylic Acid Derivatives

Because the molecule has distinct functional groups, as do many pharmacological compounds, *p*-aminobenzoic acid ($C_7H_7NO_2$, PABA) is a practicable substance for the model material. The chemical is largely utilized in the production of medicines. Perfumes, colors, and feedstock additives are some of the other applications. PABA polymorphs have sparked much curiosity among scientists. Two polymorphs are known to exist: the α- and β-forms. The majority of the reported research on this drug has been focused on crystallography. Jarchow and Banerjee use NMR to confirm the phase transition of α-form crystals at 32 °C [83,84]. When crystallized from a solvent, the transition temperature between the two forms is approximately 25 °C; β-PABA is thermodynamically stable below this temperature [85]. Yang et al. discover that when heated to 96 °C, the β-form can be transformed into the α-form [86].

According to published sources, the two polymorphs are synthesized: the α-polymorph, which is commercially accessible and resembles long, fibrous needles; and the β-polymorph, which resembles prisms. The function of high pressure on the two forms in a DAC is investigated via in situ Raman spectroscopy [87]. Experiments demonstrate that both forms keep their original structures up to a pressure of 13 GPa. For determining the variations in intermolecular interactions, the Hirshfeld surface and fingerprint plot have been applied. Upon a thorough comparison of the small structural alterations and anisotropic properties, we find that the particular dimer link is a factor in maintaining the stability of α-form crystals, while the presence of hydrogen-bonded networks with a four-membered ring structure is a factor in maintaining the stability of the β-form. For α-PABA, the molecular pairs are connected via two O–H ··· O bonds; a hydrogen-bonded bridge is formed between the two molecules. This bridge resists the effect of pressure along the direction of hydrogen bonds. This visualization is analogous to the behaviors of the α- and δ-forms in pyrazinamide polymorphs. In the tetramer structure of β-PABA, the special four-membered hydrogen-bonded networks can easily twist to release the increased intermolecular interactions, as well as maintain the balance of hydrogen bonding and van der Waals interactions; the structural stability is maintained. The crystal structures of α-PABA and β-PABA under ambient conditions are shown in Figure 7.

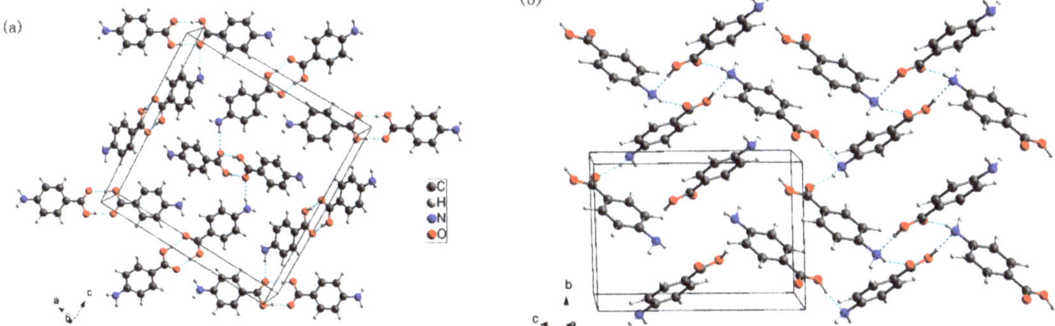

Figure 7. Crystal structures of (**a**) α-PABA and (**b**) β-PABA under ambient conditions; the hydrogen bonds are marked as dashed lines. Reprinted with permission from Ref. [87]. Copyright 2014 Royal Society of Chemistry.

Chemists and biologists have always been interested in cinchomeronic acid (CA; pyridine-3,4-dicarboxylic acid; $C_7H_5NO_4$) due to its unique structure and characteristics [88–92]. CA is commonly employed in the building of coordination networks due to the flexibility of its metal coordination modes [93,94]. Since 1971, the PDF-2 has been reported to include two CA polymorphs. Form-I and form-II can be prepared concomitantly from the recrystallization of form-II in ethanol/water solution at ambient conditions. A slurry conversion experiment converts form-II to form-I, which will disintegrate before melting, with the form-I melting at 263 °C and form-II melting at 259 °C.

The compression behavior of the two forms is investigated using DACs in conjunction with Raman spectroscopy and ADXRD [95,96]. Once a phase change is produced as the form-I is compressed to about 6.5 GPa, the new polymorph form-III is produced. The ADXRD pattern indicates that this polymorphic transition is partially reversible when the sample is brought back to ambient pressure, with a portion of the sample reverting to its original form-I structure (Figure 8a). The Raman spectra's lattices and internal modes are evaluated to determine the alterations to the CA form-I molecules' local environment. A low-symmetry triclinic structure with space group $P1$ is shaped by the indexing and refinement of form-III. In CA form-I, the phase transition may be triggered due to the reconstruction of the hydrogen-bonded networks.

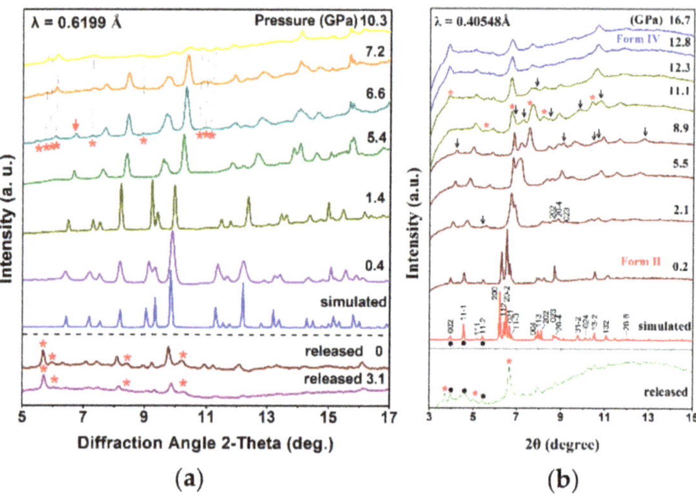

Figure 8. (a) Representative high-pressure ADXRD patterns of CA form-I. Red asterisks indicate new peaks. Down arrow indicates the disappearing diffraction peak. Reprinted with permission from Ref. [95]. Copyright 2019 Institute of Physics (b) Representative high-pressure ADXRD patterns of CA form-II. Red asterisks represent the peaks of form IV. Black dots show the peaks of form II. Down arrows show the disappearing diffraction peaks of form II. Reprinted with permission from Ref. [96]. Copyright 2021 American Chemical Society.

At ~11–13 GPa, a polymorphic transition is visible in the Raman spectra and ADXRD patterns of form-II. From form-II, a new CA polymorph named form-IV is formed. Form-IV could be a monoclinic structure with the space group $P2_1/c$, according to ab initio calculations. In the released ADXRD pattern, as shown in Figure 8b, form-IV peaks coexist with form-II peaks. The reconstruction of hydrogen-bonded networks is the most likely cause of the polymorphic transformation in both experimental and theoretical data.

Phenyl carbamate ($C_7H_7NO_2$, PC) is a widely used pharmaceutical intermediate with a wide range of pharmacological characteristics and uses. Three pure PC polymorphs have been discovered via traditional crystallization methods: form-I, form-II, and form-III. Methanol and acetonitrile are used to crystallize form-I, while ethyl acetate is used to crystallize form-II. At 25 °C, solution-mediated phase transition and solid-state phase transition produce form-I. The computed phenotype-II is 4.6% denser than phenotype-I, according to the Burgers density rule. Changes in HSM, PXRD, and DSC spectra are used to track the development of form-III.

PC form-I has been used to investigate the pressure-induced polymorphic transition and disorder in the amorphous state in polymorphic molecular systems [97]. Under high pressure, no polymorphic transition from crystalline PC-I to type-II crystals can be observed. At a pressure of 12.7 GPa, the evolution of the ADXRD and Raman spectra suggest that a reversible pressure-induced amorphization (PIA) occurs in PC type-I. The ADXRD patterns and diffraction patterns of PC form-I under different pressures are presented in Figure 9. The modifications of PC-I hydrogen bond networks and molecular configuration under pressure are computed by ab initio approach. Hirshfeld surfaces and fingerprint plots are utilized to rapidly compare changes in packing patterns and intermolecular interactions. Based on experimental and theoretical data, we predict that the PIA of crystalline PC form-I is driven by competition between close packing and long-range ordering. This competition results in the collapse of the hydrogen bonds.

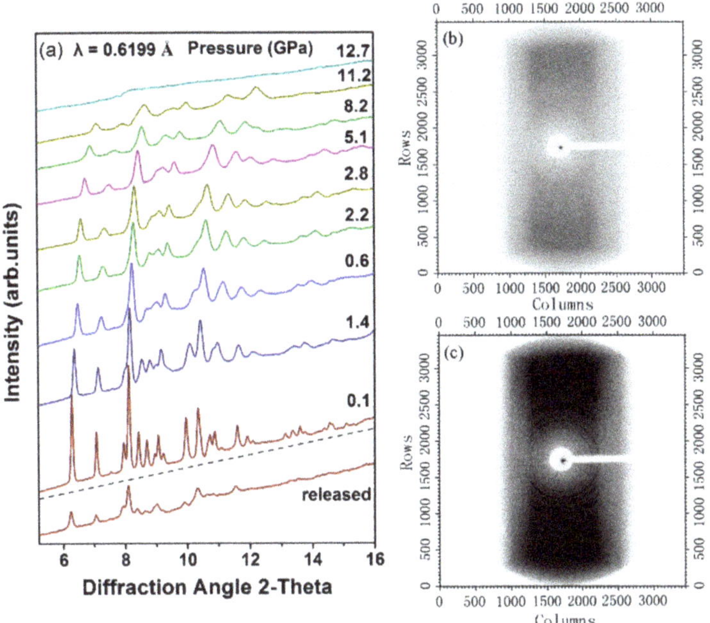

Figure 9. Synchrotron XRD patterns of PC form-I under different pressures (**a**). Diffraction images of PC form-I compressed under 12.7 GPa (**b**) and decompressed under ambient pressure (**c**). Reprinted with permission from Ref. [97]. Copyright 2017 American Chemical Society.

5. Summary

In this review, we have aimed to summarize some of the studies of our group on high-pressure polymorphism in hydrogen-bonded crystals. These crystals are divided into three categories, namely, amides, hydrazides, and carboxylic acid derivatives. These studies have shown that pressure can induce amorphization, the formation of new polymorphs, and phase transitions between different polymorphs. Experimental and computational results have revealed that pressure-induced structural distortions and structural stability can be correlated with the hydrogen bonding.

High-pressure polymorphism is a growing and rapidly expanding scientific topic with a promising future. Hydrogen-bonded crystal research will continue to evolve and improve as new discoveries in powder diffraction, high-pressure crystal formation from solution, and theoretical techniques are established (e.g., density functional theory and crystal structure prediction). Although several fundamental issues remain unsolved, significant progress is being made in several laboratories. However, more research still needs to be done by professionals using high pressures.

Author Contributions: Conceptualization, X.W.; investigation, J.W.; Writing—original draft preparation, Q.F. and Y.Z.; writing—review and editing, T.Y. and D.X. All authors have read and agreed to the published version of the manuscript.

Funding: This review was funded by the National Natural Science Foundation of China, grant numbers 11604224 and 51805336, and Liaoning Provincial Department of Education, grant number LJKZ0596.

Institutional Review Board Statement: Not applicable.

Informed Consent Statement: Not applicable.

Data Availability Statement: Not applicable.

Conflicts of Interest: The authors declare no conflict of interest.

References

1. Cruz-Cabeza, A.J.; Bernstein, J. Conformational Polymorphism. *Chem. Rev.* **2014**, *114*, 2170–2191. [CrossRef] [PubMed]
2. Johnstone, R.D.; Lennie, A.R.; Parker, S.F.; Parsons, S.; Pidcock, E.; Richardson, P.R.; Warren, J.E.; Wood, P.A. High-Pressure Polymorphism in Salicylamide. *CrystEngComm* **2010**, *12*, 1065–1078. [CrossRef]
3. Bond, A.D. Polymorphism in Molecular Crystals. *Curr. Opin. Solid State Mater. Sci.* **2009**, *13*, 91–97. [CrossRef]
4. Price, C.P.; Grzesiak, A.L.; Matzger, A.J. Crystalline Polymorph Selection and Discovery with Polymer Heteronuclei. *J. Am. Chem. Soc.* **2005**, *127*, 5512–5517. [CrossRef] [PubMed]
5. David, W.I.F.; Shankland, K.; Pulham, C.R.; Blagden, N.; Davey, R.J.; Song, M. Polymorphism in Benzamide. *Angew. Chem.* **2005**, *117*, 7194–7197. [CrossRef]
6. Gavezzotti, A.; Filippini, G. Polymorphic Forms of Organic Crystals at Room Conditions: Thermodynamic and Structural Implications. *J. Am. Chem. Soc.* **1995**, *117*, 12299–12305. [CrossRef]
7. Nangia, A. Conformational Polymorphism in Organic Crystals. *Acc. Chem. Res.* **2008**, *41*, 595–604. [CrossRef]
8. Haleblian, J.; McCrone, W. Pharmaceutical Applications of Polymorphism. *J. Pharm. Sci.* **1969**, *58*, 911–929. [CrossRef]
9. Feng, Y.; Hao, H.; Chen, Y.; Wang, N.; Wang, T.; Huang, X. Enhancement of Crystallization Process of the Organic Pharmaceutical Molecules through High Pressure. *Crystals* **2022**, *12*, 432. [CrossRef]
10. Ali, I.; Tang, J.; Han, Y.; Wei, Z.; Zhang, Y.; Li, J. A Solid-Solid Phase Transformation of Triclabendazole at High Pressures. *Crystals* **2022**, *12*, 300. [CrossRef]
11. Porter, W.W., III; Elie, S.C.; Matzger, A.J. Polymorphism in Carbamazepine Cocrystals. *Cryst. Growth Des.* **2008**, *8*, 14–16. [CrossRef] [PubMed]
12. Nauha, E.; Saxell, H.; Nissinen, M.; Kolehmainen, E.; Schäfer, A.; Schlecker, R. Polymorphism and Versatile Solvate Formation of Thiophanate-Methyl. *CrystEngComm* **2009**, *11*, 2536–2547. [CrossRef]
13. Bond, A.D.; Boese, R.; Desiraju, G.R. On the Polymorphism of Aspirin. *Angew. Chem. Int. Ed.* **2007**, *46*, 615–617. [CrossRef] [PubMed]
14. Nath, N.K.; Kumar, S.S.; Nangia, A. Neutral and Zwitterionic Polymorphs of 2-(p-tolylamino) Nicotinic Acid. *Cryst. Growth Des.* **2011**, *11*, 4594–4605. [CrossRef]
15. Yu, L. Polymorphism in Molecular Solids: An Extraordinary System of Red, Orange, and Yellow Crystals. *Acc. Chem. Res.* **2010**, *43*, 1257–1266. [CrossRef]
16. Wilding, M.C.; Wilson, M.; McMillan, P.F. Structural Studies and Polymorphism in Amorphous Solids and Liquids at High Pressure. *Chem. Soc. Rev.* **2006**, *35*, 964–986. [CrossRef]
17. Carvalho, P.H.B.; Mace, A.; Nangoi, I.M.; Leitão, A.A.; Tulk, C.A.; Molaison, J.J.; Andersson, O.; Lyubartsev, A.P.; Häussermann, U. Exploring High-Pressure Transformations in Low-Z (H_2, Ne) Hydrates at Low Temperatures. *Crystals* **2021**, *12*, 9. [CrossRef]
18. Daniel, I.; Oger, P.; Winter, R. Origins of Life and Biochemistry under High-Pressure Conditions. *Chem. Soc. Rev.* **2006**, *35*, 858–875. [CrossRef]
19. Chen, S.; Guzei, I.A.; Yu, L. New Ppolymorphs of ROY and New Record for Coexisting Polymorphs of Solved Structures. *J. Am. Chem. Soc.* **2005**, *127*, 9881–9885. [CrossRef]
20. Vasileiadis, M.; Pantelides, C.C.; Adjiman, C.S. Prediction of the Crystal Structures of Axitinib, a Polymorphic Pharmaceutical Molecule. *Chem. Eng. Sci.* **2015**, *121*, 60–76. [CrossRef]
21. Fabbiani, F.P.; Allan, D.R.; David, W.I.; Moggach, S.A.; Parsons, S.; Pulham, C.R. High-Pressure Recrystallisation—A Route to New Polymorphs and Solvates. *CrystEngComm* **2004**, *6*, 504–511. [CrossRef]
22. Fabbiani, F.P.A.; Pulham, C.R. High-Pressure Studies of Pharmaceutical Compounds and Energetic Materials. *Chem. Soc. Rev.* **2006**, *35*, 932–942. [CrossRef] [PubMed]
23. Laniel, D.; Downie, L.E.; Smith, J.S.; Savard, D.; Murugesu, M.; Desgreniers, S. High Pressure Study of a Highly Energetic Nitrogen-Rich Carbon Nitride, Cyanuric Triazide. *J. Chem. Phys.* **2014**, *141*, 234506. [CrossRef] [PubMed]
24. Fabbiani, F.P.; Allan, D.R.; Dawson, A.; David, W.I.; McGregor, P.A.; Oswald, I.D.; Parsons, S.; Pulham, C.R. Pressure-Induced Formation of a Solvate of Paracetamol. *Chem. Commun.* **2003**, *24*, 3004–3005. [CrossRef] [PubMed]
25. Fabbiani, F.P.; Allan, D.R.; Marshall, W.G.; Parsons, S.; Pulham, C.R.; Smith, R.I. High-Pressure Recrystallisation—A Route to New Polymorphs and Solvates of Acetamide and Parabanic acid. *J. Cryst. Growth* **2005**, *275*, 185–192. [CrossRef]
26. Fabbiani, F.P.; Buth, G.; Dittrich, B.; Sowa, H. Pressure-Induced Structural Changes in Wet Vitamin B12. *CrystEngComm* **2010**, *12*, 2541–2550. [CrossRef]
27. Fabbiani, F.P.; Levendis, D.C.; Buth, G.; Kuhs, W.F.; Shankland, N.; Sowa, H. Searching for Novel Crystal Forms by in situ High-Pressure Crystallisation: The Example of Gabapentin Heptahydrate. *CrystEngComm* **2010**, *12*, 2354–2360. [CrossRef]

28. Fabbiani, F.P.; Allan, D.R.; Parsons, S.; Pulham, C.R. Exploration of the High-Pressure Behaviour of Polycyclic Aromatic Hydrocarbons: Naphthalene, Phenanthrene and Pyrene. *Acta Crystallogr. Sect. B Struct. Sci.* **2006**, *62*, 826–842. [CrossRef]

29. Oswald, I.D.; Chataigner, I.; Elphick, S.; Fabbiani, F.P.; Lennie, A.R.; Maddaluno, J.; Marshall, W.G.; Prior, T.J.; Pulham, C.R.; Smith, R.I. Putting Pressure on Elusive Polymorphs and Solvates. *CrystEngComm* **2009**, *11*, 359–366. [CrossRef]

30. Neumann, M.; Van De Streek, J.; Fabbiani, F.; Hidber, P.; Grassmann, O. Combined Crystal Structure Prediction and High-Pressure Crystallization in Rational Pharmaceutical Polymorph Screening. *Nat. Commun.* **2015**, *6*, 7793. [CrossRef]

31. Fabbiani, F.P.; Pulham, C.R.; Warren, J.E. A High-Pressure Polymorph of Propionamide from in situ High-Pressure Crystallisation from Solution. *Z. Für Krist. Cryst. Mater.* **2014**, *229*, 667–675. [CrossRef]

32. Fabbiani, F.P.; Dittrich, B.; Florence, A.J.; Gelbrich, T.; Hursthouse, M.B.; Kuhs, W.F.; Shankland, N.; Sowa, H. Crystal Structures with a Challenge: High-Pressure Crystallisation of Ciprofloxacin Sodium Salts and Their Recovery to Ambient Pressure. *CrystEngComm* **2009**, *11*, 1396–1406. [CrossRef]

33. Minkov, V.S.; Goryainov, S.V.; Boldyreva, E.V.; Görbitz, C.H. Raman Study of Pressure-Induced Phase Transitions in Crystals of Orthorhombic and Monoclinic Polymorphs of L-Cysteine: Dynamics of the Side Chain. *J. Raman Spectrosc.* **2010**, *41*, 1748–1758. [CrossRef]

34. Boldyreva, E.; Ivashevskaya, S.; Sowa, H.; Ahsbahs, H.; Weber, H.-P. Effect of High Pressure on Crystalline Glycine: A New High-Pressure Polymorph. *Dokl. Phys. Chem.* **2004**, *396*, 111–114. [CrossRef]

35. Boldyreva, E.V.; Ivashevskaya, S.N.; Sowa, H.; Ahsbahs, H.; Weber, H.-P. Effect of Hydrostatic Pressure on the γ-Polymorph of Glycine.1. A polymorphic transition into a New δ-Form. *Z. Für Krist. Cryst. Mater.* **2005**, *220*, 50–57. [CrossRef]

36. Dawson, A.; Allan, D.R.; Belmonte, S.A.; Clark, S.J.; David, W.I.; McGregor, P.A.; Parsons, S.; Pulham, C.R.; Sawyer, L. Effect of High Pressure on the Crystal Structures of Polymorphs of Glycine. *Cryst. Growth Des.* **2005**, *5*, 1415–1427. [CrossRef]

37. Goryainov, S.; Kolesnik, E.; Boldyreva, E. A Reversible Pressure-Induced Phase Transition in β-Glycine at 0.76 GPa. *Phys. B Condens. Matter* **2005**, *357*, 340–347. [CrossRef]

38. Fabbiani, F.P.A.; Allan, D.R.; Parsons, S.; Pulham, C.R. An Exploration of the Polymorphism of Piracetam using High Pressure. *CrystEngComm* **2005**, *7*, 179–186. [CrossRef]

39. Fabbiani, F.P.; Allan, D.R.; David, W.I.; Davidson, A.J.; Lennie, A.R.; Parsons, S.; Pulham, C.R.; Warren, J.E. High-Pressure Studies of Pharmaceuticals: An Exploration of the Behavior of Piracetam. *Cryst. Growth Des.* **2007**, *7*, 1115–1124. [CrossRef]

40. Prins, L.J.; Reinhoudt, D.N.; Timmerman, P. Noncovalent Synthesis Using Hydrogen Bonding. *Angew. Chem. Int. Ed.* **2001**, *40*, 2382–2426. [CrossRef]

41. Vippagunta, S.R.; Brittain, H.G.; Grant, D.J. Crystalline solids. *Adv. Drug Deliv. Rev.* **2001**, *48*, 3–26. [CrossRef]

42. Steiner, T. The Hydrogen Bond in the solid state. *Angew. Chem. Int. Ed.* **2002**, *41*, 48–76. [CrossRef]

43. Abe, Y.; Harata, K.; Fujiwara, M.; Ohbu, K. Molecular Arrangement and Intermolecular Hydrogen Bonding in Crystals of Methyl 6-O-Acyl-D-Glycopyranosides. *Langmuir* **1996**, *12*, 636–640. [CrossRef]

44. Boldyreva, E.V. High-Pressure Studies of the Hydrogen Bond Networks in Molecular Crystals. *J. Mol. Struct.* **2004**, *700*, 151–155. [CrossRef]

45. Boldyreva, E.V. High-pressure Studies of the Anisotropy of Structural Distortion of Molecular Crystals. *J. Mol. Struct.* **2003**, *647*, 159–179. [CrossRef]

46. Joseph, J.; Jemmis, E.D. Red-, Blue-, or No-Shift in Hydrogen Bonds: A Unified Explanation. *J. Am. Chem. Soc.* **2007**, *129*, 4620–4632. [CrossRef]

47. Yan, T.T.; Wang, K.; Tan, X.; Liu, J.; Liu, B.B.; Zou, B. Exploration of the Hydrogen-Bonded Energetic Material Carbohydrazide at High Pressures. *J. Phys. Chem. C* **2014**, *118*, 22960–22967. [CrossRef]

48. Bi, J.; Tao, Y.; Hu, J.; Wang, H.; Zhou, M. High-Pressure Investigations on Urea Hydrogen Peroxide. *Chem. Phys. Lett.* **2022**, *787*, 139230. [CrossRef]

49. Yan, T.T.; Wang, K.; Tan, X.; Yang, K.; Liu, B.B.; Zou, B. Pressure-Induced Phase Transition in N–H⋯O Hydrogen-Bonded Molecular Crystal Biurea: Combined Raman Scattering and X-ray Diffraction Study. *J. Phys. Chem. C* **2014**, *118*, 15162–15168. [CrossRef]

50. Yan, T.T.; Li, S.U.; Wang, K.; Tan, X.; Jiang, Z.M.; Yang, K.; Liu, B.B.; Zou, G.; Zou, B. Pressure-Induced Phase Transition in N–H⋯O Hydrogen-Bonded Molecular Crystal Oxamide. *J. Phys. Chem. B* **2012**, *116*, 9796–9802. [CrossRef]

51. Moggach, S.A.; Allan, D.R.; Morrison, C.A.; Parsons, S.; Sawyer, L. Effect of Pressure on the Crystal Structure of L-Serine-I and the Crystal Structure of L-Serine-Ii at 5.4 Gpa. *Acta Crystallogr. Sect. B Struct. Sci.* **2005**, *61*, 58–68. [CrossRef] [PubMed]

52. Boldyreva, E.; Shakhtshneider, T.; Ahsbahs, H.; Sowa, H.; Uchtmann, H. Effect of High Pressure on the Polymorphs of Paracetamol. *J. Therm. Anal. Calorim.* **2002**, *68*, 437–452.

53. Allan, D.; Marshall, W.; Francis, D.; Oswald, I.; Pulham, C.; Spanswick, C. The Crystal Structures of the Low-Temperature and High-Pressure Polymorphs of Nitric Acid. *Dalton Trans.* **2010**, *39*, 3736–3743. [CrossRef] [PubMed]

54. Martins, D.M.; Spanswick, C.K.; Middlemiss, D.S.; Abbas, N.; Pulham, C.R.; Morrison, C.A. A New Polymorph of N, N'-Dimethylurea Characterized by X-Ray Diffraction and First-Principles Lattice Dynamics Calculations. *J. Phys. Chem. A* **2009**, *113*, 5998–6003. [CrossRef] [PubMed]

55. Munday, L.B.; Chung, P.W.; Rice, B.M.; Solares, S.D. Simulations of High-Pressure Phases in Rdx. *J. Phys. Chem. B* **2011**, *115*, 4378–4386. [CrossRef] [PubMed]

56. Oswald, I.D.; Urquhart, A.J. Polymorphism and Polymerisation of Acrylic and Methacrylic Acid at High Pressure. *CrystEngComm* **2011**, *13*, 4503–4507. [CrossRef]
57. Valkenburg, A.V., Jr. Visual Observations of High Pressure Transitions. *Rev. Sci. Instrum.* **1962**, *33*, 1462. [CrossRef]
58. Moggach, S.A.; Parsons, S.; Wood, P.A. High-Pressure Polymorphism in Amino Acids. *Cryst. Rev.* **2008**, *14*, 143–184. [CrossRef]
59. Bassett, W.A. Diamond Anvil Cell, 50th Birthday. *High Press. Res.* **2009**, *29*, 163–186. [CrossRef]
60. Klotz, S.; Chervin, J.; Munsch, P.; Le Marchand, G. Hydrostatic Limits of 11 Pressure Transmitting Media. *J. Phys. D Appl. Phys.* **2009**, *42*, 075413. [CrossRef]
61. Mao, H.K.; Xu, J.-A.; Bell, P.M. Calibration of the Ruby Pressure Gauge to 800 Kbar under Quasi-Hydrostatic Conditions. *J. Geophys. Res. Solid Earth* **1986**, *91*, 4673–4676. [CrossRef]
62. Becker, C.; Dressman, J.; Amidon, G.; Junginger, H.; Kopp, S.; Midha, K.; Shah, V.; Stavchansky, S.; Barends, D. Biowaiver Monographs for Immediate Release Solid Oral Dosage Forms: Pyrazinamide. *J. Pharm. Sci.* **2008**, *97*, 3709–3720. [CrossRef] [PubMed]
63. Castro, R.A.; Maria, T.M.; Évora, A.O.; Feiteira, J.C.; Silva, M.R.; Beja, A.M.; Canotilho, J.; Eusébio, M.E.S. A New Insight into Pyrazinamide Polymorphic Forms and Their Thermodynamic Relationships. *Cryst. Growth Des.* **2010**, *10*, 274–282. [CrossRef]
64. Borba, A.; Albrecht, M.; Gómez-Zavaglia, A.; Suhm, M.A.; Fausto, R. Low Temperature Infrared Spectroscopy Study of Pyrazinamide: From the Isolated Monomer to the Stable Low Temperature Crystalline Phase. *J. Phys. Chem. A* **2010**, *114*, 151–161. [CrossRef]
65. Tan, X.; Wang, K.; Li, S.U.; Yuan, H.S.; Yan, T.T.; Liu, J.; Yang, K.; Liu, B.B.; Zou, G.T.; Zou, B. Exploration of the Pyrazinamide Polymorphism at High Pressure. *J. Phys. Chem. B* **2012**, *116*, 14441–14450. [CrossRef]
66. Cherukuvada, S.; Thakuria, R.; Nangia, A. Pyrazinamide Polymorphs: Relative Stability and Vibrational Spectroscopy. *Cryst. Growth Des.* **2010**, *10*, 3931–3941. [CrossRef]
67. Dreger, Z.A.; Gupta, Y.M. High Pressure Raman Spectroscopy of Single Crystals of Hexahydro-1, 3, 5-Trinitro-1, 3, 5-Triazine (Rdx). *J. Phys. Chem. B* **2007**, *111*, 3893–3903. [CrossRef]
68. Chieh, P.C.; Subramanian, E.; Trotter, J. Crystal Structure of Malonamide. *J. Chem. Soc.* **1970**, 179–184. [CrossRef]
69. Nichol, G.S.; Clegg, W. Malonamide: An Orthorhombic Polymorph. *Acta Cryst.* **2005**, *61*, o3427–o3429. [CrossRef]
70. Nichol, G.S.; Clegg, W. Malonamide: A Tetragonal Polymorph. *Acta Cryst.* **2005**, *61*, o3424–o3426. [CrossRef]
71. Yan, T.T.; Xi, D.Y.; Ma, Z.N.; Wang, X.; Wang, Q.J.; Li, Q. Pressure-Induced Phase Transition in N–H···O Hydrogen-Bonded Crystalline Malonamide. *RSC Adv.* **2017**, *7*, 22105–22111. [CrossRef]
72. Jeffrey, G.T.; Maluszynska, H. A Survey of Hydrogen Bond Geometries in the Crystal Structures of Amino Acids. *Int. J. Biol. Macromol* **1982**, *4*, 173–185. [CrossRef]
73. Watanabe, S.; Abe, Y.; Yoshizaki, R. Tunneling Rotation of Two Inequivalent Methyl Groups in Orthorhombic Acetamide. *J. Phys. Soc. Jpn.* **1986**, *55*, 2400–2409. [CrossRef]
74. Kang, L.; Wang, K.; Li, S.R.; Li, X.; Zou, B. Pressure-Induced Phase Transition in Hydrogen-Bonded Molecular Crystal Acetamide: Combined Raman Scattering and X-Ray Diffraction Study. *RSC Adv.* **2015**, *5*, 84703–84710. [CrossRef]
75. Cradwick, P.D. Crystal Structure of the Growth Inhibitor, 'Maleic Hydrazide'(1, 2-Dihydropyridazine-3, 6-Dione). *J. Chem. Soc. Perkin Trans.* **1976**, *2*, 1386–1389. [CrossRef]
76. Katrusiak, A. A New Polymorph of Maleic Hydrazide. *Acta Crystallogr. Sect. C Cryst. Struct. Commun.* **1993**, *49*, 36–39. [CrossRef]
77. Katrusiak, A. Polymorphism of Maleic Hydrazide. I. *Acta Crystallogr. Sect. B Struct. Sci.* **2001**, *57*, 697–704. [CrossRef]
78. Wang, K.; Liu, J.; Yang, K.; Liu, B.B.; Zou, B. High-Pressure-Induced Polymorphic Transformation of Maleic Hydrazide. *J. Phys. Chem. C* **2014**, *118*, 8122–8127. [CrossRef]
79. Ahn, S.; Guo, F.; Kariuki, B.M.; Harris, K.D. Abundant Polymorphism in a System with Multiple Hydrogen-Bonding Opportunities: Oxalyl Dihydrazide. *J. Am. Chem. Soc.* **2006**, *128*, 8441–8452. [CrossRef]
80. Wen, S.; Beran, G.J. Crystal Polymorphism in Oxalyl Dihydrazide: Is Empirical Dft-D Accurate Enough? *J. Chem. Theory Comput.* **2012**, *8*, 2698–2705. [CrossRef]
81. Presti, D.; Pedone, A.; Menziani, M.C.; Civalleri, B.; Maschio, L. Oxalyl Dihydrazide Polymorphism: A Periodic Dispersion-Corrected Dft and Mp2 Investigation. *CrystEngComm* **2014**, *16*, 102–109. [CrossRef]
82. Tan, X.; Wang, K.; Yan, T.T.; Li, X.; Liu, J.; Yang, K.; Liu, B.B.; Zou, G.T.; Zou, B. Discovery of High-Pressure Polymorphs for a Typical Polymorphic System: Oxalyl Dihydrazide. *J. Phys. Chem. C* **2015**, *119*, 10178–10188. [CrossRef]
83. Jarchow, O.; Kühn, L. Die Kristallstruktur Von A-P-Aminobenzoesäure. *Acta Crystallogr. Sect. B Struct. Crystallogr. Cryst. Chem.* **1968**, *24*, 222–224. [CrossRef]
84. Banerjee, A.; Agrawal, P.; Gupta, R. Nmr Study of Solid P-Amino Benzoic Acid–Structure and Group Rotation. *J. Prakt. Chem.* **1973**, *315*, 251–257. [CrossRef]
85. Gracin, S.; Rasmuson, Å.C. Polymorphism and Crystallization of *P*-Aminobenzoic Acid. *Cryst. Growth Des.* **2004**, *4*, 1013–1023. [CrossRef]
86. Yang, X.; Wang, X.; Ching, C.B. In Situ Monitoring of Solid-State Transition of P-Aminobenzoic Acid Polymorphs Using Raman Spectroscopy. *J. Raman Spectrosc.* **2009**, *40*, 870–875. [CrossRef]
87. Yan, T.T.; Wang, K.; Duan, D.F.; Tan, X.; Liu, B.B.; Zou, B. *P*-Aminobenzoic Acid Polymorphs under High Pressures. *RSC Adv.* **2014**, *4*, 15534–15541. [CrossRef]
88. Griffiths, P. Crystallographic Data for Cinchomeronic Acid and Its Hydrochloride. *Acta Cryst.* **1963**, *16*, 1074. [CrossRef]

89. Takusagawa, F.; Hirotsu, K.; Shimada, A. The Crystal Structure of Cinchomeronic Acid. *Bull. Chem. Soc. Jpn.* **1973**, *46*, 2669–2675. [CrossRef]

90. Braga, D.; Maini, L.; Fagnano, C.; Taddei, P.; Chierotti, M.R.; Gobetto, R. Polymorphism in Crystalline Cinchomeronic Acid. *Chem. Eur. J.* **2007**, *13*, 1222–1230. [CrossRef]

91. Karabacak, M.; Bilgili, S.; Atac, A. Molecular Structure Investigation of Neutral, Dimer and Anion Forms of 3, 4-Pyridinedicarboxylic Acid: A Combined Experimental and Theoretical Study. *Spectrochim. Acta Part A Mol. Biomol. Spectrosc.* **2015**, *135*, 270–282. [CrossRef] [PubMed]

92. Evans, I.R.; Howard, J.A.; Evans, J.S.; Postlethwaite, S.R.; Johnson, M.R. Polymorphism and Hydrogen Bonding in Cinchomeronic Acid: A Variable Temperature Experimental and Computational Study. *CrystEngComm* **2008**, *10*, 1404–1409. [CrossRef]

93. Tong, M.-L.; Wang, J.; Hu, S.; Batten, S.R. A New (3, 4)-Connected Three-Dimensional Anionic Porous Coordination Net Templated by Me4n+ Cations. *Inorg. Chem. Commun.* **2005**, *8*, 48–51. [CrossRef]

94. Senevirathna, M.; Pitigala, P.; Perera, V.; Tennakone, K. Molecular Rectification: Application in Dye-Sensitized Solar Cells. *Langmuir* **2005**, *21*, 2997–3001. [CrossRef]

95. Yan, T.T.; Xi, D.Y.; Wang, J.H.; Fan, X.F.; Wan, Y.; Zhang, L.X.; Wang, K. High-Pressure-Induced Phase Transition in Cinchomeronic Acid Polycrystalline Form-I. *Chin. Phys. B* **2019**, *28*, 016104. [CrossRef]

96. Yan, T.T.; Deng, Y.Y.; Yu, Z.Q.; John, E.; Han, R.M.; Yao, Y.; Liu, Y. Exploring the Polymorphism of Cinchomeronic Acid at High Pressure. *J. Phys. Chem. C* **2021**, *125*, 8582–8588. [CrossRef]

97. Yan, T.T.; Xi, D.Y.; Ma, Z.N.; Fan, X.F.; Li, Y. Pressure-Induced Reversible Amorphization in Hydrogen-Bonded Crystalline Phenyl Carbamate Form-I. *J. Phys. Chem. C* **2017**, *121*, 19365–19372. [CrossRef]

Review

Polytypes of sp^2-Bonded Boron Nitride

Bernard Gil [1,*], Wilfried Desrat [1], Adrien Rousseau [1], Christine Elias [1], Pierre Valvin [1], Matthieu Moret [1], Jiahan Li [2], Eli Janzen [2], James Howard Edgar [2] and Guillaume Cassabois [1]

[1] Laboratoire Charles Coulomb, UMR 5221 CNRS-Université de Montpellier, F-34095 Montpellier, France; wilfried.desrat@umontpellier.fr (W.D.); adrien.rousseau@umontpellier.fr (A.R.); christine.elias@ens-paris-saclay.fr (C.E.); pierre.valvin@umontpellier.fr (P.V.); matthieu.moret@umontpellier.fr (M.M.); guillaume.cassabois@umontpellier.fr (G.C.)

[2] Tim Taylor Department of Chemical Engineering, Kansas State University, Manhattan, KS 66506, USA; jiahanli@ksu.edu (J.L.); elijanzen@ksu.edu (E.J.); edgarjh@ksu.edu (J.H.E.)

[*] Correspondence: bernard.gil@umontpellier.fr

Abstract: The sp^2-bonded layered compound boron nitride (BN) exists in more than a handful of different polytypes (i.e., different layer stacking sequences) with similar formation energies, which makes obtaining a pure monotype of single crystals extremely tricky. The co-existence of polytypes in a similar crystal leads to the formation of many interfaces and structural defects having a deleterious influence on the internal quantum efficiency of the light emission and on charge carrier mobility. However, despite this, lasing operation was reported at 215 nm, which has shifted interest in sp^2-bonded BN from basic science laboratories to optoelectronic and electrical device applications. Here, we describe some of the known physical properties of a variety of BN polytypes and their performances for deep ultraviolet emission in the specific case of second harmonic generation of light.

Keywords: boron nitride; polytypism; deep ultraviolet emission

Citation: Gil, B.; Desrat, W.; Rousseau A.; Elias C.; Valvin, P.; Moret, M.; Li, J.; Janzen, E.; Edgar, J. H.; Cassabois G. Polytypes of sp^2-Bonded Boron Nitride. *Crystals* **2022**, *12*, 782. https://doi.org/10.3390/cryst12060782

Academic Editors: Jingxiang Yang and Xin Huang

Received: 13 May 2022
Accepted: 25 May 2022
Published: 28 May 2022

Publisher's Note: MDPI stays neutral with regard to jurisdictional claims in published maps and institutional affiliations.

1. Introduction

Developing electronics capable of operating at high frequency and high power, at high temperature, and in harsh environments and optoelectronics with absorption/emission with wavelengths shorter than 400 nm requires the global control of the different branches of the technology. To date, research has focused on some group IV semiconductors such as diamond or silicon carbide and the group III element nitrides. The now fairly mature technology of nitride semiconductors extensively uses heterostructures of (Al,Ga,In)N solid solutions with atoms stacked according to the C_{6V} hexagonal (wurtzite) symmetry [1]. After some attempts, the prospect for producing commercial devices based on the cubic (zincblende) polymorphs of these alloys has been almost abandoned, because it is difficult to synthesize these materials with qualities as good as those with the wurtzite structure [2,3]. For deep UV range of emission/absorption, i.e., light with wavelengths near 200 nm, aluminum nitride and boron nitride (BN) are, among the III–V compounds, both candidates of choice [4]. Concerning the family of II–VI compounds, magnesium-rich solid solutions of (Zn,Mg)O will be probably used. Each of these three candidate semiconductor systems share similar severe difficulties, such as the control of their electronic properties by impurity doping. There are also more specific ones: a wurtzite to rock-salt phase transition versus Mg composition for the II–VI compounds alluded to earlier [4,5], or polytypism for boron nitride. All these variations in the crystal structure impact charge carrier mobilities and diffusion lengths and the interaction of electromagnetic fields with the electronic states (optical internal quantum efficiency). In this article, we focus on the polytypism in sp^2-bonded boron nitride, a layered two-dimensional compound similar to graphite, mica, MoS$_2$, or InSe. The different polytypes are associated with different space groups and different physical properties such as the occurrence or not of piezoelectricity, ferroelectricity, spontaneous polarization and second harmonic generation. The choice of this material is linked to its technologically demonstrated interest and to its optical properties: stimulated light emission at 215 nm was demonstrated in single crystals by cathodoluminescence as early as 2004 [6], and an electrically-driven field emission device was later demonstrated [7]. Boron nitride is also a developing material for quantum technologies, specifically single photon sources operating in the UV from topological defects of the crystal. Such quantum emitters have not been reported in classical nitrides, although it has been already demonstrated in hBN with operation at 4 eV [8]. It was recently reported by a bottom–up growth technique at the UCB (University of California, Berkeley, CA, USA) that slightly twisted two-layered BN crystals with reduced symmetry could dramatically enhance the intensity of this emission [9]. A various series of applications have also been proposed regarding nanophotonics with hBN, which are partly reviewed in [10].

There exists no theory about the origin of polytypism in stackings of arrays of six-ring made alternating boron and nitrogen atoms. A case by case analysis is required for the interpretation of the origin of this phenomenon in a given compound [11]. In SiC, ZnS, CdI$_2$, or micas, this has been already understood, based on a sophisticated application of advanced thermodynamic models and dislocations-assisted growth mechanisms [12–14]. These studies demonstrate that there does not exist a universal model to describing the origins of polytypism and to quantitatively anticipate its consequences on the physical properties of BN crystals out of campaigns of experimental and theoretical investigations.

2. The Many Shapes of BN: Polymorphism and Polytypism

Boron nitride was synthesized as early as 1842 [15], and it appeared as a white powder made of small flakes revealed by optical microscopy. Advantage was taken of its high melting temperature (above 2900 °C) for hot-pressing this powder in order to design specific forms (crucibles for the growth of crystals for instance) for industrial operations in high-temperature conditions. Its powders could be used as a dry solid lubricant, analogous to black graphite or MoS$_2$. The efficient interaction of neutrons with the nucleus of the ^{10}B isotope, which is present with ^{11}B with respective proportions of 20/80 in natural boron, was also evidenced [16–18]. Adding it to concrete can provide some protection to neutron irradiation. These physical properties paved the way for interesting industrial applications, including cosmetics, a couple of decades before the birth of solid-state electronics. The crystalline structure of such BN powders leads to controversial suggestions [19–22] until R.S. Pease [23] clearly proposed (for the hBN powders, he was studying the X-ray diffraction features), an sp^2-type of chemical bonding. The easy gliding of the (001) crystallographic planes under shear stress is prototypical of layer compounds including hBN, graphite and MoS$_2$. These and many other layered compounds typically have strong chemical bonds in the plane of the layers and weak van der Waals bonds orthogonal to the planes. Thus, van der Waals interactions rule the stability of the stacking in the $\langle 001 \rangle$ direction. Recently, Vuong et al. [24,25] extracted the electronic distribution around nitrogen and boron atoms by studying the X-ray diffraction features over the full reciprocal space of high-quality hBN single crystals. The electron density is significantly higher around nitrogen atoms than boron atoms as theoretically predicted [26], with in addition, some slight differences around ^{10}B or ^{11}B isotopes in case of isotopically purified hBN [25]. The six-ring stacking that is derived from these studies is a perfectly (at the scale of the sensitivity of the x-ray experiments) ordered stacking of a honeycomb two-dimensional lattice of six-ring layers, with perfectly aligned lines of alternating boron and nitrogen atoms from one plane to another, along the $\langle 001 \rangle$ direction [23]. Such three-dimensional arrangement of the boron and nitrogen atoms follows the D$_{6h}^4$ (also noted $P6_3/mmc$ or N°194) space group symmetry, and it is sketched in along the $\langle 001 \rangle$ direction of the hexagonal lattice in Figure 1a. Hexagonal boron nitride is the most common BN polytype, but it is not the only one: far from it [27–32].

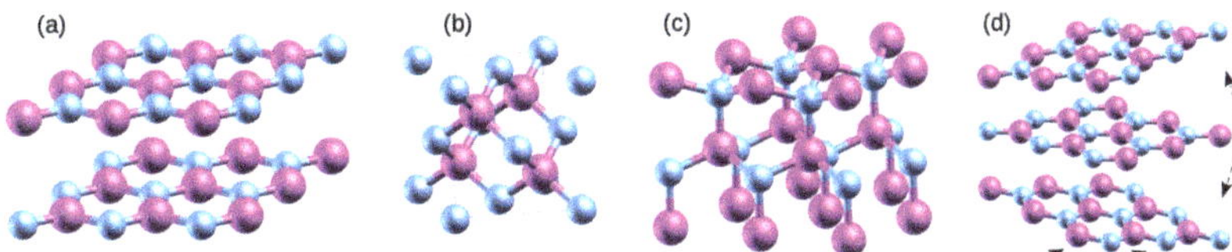

Figure 1. BN polymorphs. The stacking of boron (magenta spheres) and nitrogen (blue spheres) atoms for (**a**) the sp^2 bonded hBN, and sp^3 bonded (**b**) zinc blend, and (**c**) wurtzite polymorphs and (**d**) the turbostratic disordered stacking. The vertical axis is the $\langle 001 \rangle$ crystallographic direction for all the structures.

2.1. Polymorphisms

Allotropy (in case of a single element) and polymorphism (for a pure compound), which are frequent phenomena, are two very much documented experimental and theoretical branches of materials science. In 1788, Martin Heinrich Klaproth has revealed that rhombohedral calcite and orthorhombic aragonite were astonishingly two polymorphs of calcium carbonate $CaCO_3$. Two centuries before fullerenes and carbon nanotubes were identified, Smithson Tennant and William Hyde Wollaston demonstrated in 1796 that diamond was an allotrope of carbon with cubic symmetry. These demonstrations were very exciting for contemporary scientists, and they are the origins of the science of polymorphism and polytypism in BN crystals discussed here. It took about one century after Balmain grew sp^2-BN for achieving the growth of boron nitride crystals with four-fold coordinated atoms (sp^3 hybridizations of the atomic states) with cubic (zinc blende) symmetry. It required a pressure-induced (at about 4.5 GPa) polymorphic phase transition at 1500 °C of sp^2-bonded boron nitride powders [29,33,34]. The atomic stacking of boron and nitrogen atoms follows the T$_d^2$ (also noted $F\bar{4}3m$ or N° 216) space group symmetry [35], which is sketched in along the $\langle 001 \rangle$ direction of the cubic lattice in Figure 1b. Studies were launched to establish the complete phase diagram of boron nitride [36–39], in the exciting high-temperature (and eventually pressures) ranges required to induce the phase transitions between the different polymorphs. The cubic polymorph cBN exhibits a super high hardness (second only to diamond), thus motivating its main use in abrasive applications [40]. With its large bandgap and high thermal conductivity, cBN is a potential challenger of other wide bandgap semiconductors such as SiC and GaN, for electronic operations under extreme conditions [41]. Under high-pressure and high-temperature conditions, a further polymorphic phase transition toward wurtzite BN can be also achieved [42–46]. The number of atoms per shells of successive neighboring (first, second, …) can be very different for the polymorphic structures of a given compound: for instance, four in the case of the zincblende polytype, to six for the rock-salt, and eight for CsCl. The atomic stacking of boron and nitrogen atoms which follows the C$_{6v}^4$ (also noted $P6_3mc$ or N° 186) space group symmetry is sketched in along the $\langle 001 \rangle$ direction of the hexagonal lattice in Figure 1c. More recently, BN nanotubes were also grown [47,48].

2.2. Polytypisms

In addition to the structure variation with sp^3-bonded cBN and wBN, sp^2-bonded BN polytypes are also possible. Polytypism is a crystallographic property found with many mineral and organic crystals with compact or layered structures. Crystallographers were the first to investigate the occurrence of a large number of structural varieties in specific compounds. The best example of a material with many polytypes is silicon carbide. Its polytypism was discovered early [49,50], and this triggered pioneering theoretical studies to elucidate its origin, some based on thermodynamics [51] and others invoking specific growth mechanisms such as the defect-assisted spiral growth [52]. Chemists, physicists and engineers became interested in this, as it offers many fundamental issues to disentangle. Some physical properties such as: hardness and density remain almost unchanged between SiC polytypes. Other properties are significantly different. Piezo-electricity vanishes with the occurrence of inversion symmetry, pyroelectricity can exist or not, charge carrier mobilities and optical properties are all sensitive to the polytype. The ability to control the polytype at will would enable the tuning of these properties. If the elementary cells of these atomic stackings share two lattice parameters, whilst the third value changes from one polytype to another one, the polytypism is called monodimensional polymorphism. There exists some cases where the value of this third parameter is not changed; in that case, polytypes are distinguished by different structures of the crystalline pattern. The structures of the different polytypes of a given compound can be described using different Bravais lattices and can belong to different space groups. The structural patterns of the polytypes, whether they are compact three-dimensional compounds (such as SiC, ZnS) or layered compounds (such as mica, BN, graphite, Transition Metal Dichalcogenides, etc.) can always be described in terms of different stackings under the vertical direction of an invariant generic stacking. In the case of some compounds, there also exists a fully disordered stacking structure for which the periodicity vanishes along the normal to the layers [53]. There are many possible intermediates between aperiodic and periodic stackings. The most interesting are these based on periodic stackings flanked by a variable amount of disordered stacking faults. This is known as the turbostratic stacking, tBN, which is sketched in Figure 1d [54]. Among the polytypes, the macroscopic shapes of the crystals do not change in contrast to the case of polymorphs.

Figure 2 contains plots of the different ordered atomic stackings described in this review. In the AA stacking of Figure 2a, the six rings of alternating B and N atoms are perfectly stacked along the c direction of the crystal, a given atom of a plane facing similar ones in the planes above and below. In the AA′ stacking (also called hBN) in Figure 2b, the alternating planes experience a 60° twist from one to another, giving along the c axis, atomic lines made of alternating B and N atoms. Assuming a model of rigid atomic planes, we can conclude that AA stacking is unstable, due to the nitrogen–nitrogen and boron–boron repulsive electrostatic interactions. AA′ stacking is much more stable due to the attraction between boron and nitrogens in adjacent layers [11]. The staggered AB stacking displayed in Figure 2c, also called Bernal stacking, bBN, is obtained from the AA stacking by gliding the atomic plane of the upper layer to the center of the six rings of the lower layer in such a way that half of the boron and nitrogen atoms of adjacent planes face each other; the remaining others sit at the center of the interstitial voids of the honeycomb lattice. This structure is similar to the structure of graphite, and it is energetically more stable than AA [11]. There are two other staggered stackings. In one of them, called AB_1 here, the gliding of one plane to the other one is similar to that in the AB case, but it occurs in such a way that the boron atoms of successive planes are facing each other whilst nitrogen atoms sit at the center of the interstitial vacancy of the hexagonal lattice (Figure 2d). Finally, we indicate the AB_2 stacking where the nitrogen atoms of successive planes are facing each other whilst boron atoms sit at the center of the interstitial vacancy of the hexagonal lattice (Figure 2e). After this description of possible bilayer stacking sequences, we discuss the possibilities of longer periodic stacking with more than two layers. This is the case of the rhombohedral stacking plotted in Figure 2f. Here, the gliding vector is smaller than it is for the AB structure, and two slidings are required to recover the crystalline periodicity, which we call ABC or rBN. We restrict to these polytypes our journey into the many possibilities, since the highest probability of occurrence of polytypes generally occurs for the smaller periodicities.

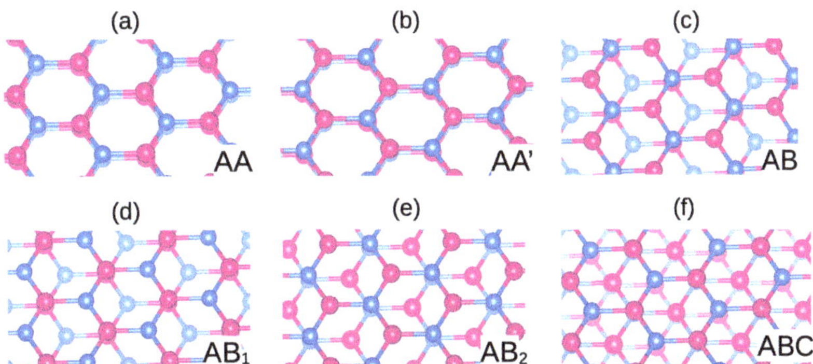

Figure 2. Top views of the sp^2 polytypes of BN with boron and nitrogen atoms indicated by magenta and blue spheres, respectively. The atomic stacking of the individual layers are: (**a**) AA, (**b**) AA' or hBN, (**c**) AB or bBN, (**d**) AB$_1$, (**e**) AB$_2$, and (**f**) ABC or rBN. The six stackings are represented as an artist view in the (001) plane slightly tilted.

2.3. Crystal Cohesive Energies and Values of the Lattice Parameters

In the previous section, we discussed the stability of the different BN stackings, founded on the rigid atom basic approach. With the aim at quantifying this stability, we have calculated the energy as a function of the volume data for nine cristallographic structures of boron nitride. The total energies of the different polymorphs have been computed with the quantum ESPRESSO code within the density functional theory, based on Perdew–Zunger local density approximation exchange–correlation potentials [55,56]. The crystallographic cells have been first relaxed before calculating the total energy as a function of the cell volume by varying the cell parameters. A k-point sampling of $10 \times 10 \times 10$ and a plane-wave cutoff of 100 Ry were used. The calculated total energies per atom are displayed as a function of the volume per atom in Figure 3. Our results are in overall agreement with the calculations of Refs. [26,57–65]. In addition, we present the result predicted for the rock-salt stacking, which matches very well to those published by Furthmüller et al. [58]. Last, many more different layered stackings are computed in ref. [61].

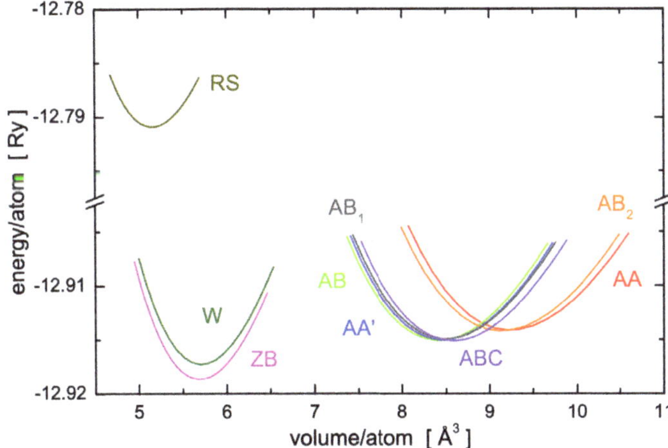

Figure 3. The energy per atom vs. volume per atom computed for nine different structures of boron nitride: rocksalt (RS), wurtzite (W), zincblende (ZB), and the six indicated polytypes.

The values of the relaxed lattice parameters are given in Table 1. In the last column, we reproduce the experimental parameters of ref. [66] that are of interest for comparing with the computed ones. Generally, the energies for AA', AB and AB_1 stackings are very close. The energies for AA and AB_2 stackings are higher. While the in-plane boron–nitrogen bond lengths do not vary drastically, that is not the case concerning the values of the c-axis parameter. The interlayer spacings are globally similar for the AA', AB, AB_1 and ABC stackings (evidently, the c parameter is 1.5 times larger for the rhombohedral stacking than for the two-layer periodic stackings), while the values of the AA and AB_2 stackings are significantly larger. Although the rigid atom model is far from exact and does not distinguish the different "sizes" of the B and N atoms as derived from quantum mechanics, it contains the spirit of the physics. The major difference between the AB_1 and AB_2 stackings relies on the larger electronic clouds of the nitrogen atoms in relation to the boron atoms. The N-N superposition in the AA and AB_2 stackings lead to strong repulsive interactions between adjacent layers and make these structures energetically unstable in relaxed conditions. The experimental values of the lattice parameters of the AA' and ABC stackings are the only ones to have been accurately determined experimentally [23–25,67–70], and their values agree with the theoretical predictions.

Table 1. Theoretical and experimental values of the lattice parameters for nine Bravais lattices of boron nitride.

Structure	Atoms facing along c (When Any)	Theory		Experiment [66]	
		a (nm)	c (nm)	a (nm)	c (nm)
Rocksalt	-	0.3451	-	-	-
Zincblende	-	0.3566	-	0.3615	-
Wurtzite	-	0.2513	0.4159	0.2551	0.4210
AA'	N-B	0.2478	0.6354	0.2504	0.6656
AB	N-B	0.2477	0.6319	-	-
AB_1	B-B	0.2476	0.6384	-	-
ABC	-	0.2476	0.9679	0.2504	0.999
AA	N-N and B-B	0.2476	0.3468	-	-
AB_2	N-N	0.2476	0.686	-	-

2.4. X-ray Diffraction Spectra for the Different Polytypes

In Figure 4, we plot the theoretical intensities of the X-ray diffraction (XRD) peaks of powders of the different ordered sp^2 stackings under discussion. In each case, given a polytype, the intensities of the XRD peaks are plotted relatively to the normalized intensity of the $(00l)$ planes, with $l = 1$ for AA, $l = 2$ for AA', AB, AB_1 and AB_2, and $l = 3$ for ABC. The diffraction angles are calculated from the values of the lattice parameters given in Table 1 for the relaxed cells, i.e., which minimize the total energy. The interesting point is that the values of the diffraction angles for the AA and AB_2 stackings are lower by approximately $2°$ than their analogs in the series of the sp^2 BN polytypes plotted in Figure 4. The vertical dashed line highlights the diffraction angle of the (002) peak in the case of the AA' stacking. We note that the angle value of the turbostratic boron nitride, in which the stacking of the layers is irregular, is shifted by about $0.5°–1°$ with respect to the (002) peak of hBN [67,71,72]. In some cases, this value may be similar to the shift observed for the AA and AB_2 stackings, which could eventually lead to a misinterpretation of the structural nature of the epilayers.

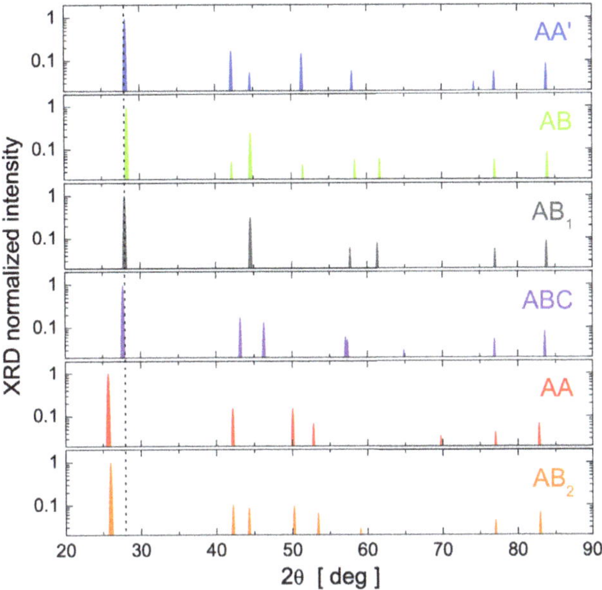

Figure 4. X-ray diffractograms computed for powders of sp^2-bonded ordered BN stackings, based on the theoretical lattice parameters. The X-ray wavelength was taken equal to 1.541842 Å(Cu-Kα line).

2.5. Symmetries and Dispersion Relations of Phonons for the Different Polytypes

The dispersion relations of phonons computed for the relaxed structures are plotted in Figure 5 along the $\Gamma - K$ path for the AA, AA' and ABC polytypes. At first sight, the three phonon band structures look similar, but significant differences exist. We have summarized in Table 2 some specificities of the phonon modes of the monolayer AA stacking (D$_{3h}$ point group), of the AA' and AB bilayer stackings (D$_{6h}$ and D$_{3h}$, respectively), and of the three-layer ABC stacking (C$_{3v}$). There are six, twelve and eighteen vibrational modes for the one, two and three-layer stackings. Thanks to the characters of the representation tables of the different groups, the modes present A-like symmetries (label A indicates vibrations of atoms that occur in the plane orthogonal to the layers), E-type symmetries (label E indicates a double degeneracy linked to vibrations in the plane of the layers), and B$_{1g}$ symmetries that correspond to silent modes in the case of the AA' stacking. Silent modes cannot be detected using an optical experiment, but they exist, and they can impact the diffusion of carriers. Of course, in all cases, there are three acoustic phonon modes. It is worthwhile noticing the lack of a low-energy mode of E type symmetry, which is either Raman active or both Raman and IR active, in the 55–60 cm^{-1} range and the existence of a Raman activity in the 800 cm^{-1} range for the ABC stacking.

Therefore, each stacking reacts specifically to incoherent light scattering, and infrared absorption experiments can (at least theoretically) permit a structural diagnosis. As usual, the linewidth of the optical features will be decisive for achieving it. Since the pioneering work of Geick et al. [73], there have been many studies dedicated to phonon properties in hBN (AA') [74–84] due to its efficient potential for thermal management [77,85–92] or nanophotonics [93–98].

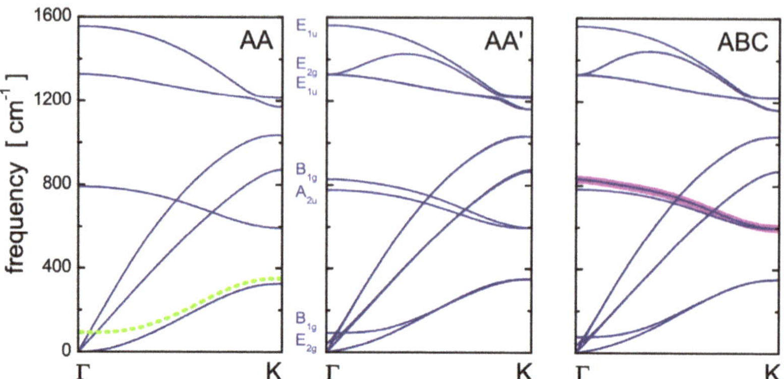

Figure 5. Dispersion of the phonons along $\Gamma - K$ for the AA (**a**), AA' (**b**) and ABC (**c**) stackings of boron nitride. The green dotted line in (**a**) shows a mode which is absent in the AA stacking. The magenta curve in (**c**) stands for the additional Raman mode in the 750–800 cm^{-1} range.

Table 2. Phonon modes specifications of four BN stackings.

Stacking	Energy Range 1370 cm^{-1}	Energy Range 750–800 cm^{-1}	Energy Range 0–120 cm^{-1}	Specific Items to Outline
AA	E' (IR + R)	A''$_2$ (IR)	A''$_2$(ac) + E'(ac)	No vibration mode for Raman activity exists in the 0–120 cm^{-1} range. Only acoustic modes.
AA'	E$_{2g}$ (R) + E$_{1u}$ (IR)	B$_{1g}$ (S) + A$_{2u}$ (IR)	A$_{2u}$(ac) + E$_{1u}$(ac) + B$_{1g}$ (S) + E$_{2g}$ (R)	There are B$_{1g}$ silent modes. Raman and IR active modes have different symmetries.
AB	2E'(IR + R)	2A''$_2$ (IR)	A''$_2$(ac) + E'(ac) + E' (IR+R) + A''$_2$ (IR)	The AA' silent modes B$_{1g}$ become IR active (A''$_2$). E$_{1u}$ and E$_{2g}$ modes become simultaneously IR and Raman vanishing of inversion symmetry.
ABC	3E (IR + R)	3A$_1$ (IR + R)	A$_1$(ac) + E(ac) + 2(E + A$_1$)(IR + R)	Raman active modes in the 750–800 cm^{-1} range.

2.6. Polytypism in Optical Experiments of the Early Days

Since the beginning of their experimental investigations, the fluorescence spectra of the sp^2-bonded crystals were measured as broad bands of about 500 meV width at half maximum, peaking in the 5.75 eV range and accompanied with many additional contributions on the low and high-energy sides of the main peak. In fact, such measurements were performed very early on powders or nanocrystals. This complicated the interpretation of light emission by layered boron nitride for decades. Representative BN spectra can be found in Refs. [63,99–104]. Fluorescence spectra, produced by electrons excited by a highly energetic photon or electron beam (cathodoluminescence), give a snapshot of the radiative recombination mechanisms of carriers with populations ruled by thermalization processes between all their possible levels in the bandgap [105]. These spectra give an image of the crystal properties in terms of its extrinsicity and promote the low-energy radiative recombination energies.

Absorption or reflectivity experiments, in contrast, probe the intrinsic and direct (in reciprocal space) transitions [106]. The oldest of them can probably be attributed to Professor Wolfgang Choyke. He never published his experimental data himself, but he offered his result to Doñi and Parravicini [107] to help them improve the tight-binding description of the band structure of BN. Since that time, there have been several studies dedicated to reproducing and improving his pioneering experiment. In Figure 6, we plotted some measurements of the dielectric constant of sp^2-bonded BN recorded using different techniques (ellipsometry, absorption, photoconductivity) at different periods, with time passing. Spectra (a), (b) and (c) are experiments taken on pyrolytic BN powders [107–109]. Note the two main resonances: one at about 6 eV and another about 1 eV higher.

Other features, which cover narrower energy scales, were collected on bulk crystals [110] (d) and epilayers [111,112] (e) and (f). The trend in going from pyrolytic BN to bulk crystals via epilayers is found when performing a photoconductivity experiment on a bulk sample, as indicated by the spectrum (g) [113]. These results were published by different groups, and the scatter in the data cannot be attributed to measurement artefacts or at the different temperatures at which experiments were performed. However, in light of what is known today, this can be attributed to the different crystallographic structures of the samples investigated by the different groups.

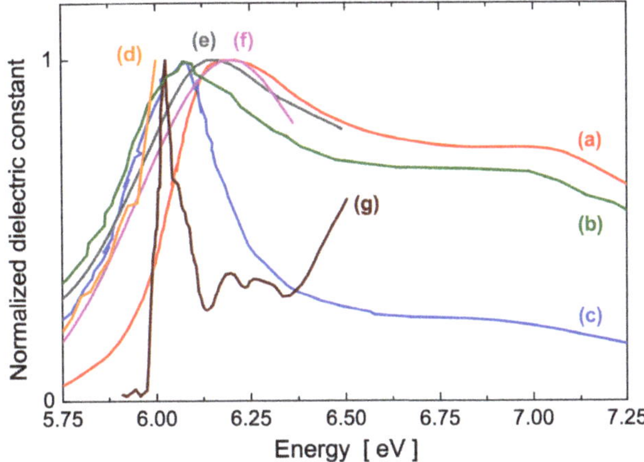

Figure 6. Dielectric constant vs. energy measured at room temperature on powders, bulk crystals and epilayers. Note the distribution of tendencies for the pyrolitic BN (**a–c**), the epilayers (**e,f**) and the bulk hBN crystals (**d,g**).

Arguably, the differences are linked to inhomogeneities of the strain states of the samples that were investigated. Under the application of a hydrostatic pressure, the direct bandgap of the AA′ stacking decreases of 26 ± 2 meV/GPa [114,115]. A compression of 1 GPa produces a variation of 2.41% of the c lattice parameter [116]. This leads to a variation of the lowest of the diffraction angles of the $(00l)$ planes of about $0.8°$ toward high angles, which was not reported experimentally. Therefore, the changes of the resonance energies cannot be attributed to a hydrostatic stress effect, at least to first order. Regarding anisotropic stress that could be operating in the plane of the layer, we consider now a uniaxial stress σ along an in-plane direction x (see Figure 7), which we represent as follows in the Voigt notation [117]:

$$\sigma' = \sigma \begin{pmatrix} 1 \\ 0 \\ 0 \\ 0 \\ 0 \\ 0 \end{pmatrix} \tag{1}$$

It generates a strain field that writes:

$$\epsilon' = \sigma \begin{pmatrix} S_{11} \\ S_{12} \\ S_{13} \\ 0 \\ 0 \\ 0 \end{pmatrix} \tag{2}$$

The signs of the components of the compliance tensor S_{11} and S_{12} or S_{13} are different, and the changes of the lengths of the vectors x, y, z are [117]:

$$\begin{pmatrix} x(\sigma) \\ y(\sigma) \\ z(\sigma) \end{pmatrix} = \begin{pmatrix} 1+\epsilon_{xx} & \epsilon_{xy} & \epsilon_{xz} \\ \epsilon_{xy} & 1+\epsilon_{yy} & \epsilon_{yz} \\ \epsilon_{xz} & \epsilon_{yz} & 1+\epsilon_{zz} \end{pmatrix} \begin{pmatrix} x(0) \\ y(0) \\ z(0) \end{pmatrix} \tag{3}$$

In that case, the orientations of the crystallographic directions x, y, and z are constant, but the unit cell lengths change. The symmetry of the crystal becomes orthorhombic and the table of character of its representative point group contains only real numbers. If a negative value is attributed to a stress compression along x, then $S_{11} > 0$ and $S_{12} < 0$. Using the notations of Figure 7, the length of x decreases; therefore, the length ΓK_1 increases and the length of ΓK_2 decreases. The direct transition at the series of K_i points of the Brillouin zone, plotted in green in Figure 7, splits into two (one for the valence and conduction states both at K_1 and K_4, another involving those of these states sitting at the remaining states sitting at K_2, K_3, K_5 and K_6). Theoretically, it results in a double structure or in a broadening or, depending on the value of the stress and of the deformation parameters of the conduction and valence band extrema at a given family of Ks, in the natural broadening of the reflectivity feature. This simple description of the effect of a uniaxial strain inside a BN layer does not explain reasonably the qualitative differences observed in the experimental curves of the dielectric constant plotted in Figure 6.

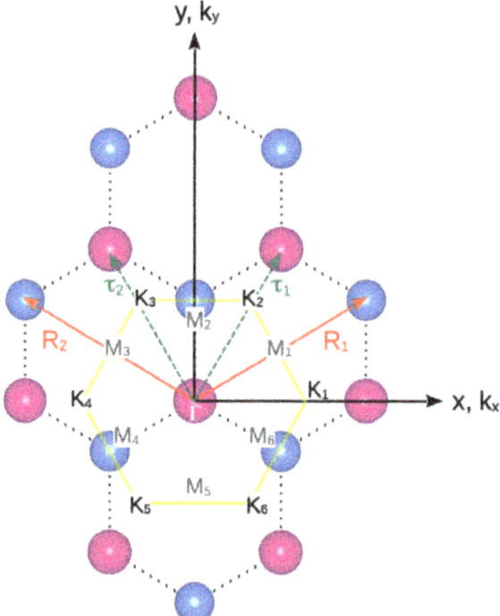

Figure 7. Atomic arrangement of boron (magenta) and nitrogen (blue) atoms in a monolayer of six-ring boron nitride in the (001) plane. The vectors of the direct lattice (τ_1 and τ_2) are plotted in dark green and the vectors of the reciprocal lattice (R_1 and R_2) are plotted in red. The edges of the Brillouin zone are shown as light green lines. The black lines represent the orientations of the vectors x and y of the international basis set representation for the direct lattice and for their analogs k_x and k_y. Note the positions of the specific points of the reciprocal lattice Γ, K_i, and M_i.

Recent ellipsometry measurements recorded worldwide at room temperature confirm a broad reflectivity feature near the value of the fundamental bandgap [71] with some puzzling discrepancies attributed to differing amounts of hBN and turbostratic BN, and/or the co-existence of sp^2 and sp^3-bonded crystallites [118].

In Figure 8, we reproduced the low-temperature reflectivity of a bulk crystal with pure AA′ stacking. This is restricted to the energy range of the direct bandgap [119], and it is complementary to the recent room temperature investigation in the broad spectral range [120]. In addition to the feature at 6.125 eV which probes the value of the fundamental direct bandgap at the K point of the BZ, phonon-assisted transitions are also evident that feature the low-energy wing of the main reflectance structure. Details about the in-depth interpretation of this reflectance experiment can be found in [119]. We believe that a uniaxial stress effect can just slightly enhance the reflectance feature of super broad intrinsic origin at the energy of the direct bandgap of hBN. When the length of y increases, the lengths ΓM_2 and ΓM_5 increase and the lengths of ΓM_1, ΓM_3, ΓM_4, and ΓM_6 decrease. Indirect transitions involving the series K_i and M_i points of the BZ of the stressed crystal will be broadened and thus will smoothen beyond the possibility to detect them.

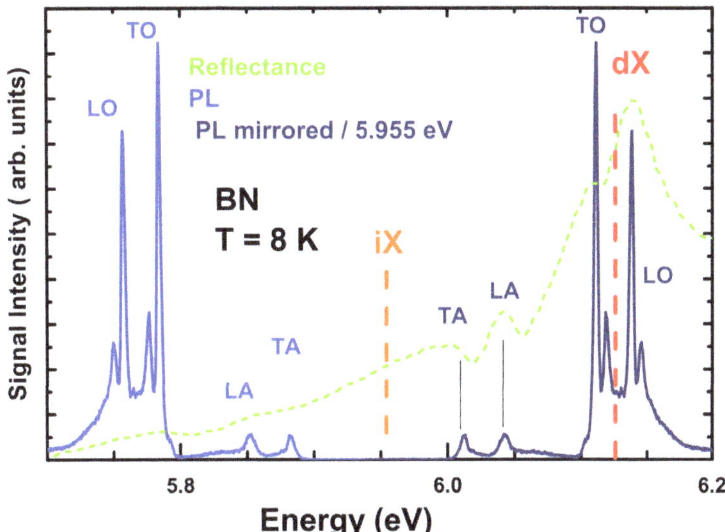

Figure 8. In blue is plotted the low-temperature photoluminescence spectrum of an ultra-pure hBN bulk crystal. In navy is plotted its mirror image relative to the energy (dashed orange line) of the indirect exciton (iX) at 5.955 eV. The reflectivity feature (green dashed line) displays a strong and broad feature at the energy (dashed red line at 6.125 eV) of the direct exciton dX. The series of lower energy features that are, versus energy of iX, one by one the mirror image of another one, is the evidence of indirect transitions that are observed thanks to the absorption of emissions of phonons [105,106,119].

3. The Optical Signatures of the Different Polytypes in the Deep Ultraviolet

3.1. Linear Optical Properties of hBN

Several authors have examined the problem of the determining of the bandgap of the different polytypes in the absence of excitonic effects or including them [26,43,57,62,104,121–128]. All studies predict different optical properties, in terms of photoluminescence energies, energies of singularities and their shapes, for the different stackings. The blue spectrum in Figure 8 reports a series of phonon-assisted transitions that are shifted from the energy of the forbidden indirect exciton relative to the energy of one phonon at the middle distance from the edge of the BZ. The band structure of hBN is indirect between the minimum of the conduction band at M and the top of the valence band in the vicinity of K [129]. It requires a phonon of the BZ at the middle of the $\Gamma - K$ direction to fulfill momentum conservation [130]. These phonon-assisted features are labeled in terms of LO, TO, LA, TA, ZO_1, and ZO_2 phonon replicas. The shape of the PL is ruled by selection rules [24,131] and the configuration of the experiment. At the low-energy side of the main peak of each phonon-assisted radiative recombination, there are overtones corresponding to a series of emissions associated with the low-energy Raman active mode E_{2g} (at about 7 meV) [132]. Observing these overtones requires high-quality material; it is not possible for the AA stacking, as this low-energy Raman active mode does not exist as discussed earlier. At even lower energy, that is to say below 5.65 eV in hBN, there is emission involving one or more complementary $TO(K)$ phonons. This is an intervalley scattering process that can be a priori observed at specific energies for all polytypes. High-quality BN material is free from such contributions, as they require specific extended defects to furnish to the Fermi golden rule the final density of states that controls the intensity of light emission at these energies [6,100,104,133]. The presence or absence of such signatures in the photoluminescence (PL) or cathodoluminescence (CL) spectra can theoretically be used to reveal whether there is a given polytype or another one as a contaminant in the host matrix.

3.2. Linear Optical Properties of bBN

To stimulate the growth of different bulk crystals of BN polytypes, we have modified our growth protocol by adding a substantial amount of graphite to the molten Cr + Ni eutectic, which is used to grow BN by a reversed solubility method. Impurities can induce non-bulk (different from AA′) stacking [134,135]. The Bernal staggered stacking of the graphitic layers is expected to impact the precipitation of a different polytype of BN out of the molten ingot saturated in B and N when cooling it down. Luckily, we obtained (001)-oriented BN crystalline flakes with slightly different X-ray diffraction features compared to high-quality hBN. Namely, broader features associated to the (001) planes and complementary diffraction peaks evidenced the inclusion of polytypisms with bBN. Unfortunately, the BN polytype cannot be determined from the morphology of the crystal. Another sample was grown in which 5% vanadium was added to the Cr + Ni eutectic solvent. Both samples grown with Cr + Ni + C and Fe + V exhibited Bernal-related PL, as we shall see later, but in contrast to the X-ray spectrum of the former, the latter did not contain misoriented polymorphic inclusions [136]. However, its X-ray diffraction spectrum is not as good as those of a high-quality hBN sample. In Figure 9, we gathered typical PL spectra recorded on BN samples, grown with Cr + Ni + C (green), Fe + V (red), and a very high-quality hBN sample grown using Fe only as a metal solvent [137] (blue). The astonishing feature is a new photoluminescence line at 6.035 eV in samples grown from Cr + Ni + C and Fe + V that we attribute to the signature of the direct bandgap of the AB stacking, i.e., bBN, and that is never detected in high-purity BN samples. To do so, we take advantage of theoretical calculations [123,124], which included an excitonic effect, and predicted that the bandgap of bBN is larger than that for hBN. It is also higher than the value for rBN.

The series of defect-related lines at energies below 6.5 eV in hBN appear slightly blue shifted in bBN compared to hBN. However, there is a strong overlap of the contributions of bBN and hBN. In other words, this indicates that the sample contains multiple polytypes. In Figure 10, we plotted in blue the PL of our test hBN sample and in red the PL of the sample containing bBN. A 1200 grooves/mm grating was used to record the experimental data. The several contributions, separated by 7 meV, in the main 6.035 eV feature reveal a not strictly homogeneous system. It is very reasonable to attribute this to strain-induced splitting of the direct transition, and/or to electronic complementary low-frequency Raman scattering, but we do not have a definitive interpretation to offer. To help understand the red PL spectrum, the spectrum of hBN was blue-shifted by 74 meV and plotted in gray. Then, it becomes possible to discriminate from the red spectrum the contributions of hBN from those of bBN. Concerning the latter, as the phonon energies are similar for these polytypes, the value of the indirect exciton in bBN is located at 6.029 eV in the low-energy wing of the PL signature of the direct bandgap, about 6 meV lower. Then, the contributions of phonon-assisted transitions, associated with the indirect bandgap noted iXbBN-PT, where the PTs are phonons at the center of the Brillouin zone, can be identified. In addition, the emission of a zone center LO_Γ is detected at energy $dXBN-LO_\Gamma$ at an energy at about 5.838 eV. Note that it is worthwhile attributing the 6.035 eV luminescence to the recombination at the bottleneck of the lowest polariton branch of the direct exciton polariton systems; thus, the value of the excitonic gap is higher than this value by an amount dictated by the magnitude of the longitudinal–transverse (LT) splitting between the branches of dispersion of the exciton–polariton. This LT splitting is huge in hBN [119], about 400 meV, and from the band structure calculations referenced above [57,62,121–125,127–133,136,137], there is no reason for it to be drastically different in bBN.

A more accurate determination requires reflectivity measurements similar to the one shown in Figure 9 with lineshape fitting of the spectrum to obtain the relevant parameters of this direct exciton. Pelini et al. [138] reported in the 4 eV range the existence of a few PL lines at 4.12 eV and 4.14 eV, energies higher than the energy of the line detected in hBN at 4.097 eV [139], implying a bBN crystalline matrix. This indicates that the description of the microscopic nature of the colored center acting as an efficient light emitter at 4 eV cannot ignore the layer stacking. This is especially true, as single-photon emitters operating at 4 eV have been demonstrated [8] in hBN. Nothing is known regarding the performances of their bBN analogs.

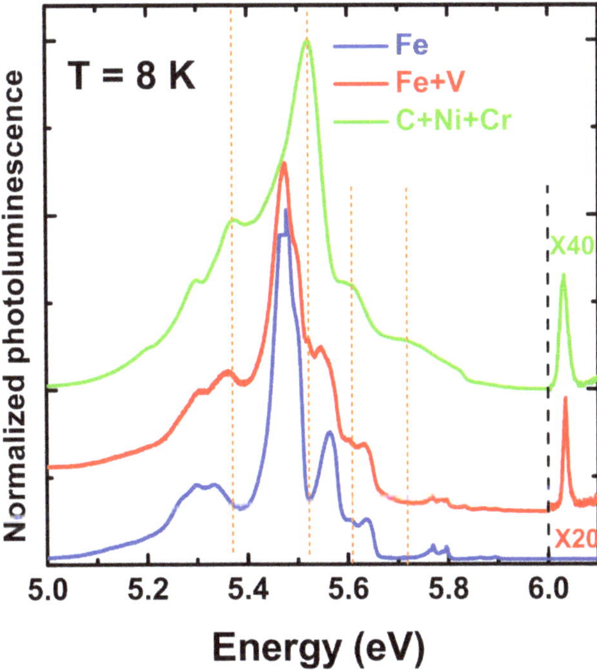

Figure 9. The 8 K macro-photoluminescence spectra in the 5–6.1 eV range of several samples showing the phonon-assisted transitions typical of hexagonal boron nitride (hBN; blue spectrum), mixed with the undulations linked to rBN and bBN inclusions impurities (green and red spectra). The signature of the direct bandgap of bBN is detected by a peak at 6.035 eV, following [136].

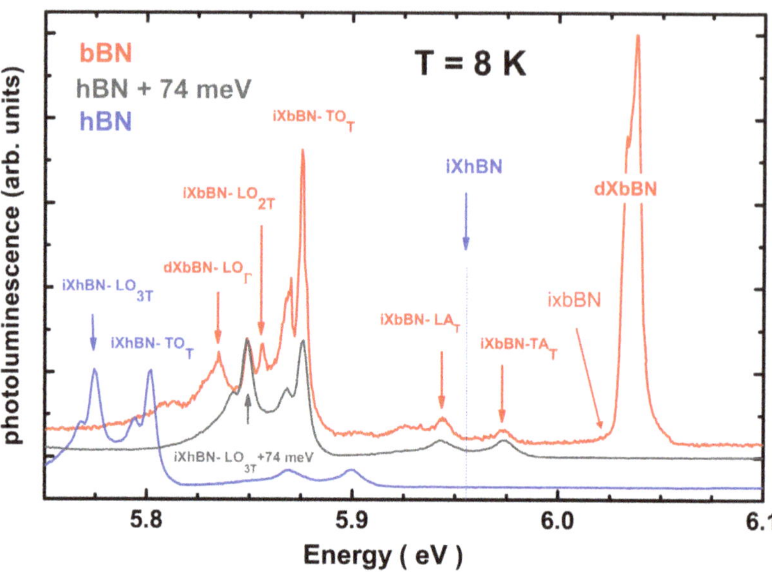

Figure 10. High resolution PL of a high-quality hBN crystal (blue) and of a piece of sp^2-bonded BN with highly dominating bBN (red). The hBN PL is re-plotted in gray after being blue-shifted 74 meV. The energies of the indirect iXbBN and direct exciton dXbBN in bBN are indicated as well as the indirect exciton iXhBN in hBN. The different phonon replicas are also identified.

3.3. Linear Optical Properties of rBN

The optical properties of rBN have not yet been well documented except in Xu et al. [140]. X-ray diffraction experiments have revealed the presence of rhombohedral stacking in BN epilayers grown by chemical vapor deposition on several substrates, namely SiC, *c*-plane sapphire capped by GaN or AlN, and various orientations of the sapphire substrate [67–70,141]. To detect rBN stacking requires completing $\theta/2\theta$ Bragg–Brentano diffraction experiments by in-plane measurements, azimuthal scans (ϕ-scans) and grazing incidence diffraction (GID). Rhombohedral BN films with twinned crystals that are rotated by 60° have been detected [67–70,72]. The photoluminescence of these deposited layers are dominated by two bands centered at 5.35 eV and 5.55 eV, respectively, and no analogs of the sharp phonon-assisted series of lines traditionally recorded can be measured. It is probably due to crystal twinning, which generates inhomogeneities and broadenings as well as transfers carriers to low energy traps. Interestingly, tBN has a PL signature that occurs as a complementary band at 5.4 eV, sandwiched between the preceding two, and with an intensity that follows the proportion of tBN relative to rBN in the crystals. The PL spectra measured in a couple of these rBN-rich epilayers are plotted in Figure 11 (red and wine) in addition to those typical of AA′ (blue) and AB stackings (green) (after [72]). Similarly to what occurs for InSe crystals containing polytypism, there is a substantial overlap of the different optical signatures that may complicate the interpretation [142].

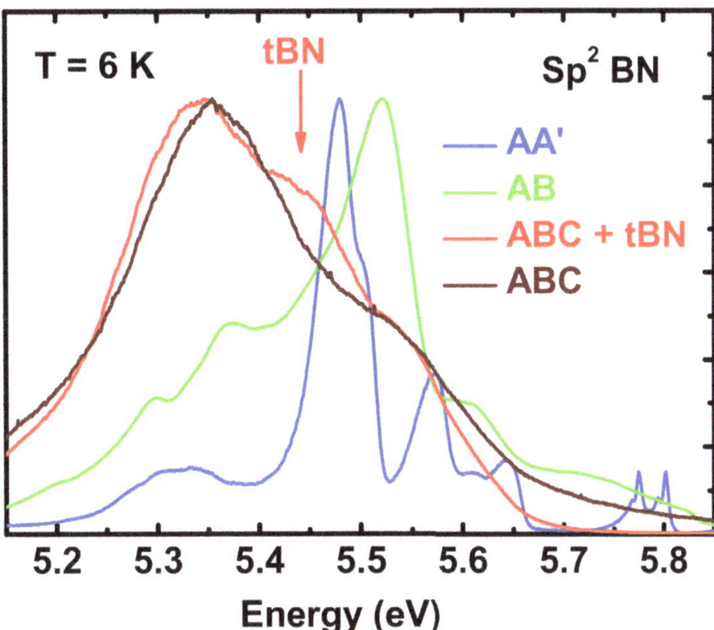

Figure 11. Photoluminescence spectra recorded on two rBN-rich epilayers (red and wine) in comparison with those typical AA' (blue) and AB (green) stackings. Note the signature of tBN. See reference [72] for more details.

3.4. Linear Optical Properties of the BN Monolayer

There has been a tremendous interest, soon after the demonstration of an indirect to direct bandgap crossover by Mak et al. [143], when decreasing the thickness of MoS_2 down to the monolayer, to reproduce this experiment in other 2D materials. There are several properties that changed from the bulk to the monolayer: the band structure itself and a strong renormalization of the exciton binding energy up to a huge value. This renormalization can be comparable in size to the value of the bandgap of the material in the absence of coupling of its electronic states with the electromagnetic field. This was established long ago through the theoretical predictions of Rytova [144] and Keldysh [145]. A more recent and detailed description of these mechanisms and their extension under the presence of disorder was published by E. V. Kirichenko and V. A. Stephanovich [146]. In BN, theoretical calculations including the Coulomb correlation predict a direct bandgap at K for the monolayer and an indirect or marginally direct bandgap for stackings of two or more layers [124,147–149]. Photoluminescence experiments [150–152] combined with reflectance measurements [150,151] have jointly confirmed using monolayers, deposited by molecular beam epitaxy on highly ordered pyrolytic graphite (HOPG) [150,151] and on suspended membranes [152], that it is a direct bandgap. The photoluminescence of the BN monolayer emits at 6.085 eV at low temperature, which was recently confirmed by the direct measurement of the density of states of a single monolayer of h-BN epitaxially grown on HOPG. This was achieved by performing low-temperature scanning tunneling microscopy (STM) and spectroscopy (STS) [153]. According to group theory, the D_{3h} point group authorizes piezoelectric behavior, which was observed experimentally [154].

3.5. Second Harmonic Generation in Some Polytypes

Among layered BN's many interesting physical properties is its ability to demonstrate second harmonic generation in the polytypes without inversion symmetry [117]. The choice of ultraviolet emission by second harmonic generation (SHG) nicely integrates in the context of the quest for compact solid state deep ultraviolet emitters. The demonstration of SHG emission was achieved by shining light on a Cr + Ni + C sample, which had the Bernal stacking of layers (bBN or AB of spatial symmetry $P\bar{6}2m$ or D^3_{3h}), with a laser operating at 400 nm and observing light emission at 200 nm. To do so, we have examined a small piece of the sample described above by micro-photoluminescence spectroscopy at $T = 8$ K, which has a spatial resolution of 200 nm [133]. A photograph of a piece of BN unambiguously containing polytypism as can be evidenced by just using an optical microscope is shown in Figure 12a. Figure 12b is a map of the PL emission by this piece of sample, when illuminated with a laser wavelength of 196 nm. This is a composite containing a region of bBN (it emits light at 6.035 eV whilst hBN does not) sandwiched between two regions of hBN (for which the light emission does not occur at this energy, as is typical of hBN). We then shifted the laser light to 400 nm and recorded the light emitted with a wavelength of 200 nm at the strict position of the Bernal polytype (see Figure 12c). This experiment demonstrates second harmonic generation at 200 nm and makes Bernal BN a good candidate for devices susceptible of deep ultraviolet emission by up-conversion of a UVA light [155].

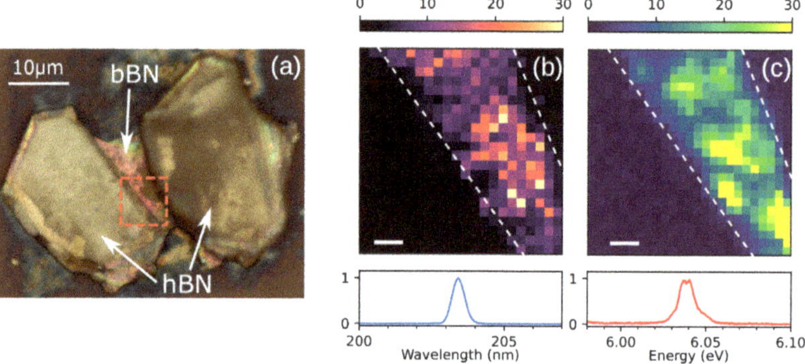

Figure 12. (**a**) Photograph of an exfoliated sample composed of a bBN region sandwiched between two hBN regions. (**b**) Mapping of the PL emission from the square area indicated in (**a**). (**c**) Mapping of the SHG emission. After [155].

4. Conclusions

Different layered boron nitride polytype crystals can be synthesized by tuning the growth conditions. In this review article, we emphasized that second harmonic generation of UVA light can produce an UVC emitter. There are not so many studies about the control of growth of layered BN with controlled polytypism, but it is obvious, having in mind the different applications of the different polytypes of silicon carbide, that such crystals can demonstrate very specific properties compared to hBN. The efficiency of SHG in the challenging area of the deep ultraviolet using Bernal boron nitride paves the way for its use in short-wavelength optoelectronics. Having in mind the SiC story, extension of our interest further than optoelectronics at short wavelengths might soon lead to other polytypes knocking at the doors of the device arena.

Author Contributions: B.G. wrote the paper. W.D. made all the theoretical calculations. M.M. performed the X-ray diffraction experiment. P.V. and G.C. supervised the photoluminescence measurements recorded by A.R. and the reflectivity measurements recorded by C.E. J.H.E. supervised the growth of boron nitride by J.L. and E.J. at KSU. All authors contributed to the interpretation of the experimental data. All authors have read and agreed to the published version of the manuscript.

Funding: This paper was financially supported by the network GaNeX (ANR-11-LABX-424 0014), the BONASPES project (ANR-19-CE30 0007), the ZEOLIGHT project (ANR-19-CE08-0016), and the Université de Montpellier. Support for BN crystal growth came from the Office of Naval Research, Award No. N00014-20-1-2474, and the National Science Foundation, Award No. CMMI 429 #1538127.

Institutional Review Board Statement: Not applicable.

Informed Consent Statement: Not applicable.

Data Availability Statement: Not applicable.

Acknowledgments: We are grateful to Sachin Sharma, Laurent Souqui, Henrik Pedersen and Hans Högberg for their epilayers with rBN polytypes. The photoluminescence spectra, which are published together with them in Moret et al. [72], are reproduced in Figure 11 and have been used to conceive the graphical abstract. We gratefully acknowledge T. Cohen and C. L'Henoret for their technical supports at the mechanics workshop.

Conflicts of Interest: The authors declare no conflict of interest.

References

1. Kneissl, M.; Seong, T.Y.; Han, J.; Amano, H. The emergence and prospects of deep-ultraviolet light-emitting diode technologies. *Nat. Photonics* **2019**, *13*, 233–244. [CrossRef]
2. Nanishi, Y. The birth of the blue LED. *Nat. Photonics* **2014**, *8*, 884–886. [CrossRef]
3. Kneissl, M.; Rass, J. (Eds.) *III-Nitride Ultraviolet Emitters: Technology and Applications*; Springer Series in Materials Science; Springer International Publishing: Cham, Switzerland, 2015; Volume 227. [CrossRef]
4. Ueta, M.; Kanzaki, H.; Toyozawa, Y.; Hanamura, E. *Handbook of Zinc Oxide and Related Materials*; Springer: Berlin/Heidelberg, Germany, 1986; Volume 60. [CrossRef]
5. Vuong, T.Q.P.; Lair, V.; Lacoste, F.R.; Halloumi, S.; Coindeau, S.; Thiel, J.; Shubina, T.V.; Pauthe, M.; Gil, B. Structural and Optical Properties of $Zn_{1-x}Mg_xO$ Prepared by Calcination of $ZnO + Mg(OH)_2$ after Hydro Micro Mechanical Activation. *Ann. Der Phys.* **2019**, *531*, 1800379.
6. Watanabe, K.; Taniguchi, T.; Kanda, H. Direct-bandgap properties and evidence for ultraviolet lasing of hexagonal boron nitride single crystal. *Nat. Mater.* **2004**, *3*, 404–409. [CrossRef]
7. Watanabe, K.; Taniguchi, T.; Niiyama, T.; Miya, K.; Taniguchi, M. Far-ultraviolet plane-emission handheld device based on hexagonal boron nitride. *Nat. Photonics* **2009**, *3*, 591–594. [CrossRef]
8. Bourrellier, R.; Meuret, S.; Tararan, A.; Stéphan, O.; Kociak, M.; Tizei, L.H.G.; Zobelli, A. Bright UV Single Photon Emission at Point Defects in h-BN. *Nano Lett.* **2016**, *16*, 4317–4321. [CrossRef]
9. Su, C.; Zhang, F.; Kahn, S.; Shevitski, B.; Jiang, J.; Dai, C.; Ungar, A.; Park, J.H.; Watanabe, K.; Taniguchi, T.; et al. Tuning Color Centers at a Twisted Interface. *arXiv* **2021**, arXiv:2108.04747v1.
10. Caldwell, J.D.; Aharonovich, I.; Cassabois, G.; Edgar, J.H.; Gil, B.; Basov, D.N. Photonics with hexagonal boron nitride. *Nat. Rev. Mater.* **2019**, *4*, 552–567. [CrossRef]
11. Baronnet, A. Some aspects of polytypism in crystals. *Prog. Cryst. Growth Charact.* **1978**, *1*, 151–211. [CrossRef]
12. Pandey, D.; Krishna, P. The origin of polytype structures. *Prog. Cryst. Growth Charact.* **1983**, *7*, 213–258. [CrossRef]
13. Mardix, S. Polytypism: A controlled thermodynamic phenomenon. *Phys. Rev. B* **1986**, *33*, 8677–8684. [CrossRef] [PubMed]
14. Verma, A.; Krishna, P. *Polymorphism and Polytypism in Crystals*; John Wiley & Sons Inc.: New York, NY, USA, 1962.
15. Balmain, W.H. Bemerkungen über die Bildung von Verbindungen des Bors und Siliciums mit Stickstoff und gewissen Metallen. *J. Prakt. Chem.* **1842**, *27*, 422–430. [CrossRef]
16. Walther Bothe's Biography. Available online: https://en.wikipedia.org/wiki/Walther_Bothe (accessed on 17 March 2022).
17. James Chadwick's Biography. Available online: https://en.wikipedia.org/wiki/James_Chadwick (accessed on 4 May 2022).
18. Henske, M.; Klein, M.; Köhli, M.; Lennert, P.; Modzel, G.; Schmidt, C.J.; Schmidt, U. The ^{10}B based Jalousie neutron detector—An alternative for ^{3}He filled position sensitive counter tubes. *Nucl. Instruments Methods Phys. Res. Sect. Accel. Spectrometers Detect. Assoc. Equip.* **2012**, *686*, 151–155. [CrossRef]
19. Hassel, O. Die Krystallstruktur des bornitrides, BN. *Norsk. Geol. Tidskr.* **1926**, *9*, 266.
20. Brager, A. An x-ray examination of the structure of boron nitride. *Acta Physicochim. URSS* **1937**, *7*, 699.
21. Brager, A. An X-ray examination of titanium nitride III. Investigation by the powder method. *Acta Physicochim. URSS* **1939**, *11*, 617.
22. Pease, R.S. Crystal Structure of Boron Nitride. *Nature* **1950**, *165*, 722–723. [CrossRef]
23. Pease, R.S. An X-ray study of boron nitride. *Acta Crystallogr.* **1952**, *5*, 356–361. [CrossRef]
24. Vuong, T.Q.P.; Cassabois, G.; Valvin, P.; Jacques, V.; Lee, A.V.D.; Zobelli, A.; Watanabe, K.; Taniguchi, T.; Gil, B. Phonon symmetries in hexagonal boron nitride probed by incoherent light emission. *2D Mater.* **2016**, *4*, 11004. [CrossRef]
25. Vuong, T.Q.P.; Liu, S.; Van der Lee, A.; Cuscó, R.; Artús, L.; Michel, T.; Valvin, P.; Edgar, J.H.; Cassabois, G.; Gil, B. Isotope engineering of van der Waals interactions in hexagonal boron nitride. *Nat. Mater.* **2018**, *17*, 152–158. [CrossRef]
26. Liu, L.; Feng, Y.P.; Shen, Z.X. Structural and electronic properties of h-BN. *Phys. Rev. B* **2003**, *68*, 104102. [CrossRef]
27. Coulson, C.A.; Taylor, R. Studies in Graphite and Related Compounds I: Electronic Band Structure in Graphite. *Proc. Phys. Soc. Sect. A* **1952**, *65*, 815–825. [CrossRef]
28. Wentorf, R.H. Cubic Form of Boron Nitride. *J. Chem. Phys.* **1957**, *26*, 956. [CrossRef]
29. Milledge, H.J.; Nave, E.; Weller, F.H. Transformation of Cubic Boron Nitride to a Graphitic Form of Hexagonal Boron Nitride. *Nature* **1959**, *184*, 715. [CrossRef]

30. Zagyansky, L.; Samsonov, G.V. On the question of the oxidizability of boron nitride. *J. Appl. Chem. USSR* **1952**, *25*, 629.
31. Corrigan, F.R.; Bundy, F.P. Direct transitions among the allotropic forms of boron nitride at high pressures and temperatures. *J. Chem. Phys.* **1975**, *63*, 3812–3820. [CrossRef]
32. Xu, F.F.; Bando, Y.; Hasegawa, M. New phases of sp2-bonded boron nitride: The 12R and 24R polytypes. *Chem. Commun.* **2002**, *14*, 1490–1491. [CrossRef]
33. Wentorf, R.H. Synthesis of the Cubic Form of Boron Nitride. *J. Chem. Phys.* **1961**, *34*, 809–812. [CrossRef]
34. Narayan, J.; Bhaumik, A. Research Update: Direct conversion of h-BN into pure c-BN at ambient temperatures and pressures in air. *APL Mater.* **2016**, *4*, 20701. [CrossRef]
35. Meng, Y.; Mao, H.K.; Eng, P.J.; Trainor, T.P.; Newville, M.; Hu, M.Y.; Kao, C.; Shu, J.; Hausermann, D.; Hemley, R.J. The formation of sp^3 bonding in compressed BN. *Nat. Mater.* **2004**, *3*, 111–114. [CrossRef]
36. Solozhenko, V.L. Thermodynamics of dense boron nitride modifications and a new phase P,T diagram for BN. *Thermochim. Acta* **1993**, *218*, 221–227. [CrossRef]
37. Eremets, M.I.; Takemura, K.; Yusa, H.; Golberg, D.; Bando, Y.; Blank, V.D.; Sato, Y.; Watanabe, K. Disordered state in first-order phase transitions: Hexagonal-to-cubic and cubic-to-hexagonal transitions in boron nitride. *Phys. Rev. B* **1998**, *57*, 5655–5660. [CrossRef]
38. Solozhenko, V.L.; Turkevich, V.Z.; Holzapfel, W.B. Refined Phase Diagram of Boron Nitride. *J. Phys. Chem. B* **1999**, *103*, 2903–2905. [CrossRef]
39. Petrescu, M.I.; Balint, M.G. Structure and properties modifications in boron nitride. Part I: Direct polymorphic transformations mechanisms. *UPB Sci. Bull. Ser. B* **2007**, *69*, 35.
40. Edgar, J.H.; Strite, S.; Akasaki, I.; Amano, H.; Wetzel, C. (Eds.) *Properties of Group III Nitrides*; EMIS Datareviews Series; INSPEC, Institution of Electrical Engineers: London, UK, 1994; Volume 23.
41. Chilleri, J.; Siddiqua, P.; Shur, M.S.; O'Leary, S.K. Cubic boron nitride as a material for future electron device applications: A comparative analysis. *Appl. Phys. Lett.* **2022**, *120*, 122105. [CrossRef]
42. Riter, J.R. Discussion of shock-induced graphite →wurtzite phase transformation in BN and implications for stacking in graphitic BN. *J. Chem. Phys.* **1973**, *59*, 1538. [CrossRef]
43. Yixi, S.; Xin, J.; Kun, W.; Chaoshu, S.; Zhengfu, H.; Junyan, S.; Jie, D.; Sheng, Z.; Yuanbin, C. Vacuum-ultraviolet reflectance spectra and optical properties of nanoscale wurtzite boron nitride. *Phys. Rev. B* **1994**, *50*, 18637–18639. [CrossRef]
44. Dubrovinskaia, N.; Solozhenko, V.L.; Miyajima, N.; Dmitriev, V.; Kurakevych, O.O.; Dubrovinsky, L. Superhard nanocomposite of dense polymorphs of boron nitride: Noncarbon material has reached diamond hardness. *Appl. Phys. Lett.* **2007**, *90*, 101912. [CrossRef]
45. Wills, R.R. Wurtzitic boron nitride—A review. *Int. J. High Technol. Ceram.* **1985**, *1*, 139–153. [CrossRef]
46. Kudrawiec, R.; Hommel, D. Bandgap engineering in III-nitrides with boron and group V elements: Toward applications in ultraviolet emitters. *Appl. Phys. Rev.* **2020**, *7*, 041314. [CrossRef]
47. Chopra, N.G.; Luyken, R.J.; Cherrey, K.; Crespi, V.H.; Cohen, M.L.; Louie, S.G.; Zettl, A. Boron Nitride Nanotubes. *Science* **1995**, *269*, 966–967. [CrossRef] [PubMed]
48. Chen, Y.I. (Ed.) *Nanotubes and Nanosheets: Functionalization and Applications of Boron Nitride and Other Nanomaterials*; CRC Press: Boca Raton, FL, USA, 2015.
49. Baumhauer, H. Uber die Krystalle des Corborundums. *Zeitshrift Kryst. Mineral.* **1912**, *50*, 33.
50. Baumhauer, H. Uber die verschiedenen Modifikationen des Carborundums und die Erscheinung der Polytypie. *Zeitshrift Kryst. Mineral.* **1915**, *55*, 249. [CrossRef]
51. Jagodzinski, H. Polytypism in SiC crystals. *Acta Crystallogr.* **1954**, *7*, 300. [CrossRef]
52. Frank, F.C. LXXXIII. Crystal dislocations—Elementary concepts and definitions. *Lond. Edinb. Dublin Philos. Mag. J. Sci.* **1951**, *42*, 809–819. [CrossRef]
53. Moore, A.W. Characterization of pyrolytic boron nitride for semiconductor materials processing. *J. Cryst. Growth* **1990**, *106*, 6–15. [CrossRef]
54. Thomas, J.; Weston, N.E.; O'Connor, T.E. Turbostratic Boron Nitride, Thermal Transformation to Ordered-layer-lattice Boron Nitride. *J. Am. Chem. Soc.* **1962**, *84*, 4619–4622. [CrossRef]
55. Giannozzi, P.; Baroni, S.; Bonini, N.; Calandra, M.; Car, R.; Cavazzoni, C.; Ceresoli, D.; Chiarotti, G.L.; Cococcioni, M.; Dabo, I.; et al. QUANTUM ESPRESSO: A modular and open-source software project for quantum simulations of materials. *J. Phys. Condens. Matter* **2009**, *21*, 395502. [CrossRef]
56. Giannozzi, P.; Andreussi, O.; Brumme, T.; Bunau, O.; Nardelli, M.B.; Calandra, M.; Car, R.; Cavazzoni, C.; Ceresoli, D.; Cococcioni, M.; et al. Advanced capabilities for materials modelling with Quantum ESPRESSO. *J. Phys. Condens. Matter* **2017**, *29*, 465901. [CrossRef]
57. Xu, Y.N.; Ching, W.Y. Calculation of ground-state and optical properties of boron nitrides in the hexagonal, cubic, and wurtzite structures. *Phys. Rev. B* **1991**, *44*, 7787–7798. [CrossRef]
58. Furthmuller, J.; Hafner, J.; Kresse, G. Ab initio calculation of the structural and electronic properties of carbon and boron nitride using ultrasoft pseudopotentials. *Phys. Rev. B* **1994**, *50*, 15606–15622. [CrossRef] [PubMed]
59. Albe, K. Theoretical study of boron nitride modifications at hydrostatic pressures. *Phys. Rev. B* **1997**, *55*, 6203–6210. [CrossRef]

60. Ohba, N.; Miwa, K.; Nagasako, N.; Fukumoto, A. First-principles study on structural, dielectric, and dynamical properties for three BN polytypes. *Phys. Rev. B* **2001**, *63*, 115207. [CrossRef]
61. Ooi, N.; Rairkar, A.; Lindsley, L.; Adams, J.B. Electronic structure and bonding in hexagonal boron nitride. *J. Phys. Condens. Matter.* **2005**, *18*, 97–115. [CrossRef]
62. Constantinescu, G.; Kuc, A.; Heine, T. Stacking in Bulk and Bilayer Hexagonal Boron Nitride. *Phys. Rev. Lett.* **2013**, *111*, 36104. [CrossRef]
63. Silly, M.G.; Jaffrennou, P.; Barjon, J.; Lauret, J.S.; Ducastelle, F.; Loiseau, A.; Obraztsova, E.; Attal-Tretout, B.; Rosencher, E. Luminescence properties of hexagonal boron nitride: Cathodoluminescence and photoluminescence spectroscopy measurements. *Phys. Rev. B* **2007**, *75*, 85205. [CrossRef]
64. Gao, S.P. Crystal structures and band gap characters of h-BN polytypes predicted by the dispersion corrected DFT and GW method. *Solid State Commun.* **2012**, *152*, 1817–1820. [CrossRef]
65. Haga, T.; Matsuura, Y.; Fujimoto, Y.; Saito, S. Electronic states and modulation doping of hexagonal boron nitride trilayers. *Phys. Rev. Mater.* **2021**, *5*, 94003. [CrossRef]
66. Kurdyumov, A.V.; Solozhenko, V.L.; Zelyavski, W.B. Lattice Parameters of Boron Nitride Polymorphous Modifications as a Function of Their Crystal-Structure Perfection. *J. Appl. Crystallogr.* **1995**, *28*, 540–545. [CrossRef]
67. Chubarov, M.; Pedersen, H.; Hogberg, H.; Jensen, J.; Henry, A. Growth of High Quality Epitaxial Rhombohedral Boron Nitride. *Cryst. Growth Des.* **2012**, *12*, 3215–3220. [CrossRef]
68. Chubarov, M.; Pedersen, H.; Hogberg, H.; Henry, A.; Czigány, Z. Initial stages of growth and the influence of temperature during chemical vapor deposition of sp^2-BN films. *J. Vac. Sci. Technol. A* **2015**, *33*, 61520. [CrossRef]
69. Chubarov, M.; Hogberg, H.; Henry, A.; Pedersen, H. Review Article: Challenge in determining the crystal structure of epitaxial 0001 oriented sp^2-BN films. *J. Vac. Sci. Technol. A* **2018**, *36*, 30801. [CrossRef]
70. Souqui, L.; Pedersen, H.; Hogberg, H. Thermal chemical vapor deposition of epitaxial rhombohedral boron nitride from trimethyl-boron and ammonia. *J. Vac. Sci. Technol. A* **2019**, *37*, 20603. [CrossRef]
71. Yamada, H.; Kumagai, N.; Yamada, T.; Yamamoto, T. Dielectric functions of CVD–grown boron nitride from 1.1 to 9.0 eV by spectroscopic ellipsometry. *Appl. Phys. Lett.* **2021**, *118*, 112101. [CrossRef]
72. Moret, M.; Rousseau, A.; Valvin, P.; Sharma, S.; Souqui, L.; Pedersen, H.; Hogberg, H.; Cassabois, G.; Li, J.; Edgar, J.H.; et al. Rhombohedral and turbostratic boron nitride: X-ray diffraction and photoluminescence signatures. *Appl. Phys. Lett.* **2021**, *119*, 262102. [CrossRef]
73. Geick, R.; Perry, C.H.; Rupprecht, G. Normal Modes in Hexagonal Boron Nitride. *Phys. Rev.* **1966**, *146*, 543–547. [CrossRef]
74. Babich, I.L. Raman spectrum of hexagonal boron nitride. *Theor. Exp. Chem.* **1974**, *8*, 594–595. [CrossRef]
75. Kuzuba, T.; Era, K.; Ishii, T.; Sato, T. A low frequency Raman-active vibration of hexagonal boron nitride. *Solid State Commun.* **1978**, *25*, 863–865. [CrossRef]
76. Kuzuba, T.; Sato, Y.; Yamaoka, S.; Era, K. Raman-scattering study of high-pressure effects on the anisotropy of force constants of hexagonal boron nitride. *Phys. Rev. B* **1978**, *18*, 4440–4443. [CrossRef]
77. Nemanich, R.J.; Solin, S.A.; Martin, R.M. Light scattering study of boron nitride microcrystals. *Phys. Rev. B* **1981**, *23*, 6348–6356. [CrossRef]
78. Cuscó, R.; Artús, L.; Edgar, J.H.; Liu, S.; Cassabois, G.; Gil, B. Isotopic effects on phonon anharmonicity in layered van der Waals crystals: Isotopically pure hexagonal boron nitride. *Phys. Rev. B* **2018**, *97*, 155435. [CrossRef]
79. Cuscó, R.; Edgar, J.H.; Liu, S.; Cassabois, G.; Gil, B.; Artús, L. Influence of isotopic substitution on the anharmonicity of the interlayer shear mode of *h*-BN. *Phys. Rev. B* **2019**, *99*, 85428. [CrossRef]
80. Cuscó, R.; Edgar, J.H.; Liu, S.; Li, J.; Artús, L. Isotopic Disorder: The Prevailing Mechanism in Limiting the Phonon Lifetime in Hexagonal BN. *Phys. Rev. Lett.* **2020**, *124*, 167402. [CrossRef] [PubMed]
81. Segura, A.; Cuscó, R.; Taniguchi, T.; Watanabe, K.; Artús, L. Long lifetime of the E_{1u} in-plane infrared-active modes of *h*-BN. *Phys. Rev. B* **2020**, *101*, 235203. [CrossRef]
82. Cuscó, R.; Pellicer-Porres, J.; Edgar, J.H.; Li, J.; Segura, A.; Artús, L. Pressure dependence of the interlayer and intralayer E_{2g} Raman-active modes of hexagonal BN up to the wurtzite phase transition. *Phys. Rev. B* **2020**, *102*, 75206. [CrossRef]
83. Cuscó, R.; Pellicer-Porres, J.; Edgar, J.H.; Li, J.; Segura, A.; Artús, L. Phonons of hexagonal BN under pressure: Effects of isotopic composition. *Phys. Rev. B* **2021**, *103*, 85204. [CrossRef]
84. Cigarini, L.; Novotný, M.; Karlický, F. Lattice dynamics in the conformational environment of multilayered hexagonal boron nitride (h-BN) results in peculiar infrared optical responses. *Phys. Chem. Chem. Phys.* **2021**, *23*, 7247–7260. [CrossRef]
85. Simpson, A.; Stuckes, A.D. The thermal conductivity of highly oriented pyrolytic boron nitride. *J. Phys. C Solid State Phys.* **1971**, *4*, 1710–1718. [CrossRef]
86. Sichel, E.K.; Miller, R.E.; Abrahams, M.S.; Buiocchi, C.J. Heat capacity and thermal conductivity of hexagonal pyrolytic boron nitride. *Phys. Rev. B* **1976**, *13*, 4607–4611. [CrossRef]
87. Mortazavi, B.; Pereira, L.F.C.; Jiang, J.W.; Rabczuk, T. Modelling heat conduction in polycrystalline hexagonal boron-nitride films. *Sci. Rep.* **2015**, *5*, 13228. [CrossRef]
88. Zhang, Z.; Hu, S.; Chen, J.; Li, B. Hexagonal boron nitride: A promising substrate for graphene with high heat dissipation. *Nanotechnology* **2017**, *28*, 225704. [CrossRef] [PubMed]

89. Lindsay, L.; Broido, D.A. Enhanced thermal conductivity and isotope effect in single-layer hexagonal boron nitride. *Phys. Rev. B* **2011**, *84*, 155421. [CrossRef]

90. Jiang, P.; Qian, X.; Yang, R.; Lindsay, L. Anisotropic thermal transport in bulk hexagonal boron nitride. *Phys. Rev. Mater.* **2018**, *2*, 64005. [CrossRef]

91. Cai, Q.; Scullion, D.; Gan, W.; Falin, A.; Zhang, S.; Watanabe, K.; Taniguchi, T.; Chen, Y.; Santos, E.J.G.; Li, L.H. High thermal conductivity of high-quality monolayer boron nitride and its thermal expansion. *Sci. Adv.* **2019**, *5*, eaav0129. [CrossRef]

92. Sharma, V.; Kagdada, H.L.; Jha, P.K.; Śpiewak, P.; Kurzydłowski, K.J. Thermal transport properties of boron nitride based materials: A review. *Renew. Sustain. Energy Rev.* **2020**, *120*, 109622. [CrossRef]

93. Cortes, C.L.; Newman, W.; Molesky, S.; Jacob, Z. Quantum nanophotonics using hyperbolic metamaterials. *J. Opt.* **2012**, *14*, 63001. [CrossRef]

94. Jacob, Z. Hyperbolic phonon–polaritons. *Nat. Mater.* **2014**, *13*, 1081–1083. [CrossRef]

95. Giles, A.J.; Dai, S.; Vurgaftman, I.; Hoffman, T.; Liu, S.; Lindsay, L.; Ellis, C.T.; Assefa, N.; Chatzakis, I.; Reinecke, T.L.; et al. Ultralow-loss polaritons in isotopically pure boron nitride. *Nat. Mater.* **2018**, *17*, 134–139. [CrossRef]

96. Poddubny, A.; Iorsh, I.; Belov, P.; Kivshar, Y. Hyperbolic metamaterials. *Nat. Photonics* **2013**, *7*, 948–957. [CrossRef]

97. Basov, D.N.; Fogler, M.M.; García de Abajo, F.J. Polaritons in van der Waals materials. *Science* **2016**, *354*.

98. Kinsey, N.; DeVault, C.; Boltasseva, A.; Shalaev, V.M. Near-zero-index materials for photonics. *Nat. Rev. Mater.* **2019**, *4*, 742–760. [CrossRef]

99. Zunger, A.; Katzir, A.; Halperin, A. Optical properties of hexagonal boron nitride. *Phys. Rev. B* **1976**, *13*, 5560–5573. [CrossRef]

100. Jaffrennou, P.; Barjon, J.; Lauret, J.S.; Attal-Trétout, B.; Ducastelle, F.; Loiseau, A. Origin of the excitonic recombinations in hexagonal boron nitride by spatially resolved cathodoluminescence spectroscopy. *J. Appl. Phys.* **2007**, *102*, 116102. [CrossRef]

101. Museur, L.; Kanaev, A. Near band-gap photoluminescence properties of hexagonal boron nitride. *J. Appl. Phys.* **2008**, *103*, 103520. [CrossRef]

102. Museur, L.; Feldbach, E.; Kanaev, A. Defect-related photoluminescence of hexagonal boron nitride. *Phys. Rev. B* **2008**, *78*, 155204. [CrossRef]

103. Pierret, A.; Loayza, J.; Berini, B.; Betz, A.; Plaçais, B.; Ducastelle, F.; Barjon, J.; Loiseau, A. Excitonic recombinations in h-BN: From bulk to exfoliated layers. *Phys. Rev. B* **2014**, *89*, 35414. [CrossRef]

104. Bourrellier, R.; Amato, M.; Galvão Tizei, L.H.; Giorgetti, C.; Gloter, A.; Heggie, M.I.; March, K.; Stéphan, O.; Reining, L.; Kociak, M.; et al. Nanometric Resolved Luminescence in h-BN Flakes: Excitons and Stacking Order. *ACS Photonics* **2014**, *1*, 857–862. [CrossRef]

105. Pankove, J.I.; Kiewit, D.A. Optical Processes in Semiconductors. *J. Electrochem. Soc.* **1972**, *119*, 156C. [CrossRef]

106. Yu, P.; Cardona, M. *Fundamentals of Semiconductors: Physics and Materials Properties*; Springer: Berlin/Heiderlberg, Germany, 2010.

107. Doni, E.; Parravicini, G.P. Energy bands and optical properties of hexagonal boron nitride and graphite. *Il Nuovo Cim. B (1965–1970)* **1969**, *64*, 117–144. [CrossRef]

108. Hoffman, D.M.; Doll, G.L.; Eklund, P.C. Optical properties of pyrolytic boron nitride in the energy range 0.05–10 eV. *Phys. Rev. B* **1984**, *30*, 6051–6056. [CrossRef]

109. Tarrio, C.; Schnatterly, S.E. Interband transitions, plasmons, and dispersion in hexagonal boron nitride. *Phys. Rev. B* **1989**, *40*, 7852–7859. [CrossRef]

110. Watanabe, K.; Taniguchi, T. Jahn-Teller effect on exciton states in hexagonal boron nitride single crystal. *Phys. Rev. B* **2009**, *79*, 193104. [CrossRef]

111. Li, X.; Jordan, M.B.; Ayari, T.; Sundaram, S.; El Gmili, Y.; Alam, S.; Alam, M.; Patriarche, G.; Voss, P.L.; Paul Salvestrini, J.; et al. Flexible metal-semiconductor-metal device prototype on wafer-scale thick boron nitride layers grown by MOVPE. *Sci. Rep.* **2017**, *7*, 786. [CrossRef] [PubMed]

112. Cheng, T.S.; Summerfield, A.; Mellor, C.J.; Davies, A.; Khlobystov, A.N.; Eaves, L.; Foxon, C.T.; Beton, P.H.; Novikov, S.V. High-temperature molecular beam epitaxy of hexagonal boron nitride layers. *J. Vac. Sci. Technol. B* **2018**, *36*, 2D103. [CrossRef]

113. Museur, L.; Brasse, G.; Pierret, A.; Maine, S.; Attal-Tretout, B.; Ducastelle, F.; Loiseau, A.; Barjon, J.; Watanabe, K.; Taniguchi, T.; et al. Exciton optical transitions in a hexagonal boron nitride single crystal. *Phys. Status Solidi (Rrl) Rapid Res. Lett.* **2011**, *5*, 214–216. [CrossRef]

114. Akamaru, H.; Onodera, A.; Endo, T.; Mishima, O. Pressure dependence of the optical-absorption edge of AlN and graphite-type BN. *J. Phys. Chem. Solids* **2002**, *63*, 887–894. [CrossRef]

115. Segura, A.; Cuscó, R.; Attaccalite, C.; Taniguchi, T.; Watanabe, K.; Artús, L. Tuning the Direct and Indirect Excitonic Transitions of h-BN by Hydrostatic Pressure. *J. Phys. Chem. C Nanomater. Interfaces* **2021**, *125*, 12880–12885. [CrossRef]

116. Solozhenko, V.; Will, G.; Elf, F. Isothermal compression of hexagonal graphite-like boron nitride up to 12 GPa. *Solid State Commun.* **1995**, *96*, 1–3. [CrossRef]

117. Nye, J.; Nye, P. *Physical Properties of Crystals: Their Representation by Tensors and Matrices*; Oxford Science Publications, Clarendon Press: Oxford, UK, 1985.

118. Hamano, T.; Matsuda, T.; Asamoto, Y.; Noma, M.; Hasegawa, S.; Yamashita, M.; Urabe, K.; Eriguchi, K. Spectroscopic ellipsometry characterization of boron nitride films synthesized by a reactive plasma-assisted coating method. *Appl. Phys. Lett.* **2022**, *120*, 031904. [CrossRef]

119. Elias, C.; Fugallo, G.; Valvin, P.; L'Henoret, C.; Li, J.; Edgar, J.H.; Sottile, F.; Lazzeri, M.; Ouerghi, A.; Gil, B.; et al. Flat Bands and Giant Light-Matter Interaction in Hexagonal Boron Nitride. *Phys. Rev. Lett.* **2021**, *127*, 137401. [CrossRef]

120. Artús, L.; Feneberg, M.; Attaccalite, C.; Edgar, J.H.; Li, J.; Goldhahn, R.; Cuscó, R. Ellipsometry Study of Hexagonal Boron Nitride Using Synchrotron Radiation: Transparency Window in the Far-UVC. *Adv. Photonics Res.* **2021**, *2*, 2000101. [CrossRef]

121. Ribeiro, R.M.; Peres, N.M.R. Stability of boron nitride bilayers: Ground-state energies, interlayer distances, and tight-binding description. *Phys. Rev. B* **2011**, *83*, 235312. [CrossRef]

122. Wickramaratne, D.; Weston, L.; Van de Walle, C.G. Monolayer to Bulk Properties of Hexagonal Boron Nitride. *J. Phys. Chem. C* **2018**, *122*, 25524–25529. [CrossRef]

123. Aggoune, W.; Cocchi, C.; Nabok, D.; Rezouali, K.; Belkhir, M.A.; Draxl, C. Dimensionality of excitons in stacked van der Waals materials: The example of hexagonal boron nitride. *Phys. Rev. B* **2018**, *97*, 241114. [CrossRef]

124. Sponza, L.; Amara, H.; Attaccalite, C.; Latil, S.; Galvani, T.; Paleari, F.; Wirtz, L.; Ducastelle, F.m.c. Direct and indirect excitons in boron nitride polymorphs: A story of atomic configuration and electronic correlation. *Phys. Rev. B* **2018**, *98*, 125206. [CrossRef]

125. Mengle, K.A.; Kioupakis, E. Impact of the stacking sequence on the bandgap and luminescence properties of bulk, bilayer, and monolayer hexagonal boron nitride. *APL Mater.* **2019**, *7*, 21106. [CrossRef]

126. Kobayashi, K.; Watanabe, K.; Taniguchi, T. First-Principles Study of Various Hexagonal BN Phases. *J. Phys. Soc. Jpn.* **2007**, *76*, 104707. [CrossRef]

127. Gilbert, S.M.; Pham, T.; Dogan, M.; Oh, S.; Shevitski, B.; Schumm, G.; Liu, S.; Ercius, P.; Aloni, S.; Cohen, M.L.; et al. Alternative stacking sequences in hexagonal boron nitride. *2D Mater.* **2019**, *6*, 21006. [CrossRef]

128. Claudio, C.; Tim, G. Polymorphism of bulk boron nitride. *Sci. Adv.* **2019**, *5*, eaau5832.

129. Arnaud, B.; Lebègue, S.; Rabiller, P.; Alouani, M. Huge Excitonic Effects in Layered Hexagonal Boron Nitride. *Phys. Rev. Lett.* **2006**, *96*, 26402. [CrossRef]

130. Cassabois, G.; Valvin, P.; Gil, B. Hexagonal boron nitride is an indirect bandgap semiconductor. *Nat. Photonics* **2016**, *10*, 262–266. [CrossRef]

131. Cassabois, G.; Valvin, P.; Gil, B. Intervalley scattering in hexagonal boron nitride. *Phys. Rev. B* **2016**, *93*, 35207. [CrossRef]

132. Vuong, T.Q.P.; Cassabois, G.; Valvin, P.; Jacques, V.; Cuscó, R.; Artús, L.; Gil, B. Overtones of interlayer shear modes in the phonon-assisted emission spectrum of hexagonal boron nitride. *Phys. Rev. B* **2017**, *95*, 45207. [CrossRef]

133. Valvin, P.; Pelini, T.; Cassabois, G.; Zobelli, A.; Li, J.; Edgar, J.H.; Gil, B. Deep ultraviolet hyperspectral cryomicroscopy in boron nitride: Photoluminescence in crystals with an ultra-low defect density. *AIP Adv.* **2020**, *10*, 75025. [CrossRef]

134. Zan, R.; Bangert, U.; Ramasse, Q.; Novoselov, K.S. Imaging of Bernal stacked and misoriented graphene and boron nitride: Experiment and simulation. *J. Microsc.* **2011**, *244*, 152–158. [CrossRef] [PubMed]

135. Shmeliov, A.; Kim, J.S.; Borisenko, K.B.; Wang, P.; Okunishi, E.; Shannon, M.; Kirkland, A.I.; Nellist, P.D.; Nicolosi, V. Impurity induced non-bulk stacking in chemically exfoliated h-BN nanosheets. *Nanoscale* **2013**, *5*, 2290–2294. [CrossRef]

136. Rousseau, A.; Moret, M.; Valvin, P.; Desrat, W.; Li, J.; Janzen, E.; Xue, L.; Edgar, J.H.; Cassabois, G.; Gil, B. Determination of the optical bandgap of the Bernal and rhombohedral boron nitride polymorphs. *Phys. Rev. Mater.* **2021**, *5*, 64602. [CrossRef]

137. Li, J.; Wang, J.; Zhang, X.; Elias, C.; Ye, G.; Evans, D.; Eda, G.; Redwing, J.M.; Cassabois, G.; Gil, B.; et al. Hexagonal Boron Nitride Crystal Growth from Iron, a Single Component Flux. *ACS Nano* **2021**, *15*, 7032–7039. [CrossRef]

138. Pelini, T.; Elias, C.; Page, R.; Xue, L.; Liu, S.; Li, J.; Edgar, J.H.; Dréau, A.; Jacques, V.; Valvin, P.; et al. Shallow and deep levels in carbon-doped hexagonal boron nitride crystals. *Phys. Rev. Mater.* **2019**, *3*, 94001. [CrossRef]

139. Vuong, T.Q.P.; Cassabois, G.; Valvin, P.; Ouerghi, A.; Chassagneux, Y.; Voisin, C.; Gil, B. Phonon-Photon Mapping in a Color Center in Hexagonal Boron Nitride. *Phys. Rev. Lett.* **2016**, *117*, 97402. [CrossRef]

140. Xu, L.; Zhan, J.; Hu, J.; Bando, Y.; Yuan, X.; Sekiguchi, T.; Mitome, M.; Golberg, D. High-Yield Synthesis of Rhombohedral Boron Nitride Triangular Nanoplates. *ChemInform* **2007**, *38*, 2141–2144. [CrossRef]

141. Sharma, S.; Souqui, L.; Pedersen, H.; Hogberg, H. Chemical vapor deposition of sp^2-boron nitride films on Al$_2$O$_3$ (0001), (11$\bar{2}$0), (1$\bar{1}$02), and (10$\bar{1}$0) substrates. *J. Vac. Sci. Technol. A* **2022**, *40*, 33404. [CrossRef]

142. de Brucker, L.; Moret, M.; Gil, B.; Desrat, W. Hexagonal and rhombohedral polytypes in indium selenide films grown on c-plane sapphire. *AIP Adv.* **2022**, *12*, 55308. [CrossRef]

143. Mak, K.F.; Lee, C.; Hone, J.; Shan, J.; Heinz, T.F. Atomically Thin MoS$_2$: A New Direct-Gap Semiconductor. *Phys. Rev. Lett.* **2010**, *105*, 136805. [CrossRef]

144. Rytova, N.S. The screened potential of a point charge in a thin film. *Mosc. Univ. Phys. Bull.* **1967**, *3*, 18.

145. Keldysh, L.V. Coulomb interaction in thin semiconductor and semimetal films. *JETP Lett.* **1979**, *29*, 658.

146. Kirichenko, E.V.; Stephanovich, V.A. The influence of Coulomb interaction screening on the excitons in disordered two-dimensional insulators. *Sci. Rep.* **2021**, *11*, 11956. [CrossRef]

147. Cudazzo, P.; Sponza, L.; Giorgetti, C.; Reining, L.; Sottile, F.; Gatti, M. Exciton Band Structure in Two-Dimensional Materials. *Phys. Rev. Lett.* **2016**, *116*, 66803. [CrossRef]

148. Koskelo, J.; Fugallo, G.; Hakala, M.; Gatti, M.; Sottile, F.; Cudazzo, P. Excitons in van der Waals materials: From monolayer to bulk hexagonal boron nitride. *Phys. Rev. B* **2017**, *95*, 35125. [CrossRef]

149. Henriques, J.C.G.; Ventura, G.B.; Fernandes, C.D.M.; Peres, N.M.R. Optical absorption of single-layer hexagonal boron nitride in the ultraviolet. *J. Phys. Condens. Matter.* **2020**, *32*, 25304. [CrossRef] [PubMed]

150. Elias, C.; Valvin, P.; Pelini, T.; Summerfield, A.; Mellor, C.J.; Cheng, T.S.; Eaves, L.; Foxon, C.T.; Beton, P.H.; Novikov, S.V.; et al. Direct band-gap crossover in epitaxial monolayer boron nitride. *Nat. Commun.* **2019**, *10*, 2639. [CrossRef] [PubMed]
151. Wang, P.; Lee, W.; Corbett, J.P.; Koll, W.H.; Vu, N.M.; Laleyan, D.A.; Wen, Q.; Wu, Y.; Pandey, A.; Gim, J.; et al. Scalable Synthesis of Monolayer Hexagonal Boron Nitride on Graphene with Giant Bandgap Renormalization. *Adv. Mater.* **2022**, *34*, 2201387. [CrossRef] [PubMed]
152. Rousseau, A.; Ren, L.; Durand, A.; Valvin, P.; Gil, B.; Watanabe, K.; Taniguchi, T.; Urbaszek, B.; Marie, X.; Robert, C.; et al. Monolayer Boron Nitride: Hyperspectral Imaging in the Deep Ultraviolet. *Nano Lett.* **2021**, *21*, 10133–10138. [CrossRef] [PubMed]
153. Román, R.J.P.; Costa, F.J.R.C.; Zobelli, A.; Elias, C.; Valvin, P.; Cassabois, G.; Gil, B.; Summerfield, A.; Cheng, T.S.; Mellor, C.J.; et al. Band gap measurements of monolayer h-BN and insights into carbon-related point defects. *2D Mater.* **2021**, *8*, 44001. [CrossRef]
154. Ares, P.; Cea, T.; Holwill, M.; Wang, Y.B.; Roldán, R.; Guinea, F.; Andreeva, D.V.; Fumagalli, L.; Novoselov, K.S.; Woods, C.R. Piezoelectricity in Monolayer Hexagonal Boron Nitride. *Adv. Mater.* **2020**, *32*, 1905504. [CrossRef]
155. Rousseau, A.; Valvin, P.; Desrat, W.; Xue, L.; Li, J.; Edgar, J.H.; Cassabois, G.; Gil, B. Bernal Boron Nitride Crystals Identified by Deep-Ultraviolet Cryomicroscopy. *ACS Nano* **2022**, *16*, 2756–2761. [CrossRef]

Article

Thermodynamic Properties of 1,5-Pentanediamine Adipate Dihydrate in Three Binary Solvent Systems from 278.15 K to 313.15 K

Liang Li [1,†], Yihan Zhao [1,†], Baohong Hou [1,2], Han Feng [1], Na Wang [1,2,*], Dong Liu [3], Yingjie Ma [1], Ting Wang [1,2,*] and Hongxun Hao [1,2]

[1] National Engineering Research Center of Industrial Crystallization Technology, School of Chemical Engineering and Technology, Tianjin University, Tianjin 300072, China; 2018207068@tju.edu.cn (L.L.); yihan@tju.edu.cn (Y.Z.); houbaohong@tju.edu.cn (B.H.); fhan@tju.edu.cn (H.F.); yingjiema207@tju.edu.cn (Y.M.); hongxunhao@tju.edu.cn (H.H.)

[2] Collaborative Innovation Center of Chemical Science and Engineering (Tianjin), Tianjin 300072, China

[3] Hebei Meibang Engineering & Technology Co., Ltd., Shijiazhuang 050000, China; 13933883607@163.com

* Correspondence: wangna224@tju.edu.cn (N.W.); wang_ting@tju.edu.cn (T.W.); Tel.: +86-22-27405754 (N.W.); Fax: +86-22-27374971 (N.W.)

† These authors contributed equally to this work.

Abstract: In this work, solubility data of 1,5-pentanediamine adipate dihydrate in binary solvent systems of water + methanol, water + ethanol and water + N,N-dimethylformamide were experimentally measured via a static gravimetric method in the temperature range from 278.15 K to 313.15 K under atmospheric pressure. The results indicated that the solubility of 1,5-pentanediamine adipate dihydrate increased with the rising of temperature in all the selected binary solvent systems. For water + N,N-dimethylformamide, solubility increased as the mole fraction of water increased. However, the rising tendency changed when the temperature was higher than 303.15 K for water + methanol, and it would show a cosolvency phenomenon for water + ethanol. Furthermore, the solubility data were fitted with modified an Apelblat equation, NRTL model, combined nearly ideal binary solvent/Redlich Kister (CNIBS/R-K) model and Jouban–Acree model. The calculation results agreed well with the experimental data. Finally, the mixing thermodynamic properties of 1,5-pentanediamine adipate dihydrate in all tested solvents were calculated based on the experimental data and NRTL model.

Keywords: 1,5-pentanediamine adipate dihydrate; solubility; mixing thermodynamics

Citation: Li, L.; Zhao, Y.; Hou, B.; Feng, H.; Wang, N.; Liu, D.; Ma, Y.; Wang, T.; Hao, H. Thermodynamic Properties of 1,5-Pentanediamine Adipate Dihydrate in Three Binary Solvent Systems from 278.15 K to 313.15 K. *Crystals* **2022**, *12*, 877. https://doi.org/10.3390/cryst 12060877

Academic Editor: George D. Verros

Received: 23 May 2022
Accepted: 18 June 2022
Published: 20 June 2022

Publisher's Note: MDPI stays neutral with regard to jurisdictional claims in published maps and institutional affiliations.

1. Introduction

1,5-pentanediamine adipate dihydrate ($C_{11}H_{24}O_4N_2 \cdot 2H_2O$, CAS registry No. 2156592-12-8, Figure 1), is major monomer for the synthesis of bio-based polyamide 56 fiber [1]. Due to its excellent mechanical properties, heat resistance, abrasion resistance and self-lubricity, polyamide 56 is not only a renewable substitute for traditional materials, but also a unique and novel polymer applied in the textile and industrial plastics [2].

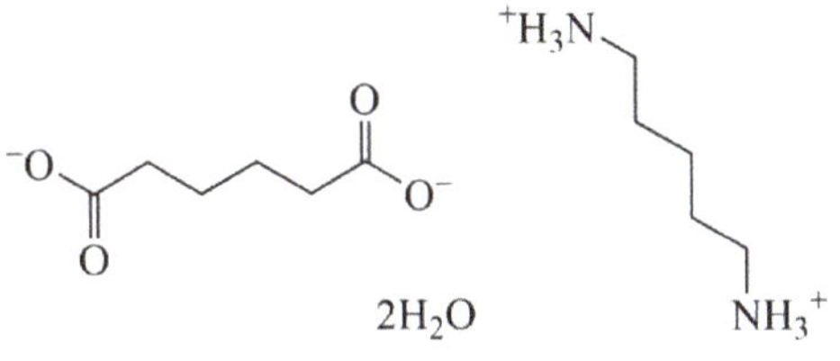

Figure 1. Molecular structure of 1,5-pentanediamine adipate dihydrate.

The performance of the polymer material is directly related to the purity of the polymerized monomer since it is greatly sensitive to impurities [3]. Therefore, the purification stage of 1,5-pentanediamine adipate dihydrate plays a critical role during the manufacturing process [4,5]. Crystallization, as a vital separation and purification process, is widely used in the pharmaceutical and food industries. As such, it is essential to know the physicochemical and thermodynamic properties of 1,5-pentanediamine adipate dihydrate to design and optimize the crystallization process [6]. Considering the high solubility of 1,5-pentanediamine adipate dihydrate in water, anti-solvent crystallization is a reasonable and efficient crystallization method for purifying it [7,8]. Besides, the temperature range from 278.15 K to 313.15 K is commonly used in industrial manufacturing and the product contains two molecules of water. However, no reports about the solubility of 1,5-pentanediamine adipate dihydrate in this temperature range have been published.

In this work, the solubility data of 1,5-pentanediamine adipate dihydrate in water + methanol, water + ethanol and water + N,N-dimethylformamide were determined in the range of 278.15 K to 313.15 K under 0.1 MPa. Besides, the modified Apelblat equation model, NRTL model, combined nearly ideal binary solvent/Redlich Kister (CNIBS/R-K) model, and Jouyban–Acree model were applied to the fitting of the experimental solubility data, respectively. Finally, the mixing thermodynamic properties, including the mixing enthalpy, the mixing entropy and the Gibbs free energy of 1,5-pentanediamine adipate dihydrate in different binary solvents were calculated and discussed.

2. Experimental Section

2.1. Materials

1,5-pentanediamine adipate dihydrate ($\geq$0.99 mass fraction) was offered by Hebei Meibang Engineering Technology Co., Ltd. (Hebei, China). Deionized water was prepared by an ultrapure water system (YPYD Co., Nanjing, China) in the laboratory. Methanol and ethanol were purchased from Guangda Pharmaceutical Co., Ltd. (Tianjin, China), and N,N-dimethylformamide was purchased from Benchmark Chemical Reagent Co., Ltd. (Tianjin, China). More details about the materials are listed in Table 1. All chemicals were used without further purification.

Table 1. Sources and mass fraction purity of chemicals.

Chemical Name	CAS Reg. No.	Mass Fraction Purity	Source	Analysis Method
1,5-pentanediamine adipate dihydrate	2156592-12-8	$\geq$0.99	Hebei Meibang Engineering Technology Co., Ltd., of China	HPLC [a]
1,5-pentanediamine adipate	13534-23-1	$\geq$0.99	Hebei Meibang Engineering Technology Co., Ltd., of China	HPLC [a]
methanol	67-56-1	$\geq$0.995	Tianjin Guangda Pharmaceutical Co., Ltd., China	GC [b]
ethanol	64-17-5	$\geq$0.997	Tianjin Guangda Pharmaceutical Co., Ltd., China	GC [b]
DMF	68-12-2	$\geq$0.995	Tianjin Benchmark Chemical Reagent Co., Ltd., China	GC [c]

[a] High performance liquid chromatography, determined by Hebei Meibang Engineering Technology Co., Ltd., of China. [b] Gas chromatography, determined by Tianjin Guangda Pharmaceutical Co., Ltd., China. [c] Gas chromatography, determined by Tianjin Benchmark Chemical Reagent Co., Ltd., China.

2.2. Apparatus and Methods

The gravimetric method was used to determine the solubility of 1,5-pentanediamine adipate dihydrate in different binary solvent mixtures [9]. Known mass of solvent and excess 1,5-pentanediamine adipate dihydrate were added to a 100 mL crystallizer. A constant temperature water bath (Nanjing Xianou Instrument Manufacturing Co., Ltd., China) was used to control the solution at specified temperature. The solution was agitated for 12 h to ensure that the solution reaches solid-liquid equilibrium. Then, the stirring was stopped and the solution was kept static for 3 h until the suspended particles settled down.

After that, the samples were drawn with a 5 mL syringe and filtered through a 0.45 μm nylon membrane into 50 mL pre-weighed beakers and the total weight was measured immediately. Besides, KF analyzer (C20s, Mettler Toledo, Switzerland) was used to determine the amount of water in the saturated solution after equilibration. Finally, the beakers were put into a vacuum oven and dried at temperature of 323.15 K. The quality of the beaker with the samples was recorded periodically until the total weight did not change any more, to ensure that there were no solvents in the solid after the drying treatment. The solubility data of 1,5-pentanediamine adipate dihydrate were calculated from the mass of the remaining solid. The crystal form of the dried solid was determined using powder X-ray diffraction (PXRD). All the samples were weighed by an analytic balance (AL204, Mettler Toledo, Switzerland) with an accuracy of ±0.0001 g. The procedure was repeated three times to determine the error of the experiments.

The solubility of 1,5-pentanediamine adipate dihydrate described in the mole percentage x_p was calculated by Equation (1):

$$x_p = \frac{\frac{m_p}{M_p}}{\frac{m_p}{M_p} + \frac{m_w}{M_w} + \frac{m_i}{M_i}} \tag{1}$$

where x, m and M represent the mole fraction solubility, mass and molar mass, respectively. The subscript p, w, and i represent 1,5-pentanediamine adipate dihydrate, water and the other organic solvents, respectively. The mass of 1,5-pentanediamine adipate dihydrate (m_p) was calculated by using experimental results. Besides, the masses of water and organic solvents were determined by the pre-designed mixed solvents of different compositions.

The samples were dried to completely convert into anhydrous 1,5-pentanediamine adipate and the mass of dissolved 1,5-pentanediamine adipate dihydrate was calculated by Equation (2):

$$m_p = \frac{M_p}{M_a} m_a \tag{2}$$

where m_a and M_a represent the mass and molar mass of anhydrous 1,5-pentanediamine adipate, respectively. The mole fraction solubility of anhydrous 1,5-pentanediamine adipate was calculated as follows:

$$x_a = \frac{\frac{m_p}{M_p}}{\frac{m_w}{M_w} + \frac{m_i}{M_i} + 3\frac{m_p}{M_p}} \tag{3}$$

The initial mole fraction of water x_w in the mixed solution was calculated as follows:

$$x_w = \frac{\frac{m_w}{M_w}}{\frac{m_w}{M_w} + \frac{m_i}{M_i}} \tag{4}$$

where m_w and m_i were determined by an analytic balance during the process of preparing mixing solvent. The final mole fraction of water x_w^0 in three binary solvent mixtures was calculated as follows.

$$x_w^0 = \frac{\frac{m_w}{M_w} + 2\frac{m_p}{M_p}}{\frac{m_w}{M_w} + \frac{m_i}{M_i} + 2\frac{m_p}{M_p}} \tag{5}$$

2.3. Characterization of 1,5-Pentanediamine Adipate Dihydrate

The crystal form of 1,5-pentanediamine adipate dihydrate, before and after solubility measurement, was determined using powder X-ray diffraction (PXRD), which was carried out on Rigaku D/max-2500 (Rigaku, Japan). The diffraction angle (2θ) ranged from $2°$ to $40°$ with a scanning rate of $8°$/min and step size of $0.02°$.

In order to characterize the melting temperature and melting enthalpy of 1,5-pentanediamine adipate dihydrate, the thermal behaviors of 1,5-pentanediamine adipate dihydrate were measured using TGA/DSC (Mettler Toledo, Zurich, Switzerland) under the protection of nitrogen. Besides, the thermal behaviors of dehydrated raw materials were also detected to eliminate the influence of dehydration peaks. The measurements were conducted within the temperature ranging from 298.15 K to 473.15 K at a heating rate of 3 K/min. The DSC instrument was calibrated by indium and zinc, of which the melting temperature was measured three times. The standard uncertainties of melting points and solid-to-solid transition temperature were estimated to be 1 K while the relative standard uncertainties of melting enthalpy and solid-to-solid transition enthalpy were estimated to be 5%.

3. Thermodynamic Models

3.1. The Modified Apelblat Equation

Combined with the Clausius–Clapeyron relationship, Apelblat et al. proposed a semi-empirical model in 1999 as Equation (6) [10,11].

$$\ln x = A + \frac{B}{T} + C \ln T \tag{6}$$

where x is the mole fraction solubility of solute at absolute temperature T. A, B and C are model constants.

3.2. NRTL Equation

The solubility of solute can be expressed by a general thermodynamic model, which is written as follows [12].

$$\ln x = -\ln \gamma_i + \frac{\Delta_f H}{R} \left(\frac{1}{T_m} - \frac{1}{T} \right) - \frac{1}{RT} \int_{T_m}^{T} \Delta C_p dT + \frac{1}{R} \int_{T_m}^{T} \frac{\Delta C_p}{T} dT \tag{7}$$

where γ_i stands for the activity coefficient, $\Delta_f H$ is the enthalpy of fusion, R is the gas constant; T and T_m refer to the solution temperature and the melting point of solute, respectively, ΔC_p refers to the difference of the molar heat capacity between melting state and solid state of the solute. When a solid undergoes a phase transition, the solubility equation should include a term for the contribution of the transition as following [13].

$$\ln x = -\ln \gamma_i + \frac{\Delta_f H}{R} \left(\frac{1}{T_m} - \frac{1}{T} \right) + \frac{\Delta_{tr} H}{R} \left(\frac{1}{T_{tr}} - \frac{1}{T} \right) - \frac{1}{RT} \int_{T_m}^{T} \Delta C_p dT + \frac{1}{R} \int_{T_m}^{T} \frac{\Delta C_p}{T} dT \tag{8}$$

where $\Delta_{tr} H$ and T_{tr} stand for the enthalpy of the transition and the transition temperature of the solute, respectively.

Considering the last two parts of the Equation (8) are less important than the first three parts due to the negligible value of ΔC_p, the solubility of the solute can be expressed by a simplified thermodynamic model, which is written as follows [14].

$$\ln x = \frac{\Delta_f H}{R} \left(\frac{1}{T_m} - \frac{1}{T} \right) + \frac{\Delta_{tr} H}{R} \left(\frac{1}{T_{tr}} - \frac{1}{T} \right) - \ln \gamma_i \tag{9}$$

The nonrandom two-liquid (NRTL) model, which is based on the theory of solid–liquid phase equilibrium can be used to calculate the activity coefficients. It can be shown as Equations (10)–(13) [15,16].

$$\ln \gamma_i = \frac{\left(G_{ji}x_j + G_{ki}x_k \right)\left(\tau_{ji}G_{ji}x_j + \tau_{ki}G_{ki}x_k \right)}{\left(x_i + G_{ji}x_j + G_{ki}x_k \right)^2} + \frac{\tau_{ij}G_{ij}x_j^2 + G_{ij}G_{kj}x_jx_k \left(\tau_{ij} - \tau_{kj} \right)}{\left(x_j + G_{ij}x_i + G_{kj}x_k \right)^2}$$
$$+ \frac{\tau_{ik}G_{ik}x_k^2 + G_{ik}G_{jk}x_jx_k \left(\tau_{ik} - \tau_{jk} \right)}{\left(x_k + G_{ik}x_i + G_{jk}x_j \right)^2} \tag{10}$$

where i, j, k are the three components of the solution system. Model parameters τ_{ij} and G_{ij} can be calculated as follows:

$$G_{ij} = \exp\left(-\alpha_{ij}\tau_{ij} \right) \tag{11}$$

$$\tau_{ij} = \frac{\Delta g_{ij}}{RT} \tag{12}$$

$$\alpha_{ij} = \alpha_{ji} \tag{13}$$

where α_{ij} represents the non-randomness between i and j components and Δg_{ij} stands for the cross-interaction energy.

3.3. The CNIBS/R-K Model

The combined nearly ideal binary solvent/Redlich–Kister (CNIBS/R–K) model, which describes the relationship between solubility and solvent composition, is suitable for binary solvent systems [17]. It describes the relationship between solubility and solvent composition as Equation (14).

$$\ln x = x_a \ln X_a + x_b \ln X_b + x_a x_b \sum_{i=0}^{N} S_i (x_a - x_b)^i \tag{14}$$

where x_a, x_b stand for the initial mole fraction of water and organic solvents in binary solvent mixtures, respectively. X_a, X_b are the mole fraction solubility of 1,5-pentanediamine adipate dihydrate in pure water and organic solvents, respectively. S_i is the model constant and N refers to the number of the solvents.

For binary solvent systems, substituting $N = 2$ and $x_a = 1 - x_b$ into Equation (14) can result in a new simplified equation as following.

$$\ln x = B_0 + B_1 x_a + B_2 x_a^2 + B_3 x_a^3 + B_4 x_a^4 \tag{15}$$

where B_0 to B_4 are empirical model parameters.

3.4. The Jouyban–Acree Model

The Jouyban–Acree model is widely used to describe the solubility by considering both the composition of the solution and the temperature [18]. The model is expressed as Equation (16):

$$\ln x = x_a \ln X_a + x_b \ln X_b + x_a x_b \sum_{i=0}^{N} \frac{J_i (x_a - x_b)^i}{T} \tag{16}$$

Subsequently, Jouyban et al. proposed to combine the Jouyban–Acree model with the van't Hoff model to obtain a seven-parameters mixed model [19] as Equation (17).

$$\ln x = A_0 + A_1 x_a + \frac{A_2 + A_3 x_a + A_4 x_a^2 + A_5 x_a^3 + A_6 x_a^4}{T} \tag{17}$$

where A_0 to A_6 are empirical model parameters.

To evaluate the fitting accuracy, the average relative deviation (ARD) and root mean-square deviations ($RMSD$) were calculated [20,21].

$$ARD = \frac{1}{N} \sum_{i=1}^{N} \left| \frac{x_i^{cal} - x_i^{exp}}{x_i^{exp}} \right| \tag{18}$$

$$RMSD = \sqrt{\frac{\sum_{i=1}^{N} \left(x_i^{cal} - x_i^{exp} \right)^2}{N}} \tag{19}$$

where N stands for the total number of experiments. x_i^{exp} and x_i^{cal} refer to the experimental mole fraction solubility and calculated mole fraction solubility of 1,5-pentanediamine adipate dihydrate, respectively.

3.5. Solution Mixing Thermodynamics

For the ideal solution, its mixing properties, including Gibbs energy, enthalpy, and entropy, can be expressed as follows [22].

$$\Delta_m G^{id} = RT \sum_{i=1}^{n} x_i \ln x_i \tag{20}$$

$$\Delta_m H^{id} = 0 \tag{21}$$

$$\Delta_m S^{id} = -R \sum_{i=1}^{n} x_i \ln x_i \tag{22}$$

where x_i represent the mole fraction of each component. The mixing thermodynamic properties of the real solution system can be calculated by using Equation (23).

$$\Delta_m M = M^E + \Delta_m M^{id} \tag{23}$$

where M refers to the Gibbs free energy (G), enthalpy (H) or entropy (S); $\Delta_m M^{id}$ and M^E represent the mixing properties of the ideal solution and excess properties, respectively. Excess mixing properties (M^E) can be calculated using the Equations (24)–(26).

$$G^E = RT \sum_{i=1}^{n} x_i \ln \gamma_i \tag{24}$$

$$H^E = -RT^2 \sum_{i=1}^{n} x_i \left(\frac{\partial \ln \gamma_i}{\partial T} \right)_{P,x} \tag{25}$$

$$S^E = \frac{H^E - G^E}{T} \tag{26}$$

where γ_i refer to the activity coefficient, which will be calculated by the NRTL model in this work.

4. Results and Discussion

4.1. Identification and Characterization of Materials

The powder X-ray diffraction (PXRD) patterns of 1,5-pentanediamine adipate dihydrate are shown in Figure 2. It can be found that the 1,5-pentanediamine adipate dihydrate crystals obtained in this work remained unchanged in each experiment, indicating no phase transformation occurred during solid–liquid equilibrium. Furthermore, the results of random KF titration experiments for the cases with a large organic solvent content at the highest studied temperatures are shown in Table 2. It can be found that the amount of water $x_w^{0,KF}$ agrees with the calculated data in Tables 3–5, indicating no additional release of water from the hydrated solid phase upon equilibration.

Table 2. The amount of water x_w^0 upon equilibration.

Sample	$x_w^{0,KF}$
water + methanol (x_w = 0.360, T = 313.15 K)	0.562
water + ethanol (x_w = 0.447, T = 313.15 K)	0.553
water + DMF (x_w = 0.648, T = 313.15 K)	0.656

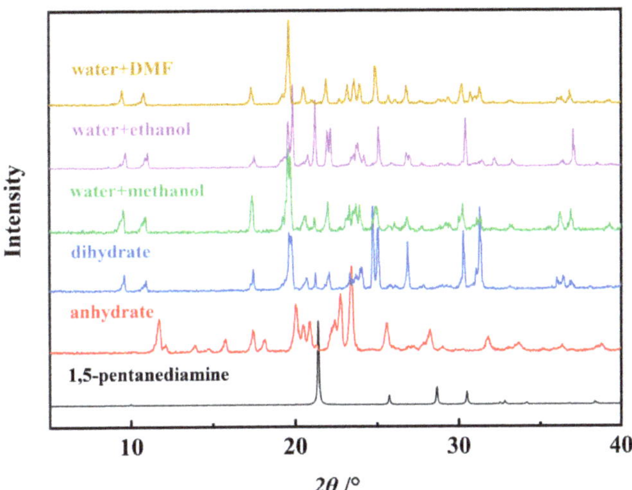

Figure 2. Powder X-ray diffraction pattern of raw material and residual solid in mixed systems: water + DMF (x_w = 0.648), water + ethanol (x_w = 0.447) and water + methanol (x_w = 0.360) at T = 298.15 K.

Table 3. Mole fraction solubility of 1,5-pentanediamine adipate dihydrate in water + methanol from 278.15 K to 313.15 K (p = 0.1 MPa) [a,b].

T/K	x_w^0	$10^2 x_a$	$10^2 x_p^{exp}$	$10^2 x_p^{Apel}$	$10^2 x_p^{NRTL}$	$10^2 x_p^{RK}$	$10^2 x_p^{JA}$
			$x_w = 0.360$				
278.15	0.432	5.412	6.029	5.738	5.934	6.036	5.795
283.15	0.440	5.896	6.641	6.837	7.027	6.641	6.944
288.15	0.455	6.976	8.055	8.132	8.117	8.062	8.270
293.15	0.474	8.233	9.793	9.654	9.439	9.793	9.789
298.15	0.488	9.158	11.14	11.43	11.27	11.14	11.52
303.15	0.520	11.17	14.29	13.53	13.05	13.74	13.49
308.15	0.535	11.74	15.85	15.97	15.93	15.87	15.71
313.15	0.569	13.86	19.01	18.82	18.98	18.89	18.21
			$x_w = 0.491$				
278.15	0.554	5.827	6.552	6.308	6.608	6.503	6.270
283.15	0.562	6.544	7.480	7.340	7.383	7.483	7.365
288.15	0.570	7.265	8.446	8.529	8.399	8.407	8.603
293.15	0.583	8.303	9.893	9.897	9.602	9.891	9.996
298.15	0.591	8.995	10.89	11.46	11.26	10.90	11.55
303.15	0.610	10.53	13.26	13.27	13.09	13.26	13.29
308.15	0.629	11.97	15.70	15.33	15.47	15.58	15.22
313.15	0.644	13.10	17.65	17.70	18.66	17.65	17.36
			$x_w = 0.600$				
278.15	0.651	6.044	6.830	6.730	7.161	6.956	6.703
283.15	0.658	6.797	7.816	7.692	7.777	7.794	7.744
288.15	0.664	7.455	8.704	8.786	8.652	8.772	8.901
293.15	0.673	8.434	10.08	10.02	9.708	10.09	10.18
298.15	0.679	9.043	10.97	11.44	11.18	10.94	11.59
303.15	0.693	10.48	13.19	13.04	12.86	13.19	13.14
308.15	0.704	11.61	15.04	14.85	15.12	15.22	14.84
313.15	0.715	12.66	16.85	16.90	18.05	16.83	16.70

Table 3. *Cont.*

T/K	x_w^0	$10^2 x_a$	$10^2 x_p^{\text{exp}}$	$10^2 x_p^{\text{Apel}}$	$10^2 x_p^{\text{NRTL}}$	$10^2 x_p^{\text{RK}}$	$10^2 x_p^{\text{JA}}$
			$x_w = 0.692$				
278.15	0.735	6.571	7.517	7.241	7.496	7.371	7.086
283.15	0.738	6.900	7.952	8.123	8.136	8.004	8.071
288.15	0.744	7.761	9.127	9.117	8.823	9.108	9.151
293.15	0.750	8.590	10.30	10.23	9.784	10.26	10.33
298.15	0.754	9.099	11.05	11.50	11.11	11.08	11.61
303.15	0.763	10.39	13.04	12.92	12.67	13.06	13.01
308.15	0.771	11.43	14.72	14.52	14.76	14.71	14.52
313.15	0.779	12.33	16.28	16.33	17.45	16.30	16.15
			$x_w = 0.771$				
278.15	0.804	6.687	7.670	7.513	7.887	7.730	7.418
283.15	0.806	7.136	8.270	8.360	8.382	8.225	8.348
288.15	0.811	7.971	9.422	9.304	8.977	9.380	9.356
293.15	0.814	8.604	10.32	10.35	9.874	10.36	10.44
298.15	0.818	9.221	11.23	11.52	11.05	11.22	11.61
303.15	0.824	10.28	12.86	12.82	12.51	12.84	12.87
308.15	0.829	11.19	14.32	14.27	14.44	14.19	14.22
313.15	0.834	12.14	16.15	15.88	16.00	15.93	15.66
			$x_w = 0.900$				
278.15	0.915	7.119	8.247	8.028	8.415	8.250	7.947
283.15	0.916	7.406	8.638	8.760	8.797	8.654	8.771
288.15	0.918	8.117	9.628	9.567	9.259	9.664	9.647
293.15	0.919	8.710	10.48	10.45	9.991	10.46	10.57
298.15	0.920	9.293	11.34	11.43	10.98	11.34	11.55
303.15	0.922	9.977	12.38	12.51	12.26	12.39	12.59
308.15	0.924	10.59	13.36	13.69	13.91	13.45	13.68
313.15	0.927	11.48	15.36	14.99	16.09	15.37	14.83
			$x_w = 1.000$				
278.15	1.000	7.313	8.512	8.263	8.843	8.509	8.334
283.15	1.000	7.506	8.869	8.945	9.107	8.865	9.060
288.15	1.000	8.106	9.613	9.691	9.490	9.603	9.821
293.15	1.000	8.774	10.57	10.50	10.07	10.57	10.61
298.15	1.000	9.152	11.18	11.39	10.94	11.18	11.44
303.15	1.000	9.869	12.27	12.37	12.06	12.27	12.31
308.15	1.000	10.62	13.40	13.43	13.53	13.37	13.21
313.15	1.000	11.50	14.78	14.59	15.46	14.78	14.14

[a] x_p^{exp} is the experimental mole fraction solubility of 1,5-pentanediamine adipate dihydrate; x_p^{Apel}, x_p^{NRTL}, x_p^{RK} and x_p^{JA} refer to the calculated mole fraction solubility according to the modified Apelblat equation, NRTL model, CNIBS/R-K model and Jouyban–Acree model, respectively. x_w is the initial mole fraction of water in three binary solvent mixtures; x_w^0 is the final mole fraction of water in three binary solvent mixtures; x_a is the mole fraction solubility of anhydrous 1,5-pentanediamine adipate. [b] Standard uncertainty is $u(T) = 0.03$ K, $u(p) = 0.3$ kPa. The relative standard uncertainty is $u_r(x_w) = 0.03$, $u_r(x_p) = 0.05$, $u_r(x_w^0) = 0.03$.

Table 4. Mole fraction solubility of 1,5-pentanediamine adipate dihydrate in water + ethanol from 278.15 K to 313.15 K ($p = 0.1$ MPa) [a,b].

T/K	x_w^0	$10^2 x_a$	$10^2 x_p^{\text{exp}}$	$10^2 x_p^{\text{Apel}}$	$10^2 x_p^{\text{NRTL}}$	$10^2 x_p^{\text{RK}}$	$10^2 x_p^{\text{JA}}$
			$x_w = 0.447$				
278.15	0.469	1.975	2.042	1.885	2.194	2.039	2.120
283.15	0.475	2.469	2.580	2.421	2.641	2.575	2.673
288.15	0.482	3.110	3.293	3.106	3.236	3.291	3.345
293.15	0.488	3.620	3.877	3.978	3.976	3.872	4.153
298.15	0.497	4.382	4.771	5.087	4.839	4.766	5.120
303.15	0.514	5.751	6.455	6.495	6.224	6.451	6.268
308.15	0.530	7.075	8.188	8.280	7.987	8.193	7.623
313.15	0.555	8.863	10.70	10.54	10.66	10.69	9.213

Table 4. *Cont.*

T/K	x_w^0	$10^2 x_a$	$10^2 x_p^{exp}$	$10^2 x_p^{Apel}$	$10^2 x_p^{NRTL}$	$10^2 x_p^{RK}$	$10^2 x_p^{JA}$
			$x_w = 0.581$				
278.15	0.613	3.790	4.074	4.175	4.364	4.114	4.041
283.15	0.621	4.678	5.126	4.951	4.912	5.198	4.874
288.15	0.626	5.189	5.751	5.852	5.636	5.783	5.840
293.15	0.634	6.055	6.844	6.892	6.573	6.899	6.954
298.15	0.644	7.110	8.235	8.090	7.761	8.297	8.233
303.15	0.654	8.069	9.561	9.466	9.328	9.605	9.693
308.15	0.662	8.940	10.81	11.04	11.29	10.77	11.35
313.15	0.677	10.33	12.94	12.84	13.94	12.99	13.22
			$x_w = 0.683$				
278.15	0.719	5.392	6.004	6.092	6.136	5.879	5.564
283.15	0.725	6.278	7.132	6.899	6.627	6.904	6.504
288.15	0.728	6.743	7.743	7.810	7.382	7.620	7.562
293.15	0.735	7.650	8.974	8.837	8.280	8.807	8.747
298.15	0.740	8.327	9.928	9.995	9.507	9.804	10.069
303.15	0.747	9.222	11.23	11.29	11.04	11.12	11.53
308.15	0.752	9.963	12.36	12.76	13.02	12.46	13.16
313.15	0.765	11.53	14.74	14.41	15.57	14.60	14.94
			$x_w = 0.764$				
278.15	0.794	6.131	6.943	6.824	7.398	7.011	6.600
283.15	0.797	6.674	7.652	7.620	7.861	7.840	7.539
288.15	0.801	7.344	8.553	8.516	8.487	8.724	8.572
293.15	0.805	8.074	9.567	9.522	9.337	9.705	9.703
298.15	0.808	8.699	10.46	10.65	10.48	10.42	10.93
303.15	0.813	9.558	11.74	11.92	11.93	11.80	12.28
308.15	0.819	10.54	13.27	13.34	13.81	13.24	13.74
313.15	0.826	11.91	15.15	14.94	16.23	15.28	15.31
			$x_w = 0.829$				
278.15	0.853	6.607	7.564	7.454	8.143	7.645	7.292
283.15	0.855	7.167	8.313	8.228	8.494	8.345	8.180
288.15	0.859	7.948	9.390	9.089	9.004	9.305	9.140
293.15	0.860	8.405	10.04	10.04	9.852	10.06	10.17
298.15	0.862	8.755	10.54	11.10	10.96	10.75	11.28
303.15	0.866	9.759	12.05	12.28	12.28	12.10	12.47
308.15	0.870	10.72	13.57	13.59	14.04	13.47	13.74
313.15	0.875	12.03	15.38	15.04	16.33	15.37	15.09
			$x_w = 0.928$				
278.15	0.939	7.163	8.307	8.151	8.558	8.205	8.152
283.15	0.940	7.588	8.889	8.855	8.833	8.781	8.904
288.15	0.941	8.140	9.662	9.628	9.274	9.660	9.696
293.15	0.942	8.632	10.36	10.47	9.956	10.29	10.52
298.15	0.943	9.255	11.28	11.40	10.86	11.12	11.39
303.15	0.944	9.959	12.35	12.42	12.06	12.29	12.30
308.15	0.945	10.56	13.31	13.53	13.62	13.39	13.25
313.15	0.947	12.13	15.05	14.76	15.65	15.01	14.24

[a] x_p^{exp} is the experimental mole fraction solubility of 1,5-pentanediamine adipate dihydrate; x_p^{Apel}, x_p^{NRTL}, x_p^{RK} and x_p^{JA} refer to the calculated mole fraction solubility according to the modified Apelblat equation, NRTL model, CNIBS/R-K model and Jouyban–Acree model, respectively. x_w is the initial mole fraction of water in three binary solvent mixtures; x_w^0 is the final mole fraction of water in three binary solvent mixtures; x_a is the mole fraction solubility of anhydrous 1,5-pentanediamine adipate. [b] Standard uncertainty is $u(T) = 0.03$ K, $u(p) = 0.3$ kPa. The relative standard uncertainty is $u_r(x_w) = 0.03$, $u_r(x_p) = 0.05$, $u_r(x_w^0) = 0.03$.

Table 5. Mole fraction solubility of 1,5-pentanediamine adipate dihydrate in water + DMF from 278.15 K to 313.15 K (p = 0.1 MPa) [a,b].

T/K	x_w^0	$10^2 x_a$	$10^2 x_p^{exp}$	$10^2 x_p^{Apel}$	$10^2 x_p^{NRTL}$	$10^2 x_p^{RK}$	$10^2 x_p^{JA}$
			$x_w = 0.648$				
278.15	0.648	0.1082	0.1084	0.1350	0.1095	0.1075	0.1237
283.15	0.649	0.2164	0.2173	0.1930	0.1630	0.2158	0.1806
288.15	0.649	0.2624	0.2637	0.2740	0.2402	0.2619	0.2602
293.15	0.650	0.3792	0.3820	0.3860	0.3571	0.3640	0.3701
298.15	0.651	0.4628	0.4671	0.5420	0.5190	0.4649	0.5204
303.15	0.652	0.6523	0.6609	0.7550	0.7410	0.6558	0.7234
308.15	0.655	1.059	1.081	1.046	1.116	1.073	0.9948
313.15	0.658	1.483	1.528	1.441	1.632	1.509	1.354
			$x_w = 0.741$				
278.15	0.745	0.8593	0.8940	1.030	1.003	0.8944	1.018
283.15	0.747	1.214	1.241	1.347	1.305	1.217	1.340
288.15	0.749	1.703	1.756	1.754	1.693	1.758	1.748
293.15	0.752	2.159	2.239	2.274	2.177	2.226	2.259
298.15	0.757	3.134	3.265	2.937	2.835	3.244	2.895
303.15	0.759	3.485	3.729	3.779	3.613	3.749	3.679
308.15	0.764	4.369	4.777	4.843	4.607	4.809	4.640
313.15	0.771	5.526	6.197	6.184	5.848	6.240	5.808
			$x_w = 0.811$				
278.15	0.823	3.087	3.291	2.997	3.079	3.108	2.677
283.15	0.823	3.292	3.465	3.535	3.625	3.553	3.298
288.15	0.826	3.835	4.090	4.167	4.242	4.107	4.034
293.15	0.828	4.498	4.918	4.911	4.955	4.950	4.900
298.15	0.830	4.949	5.711	5.785	5.839	5.665	5.914
303.15	0.835	6.023	6.762	6.810	6.724	6.726	7.093
308.15	0.839	7.030	8.026	8.013	7.787	7.980	8.457
313.15	0.843	8.013	9.374	9.422	9.078	9.295	10.02
			$x_w = 0.866$				
278.15	0.877	4.411	4.837	4.949	5.193	4.965	4.431
283.15	0.880	5.226	5.858	5.566	5.551	5.746	5.204
288.15	0.881	5.504	6.196	6.252	6.353	6.162	6.079
293.15	0.883	6.630	7.152	7.012	6.954	7.125	7.063
298.15	0.884	6.767	7.429	7.855	7.903	7.567	8.165
303.15	0.887	7.424	8.715	8.788	8.959	8.739	9.393
308.15	0.889	8.328	10.05	9.818	9.993	10.07	10.75
313.15	0.892	9.029	11.07	10.95	11.42	11.13	12.26
			$x_w = 0.9453$				
278.15	0.952	6.432	7.071	7.404	7.526	7.262	7.157
283.15	0.953	7.160	8.265	8.021	7.754	8.327	7.884
288.15	0.953	7.435	8.645	8.698	8.549	8.670	8.654
293.15	0.954	8.069	9.568	9.442	9.148	9.579	9.470
298.15	0.955	8.451	10.07	10.25	10.40	9.962	10.33
303.15	0.956	9.033	10.92	11.15	11.48	10.92	11.24
308.15	0.957	9.937	12.25	12.13	12.15	12.25	12.19
313.15	0.958	10.55	13.22	13.20	13.69	13.21	13.19

[a] x_p^{exp} is the experimental mole fraction solubility of 1,5-pentanediamine adipate dihydrate; x_p^{Apel}, x_p^{NRTL}, x_p^{RK} and x_p^{JA} refer to the calculated mole fraction solubility according to the modified Apelblat equation, NRTL model, CNIBS/R-K model and Jouyban–Acree model, respectively. x_w is the initial mole fraction of water in three binary solvent mixtures; x_w^0 is the final mole fraction of water in three binary solvent mixtures; x_a is the mole fraction solubility of anhydrous 1,5-pentanediamine adipate. [b] Standard uncertainty is $u(T)$ = 0.03 K, $u(p)$ = 0.3 kPa. The relative standard uncertainty is $u_r(x_w)$ = 0.03, $u_r(x_p)$ = 0.05, $u_r(x_w^0)$ = 0.03.

As shown in Figure 3, the dried solid was proved to completely convert into anhydrous 1,5-pentanediamine adipate by PXRD. The mass of dissolved 1,5-pentanediamine adipate dihydrate was calculated from the measured value of anhydrous 1,5-pentanediamine adipate.

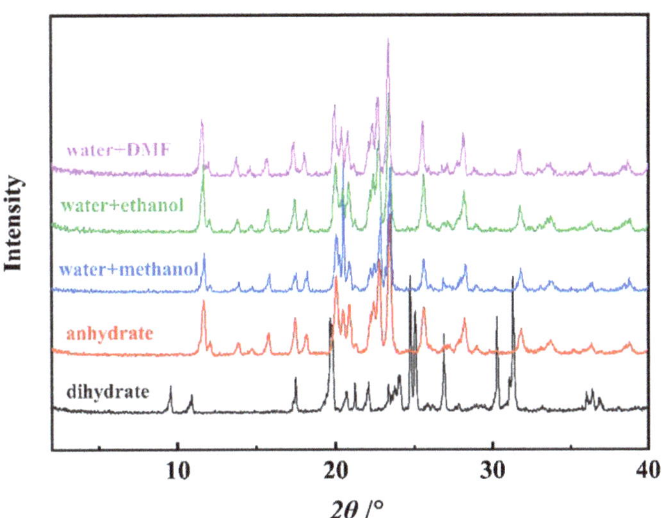

Figure 3. Powder X-ray diffraction pattern of dried solid in mixed systems: water + DMF ($x_w = 0.648$), water + ethanol ($x_w = 0.447$) and water + methanol ($x_w = 0.360$) at T = 298.15 K.

The TG/DSC results for the 1,5-pentanediamine adipate dihydrate crystal are shown in Figure 4. It was found that there was 12.49% weight loss between room temperature and 428.15 K, which is consistent with the theoretical water content of 1,5-pentanediamine adipate dihydrate (12.66 wt%). Meanwhile, there are two sharp endothermic peaks in the DSC curve, representing the dehydration process of two crystal waters in crystal. The loss of water from the lattice was divided into two stages, occurring at 341.71 K and 347.47 K, and the total dehydration enthalpy is 40.13 kJ·mol^{-1}, respectively. The broad endothermic peak at about 380 K indicated that the dehydration of polyamide 56 salt dihydrate was a slow process, accompanied by the endothermic phenomena related to the melting and dissolution in water. The endothermic peak at 456.10 K represents the decomposition process.

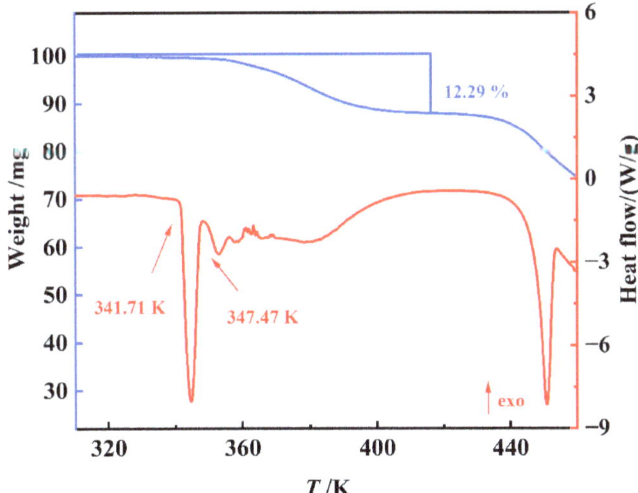

Figure 4. Thermal analysis (TG-DSC) of 1,5-pentanediamine adipate dihydrate.

Furthermore, the DSC results for anhydrous 1,5-pentanediamine adipate (Figure 5) show a sharp endothermic peak at 397.32 K, which should be the melting process, and the melting enthalpy is 22.13 kJ·mol^{-1}. As shown in Figure 6, the melting process of anhydrous 1,5-pentanediamine adipate at 397.32 K was proved through polarized optical microscopy.

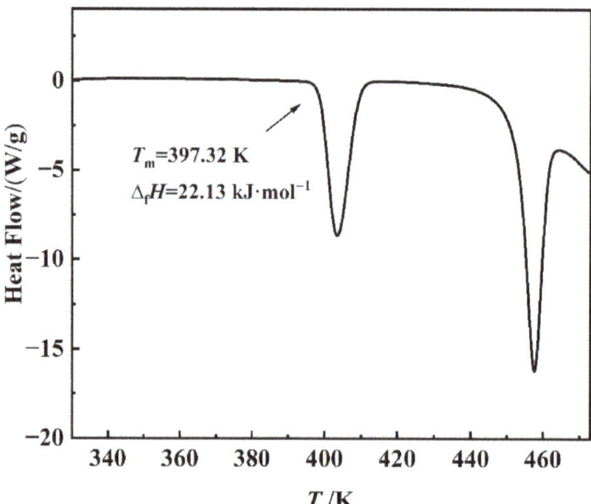

Figure 5. DSC of 1,5-pentanediamine adipate dihydrate after dehydration.

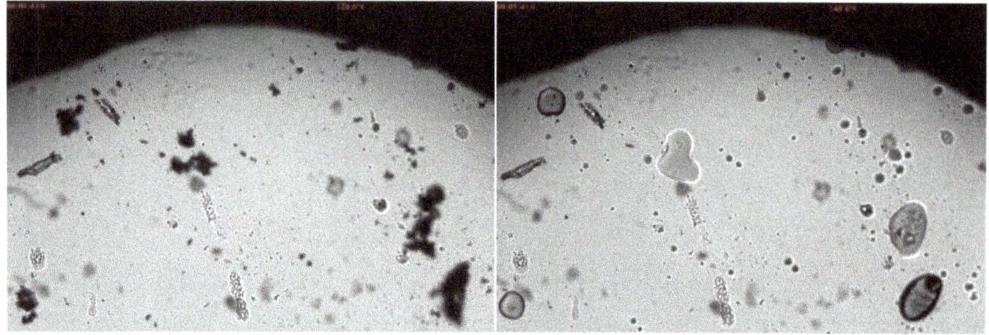

Figure 6. Polarized optical microscopy of anhydrous 1,5-pentanediamine adipate (**left**: 120.6 °C, **right**: 140.6 °C).

4.2. Solubility of 1,5-Pentanediamine Adipate Dihydrate in Binary Mixed Solvents

The experimental solubility data of 1,5-pentanediamine adipate dihydrate are listed in Tables 3–5 and are plotted in 3D mode in Figures 7–9. The results show that the solubility of 1,5-pentanediamine adipate dihydrate is positively correlated with temperature in all the tested three solvent systems. At a fixed temperature and solvent composition, the solubility order of 1,5-pentanediamine adipate dihydrate in the tested solvent systems follows the trend: (water + methanol) > (water + ethanol) > (water + DMF), which is consistent with the solvent polarity of methanol, ethanol and DMF. Taking into account the intense polarity of the molecule, the solvent effect on solubility could be explained by the "like dissolves like" rule, in which polar solute and polar solvent can result in strong interactions [23].

Solvent composition is the most important factor, which could affect the solubility of 1,5-pentanediamine adipate dihydrate in the solvent mixtures. Interestingly, the solubility values show different characteristics in the three binary mixture solvent systems. For water + methanol, a progressive rise of initial content of water result in ever-increasing solubility at lower temperature, while the trend changes when the temperature is higher than 303.15 K. Besides, cosolvency phenomenon can be observed in water + ethanol mixture, which means that there is a peak in solubility curve versus solvent composition [24]. The peak position slightly shifts with temperature and it gives the highest solubility at x_w in between 0.8 and 0.9 for water + ethanol. While, the solubility of 1,5-pentanediamine adipate dihydrate increases with the rising mole fraction of water for the binary mixed solvents water + DMF. From the above results, ethanol and DMF can be chosen as antisolvent since 1,5-pentanediamine adipate dihydrate is almost insoluble in them.

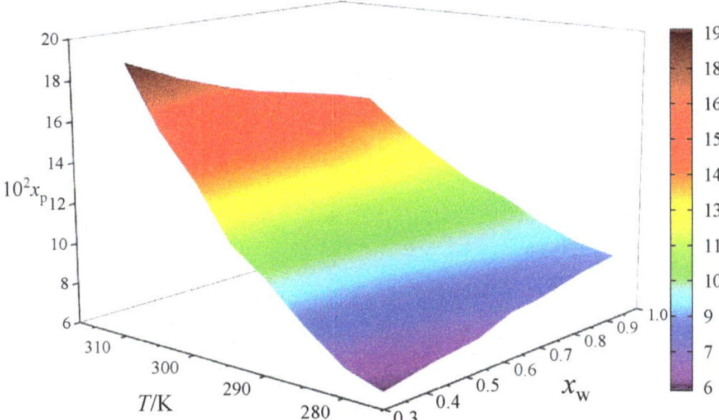

Figure 7. Mole fraction solubility of 1,5-pentanediamine adipate dihydrate in (water + methanol) mixed solvents with different mole fractions at various temperatures.

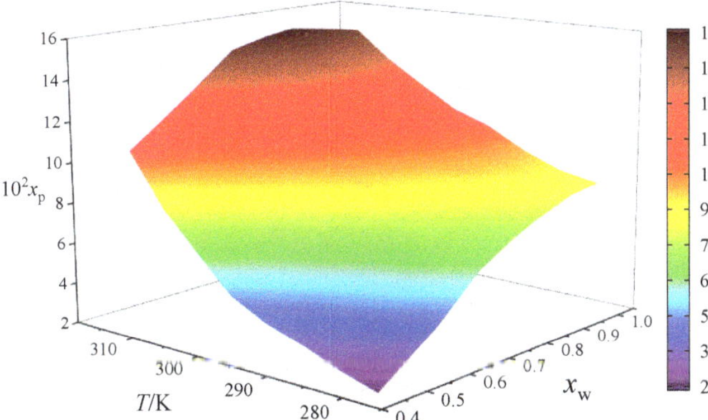

Figure 8. Mole fraction solubility of 1,5-pentanediamine adipate dihydrate in (water + ethanol) mixed solvents with different mole fractions at various temperatures.

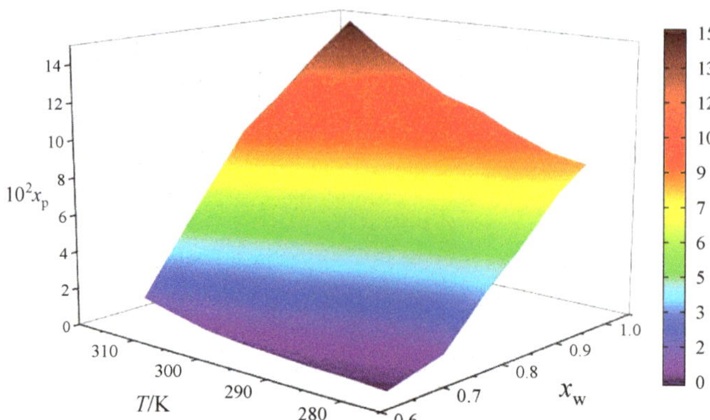

Figure 9. Mole fraction solubility of 1,5-pentanediamine adipate dihydrate in (water + DMF) mixed solvents with different mole fractions at various temperatures.

Furthermore, the experimental solubility data of this work were fitted by using the modified Apelblat Equation, NRTL model, the CNIBS/R–K model, and the Jouyban–Acree model. The values of model parameters and the calculated *ARD* and *RMSD* are shown in Tables 6–9. It can be observed the *ARD*% values of the four models used in this work and are generally lower than 5%, and the *RMSD* values are all lower than 0.006. It implies that the results fitted by these models show satisfactory consistency with the experimental values. Among them, the *ARD*% values of CNIBS/R-K model are lower than 1%, and the *RMSD* values are generally lower than 0.001. The CNIBS/R-K model shows best consistency among the four models, and as such, it was chosen to calculate the solubility of 1,5-pentanediamine adipate dihydrate in the above-mentioned solvents.

Table 6. Model parameters of modified Apelblat model for 1,5-pentanediamine adipate dihydrate in the investigated binary solvent mixtures [a,b,c].

x_w	A	B	C	$ARD\%$	$10^3 RMSD$
		water + methanol			
0.360	-82.8340	1038.99	13.5462	1.941	1.947
0.491	-77.8719	1151.40	12.6096	1.811	2.613
0.600	-82.4102	1586.55	13.1495	1.435	1.987
0.692	-95.1411	2376.78	14.9197	1.652	2.144
0.771	-80.2515	1856.47	12.6131	1.032	1.322
0.900	-81.0116	2152.54	12.5708	1.459	2.062
1.000	-75.7757	2039.99	11.7175	1.187	1.464
		water + ethanol			
0.447	-147.562	2728.03	23.7703	4.009	1.704
0.581	-44.1974	-544.343	7.63630	1.744	1.340
0.683	-81.2629	1656.53	12.8833	1.729	2.126
0.764	-94.6552	2414.27	14.7989	1.037	1.308
0.829	-85.2548	2175.21	13.2970	1.915	2.722
0.928	-78.0544	2088.14	12.0892	1.113	1.571
		water + DMF			
0.648	-103.155	-706.687	17.6053	10.73	0.553
0.741	-98.5924	393.332	16.4535	5.559	1.575
0.811	-111.074	2345.03	17.6142	2.312	1.586
0.866	-55.1677	636.236	8.86159	1.817	1.342
0.945	-81.6357	2273.41	12.5901	1.723	2.044

[a] *A, B* and *C* are the parameters of the modified Apelblat equation. [b] *ARD* is the average relative deviation. [c] *RMSD* is the root mean-square deviation.

Table 7. Model parameters of NRTL model for 1,5-pentanediamine adipate dihydrate in the investigated binary solvent mixtures [a,b,c].

Parameters	Water + Methanol	Water + Ethanol	Water + DMF
Δg_{ij}	$-345{,}527$	$-347{,}351$	$360{,}083$
Δg_{ik}	$-12{,}636.6$	-8875.18	-3418.04
Δg_{ji}	$392{,}075$	$394{,}734$	$-319{,}133$
Δg_{jk}	854.483	4455.60	$-55{,}130.0$
Δg_{ki}	$13{,}260.8$	$14{,}444.6$	8333.05
Δg_{kj}	4300.47	4725.49	$14{,}778.7$
$ARD\%$	2.678	3.652	5.387
$10^3 RMSD$	4.226	4.035	2.430

[a] Δg_{ij}, Δg_{ik}, Δg_{ji}, Δg_{jk}, Δg_{ki} and Δg_{kj} are the parameters of NRTL model. [b] ARD is the average relative deviation. [c] $RMSD$ is the root mean-square deviation.

Table 8. Model parameters of CNIBS/R-K model for 1,5-pentanediamine adipate dihydrate in the investigated binary solvent mixtures [a,b,c].

T/K	B_0	B_1	B_2	B_3	B_4	$ARD\%$	$10^3 RMSD$
			water + methanol				
278.15	-2.98656	0.535550	-0.420300	1.13613	-0.728910	0.785	0.848
283.15	-4.53891	10.7029	-22.1239	20.5480	-7.011	0.252	0.303
288.15	-2.60732	0.359600	-0.853370	1.93184	-1.17388	0.349	0.397
293.15	-1.91604	-2.84662	6.86270	-6.66656	2.32015	0.166	0.255
298.15	-1.97239	-0.888080	0.450860	1.18540	-0.966630	0.101	0.167
303.15	-0.914890	-6.90305	15.9257	-15.9672	5.76152	0.060	0.116
308.15	-1.38438	-3.35065	-9.15557	-10.9973	4.56526	0.571	1.103
313.15	-1.57444	0.361170	-2.86611	3.78493	-1.61709	0.052	0.125
			water + ethanol				
278.15	-6.85983	3.86505	15.6193	-25.9248	10.8407	0.990	0.805
283.15	-12.0539	35.5145	-50.4952	32.8953	-8.27872	1.334	1.328
288.15	-7.14881	11.4357	-6.18013	-2.53752	2.08915	0.740	0.935
293.15	-7.32735	11.6185	-1.76807	-11.6445	6.87737	0.770	0.965
298.15	-13.5664	50.5021	-86.0459	66.1292	-19.2053	0.909	1.230
303.15	-8.03959	23.0644	-34.1683	22.9691	-5.92156	0.455	0.644
308.15	-1.81365	-9.83320	31.2860	-34.0285	12.3666	0.431	0.690
313.15	-0.181820	-16.3140	42.2165	-42.3819	14.7509	0.376	0.809
			water + DMF				
278.15	219.485	-1234.93	2423.09	-2035.12	625.028	4.045	2.264
283.15	92.4160	-562.581	1120.79	-934.259	281.209	0.510	0.327
288.15	-88.0758	293.604	-381.254	223.373	-49.9886	0.423	0.294
293.15	-83.1875	272.112	-340.998	188.063	-38.2364	0.095	0.077
298.15	-330.881	1465.47	-2470.41	1859.26	-525.636	2.276	1.976
303.15	-166.220	669.528	-1035.38	716.822	-186.839	0.295	0.296
308.15	-143.599	580.109	-902.644	630.033	-165.907	0.621	0.723
313.15	-180.185	758.701	-1219.05	873.296	-234.665	0.568	0.742

[a] B_0, B_1, B_2, B_3 and B_4 are the parameters of CNIBS/R-K model. [b] ARD is the average relative deviation. [c] $RMSD$ is the root mean-square deviation.

Table 9. Model parameters of Jouyban–Acree equation for 1,5-pentanediamine adipate dihydrate in the investigated binary solvent mixtures [a,b,c].

Parameters	Water + Methanol	Water + Ethanol	Water + DMF
A_0	10.2963	15.6173	40.6254
A_1	−8.04828	−14.1420	−39.9894
A_2	−3704.93	−6730.11	−35,615.2
A_3	2326.55	8302.08	87,378.6
A_4	167.587	−3422.17	−90,712.4
A_5	−127.038	285.814	46,058.6
A_6	20.3783	475.704	−7948.44
$ARD\%$	2.022	3.270	6.128
$10^3 RMSD$	2.459	3.707	4.389

[a] A_0, A_1, A_2, A_3, A_4, A_5 and A_6 are the parameters of Jouyban-Acree equation. [b] ARD is the average relative deviation. [c] $RMSD$ is the root mean-square deviation.

4.3. The Mixing Thermodynamic Properties

The mixing properties, including mixing Gibbs free energy ($\Delta_m G$), mixing enthalpy ($\Delta_m H$), and mixing entropy ($\Delta_m S$), are essential for non-ideal binary solution mixtures and NRTL model can be adopted to calculate these data. The results are shown in Tables 10–12. It can be found that the values of $\Delta_m G$ are negative, indicating that the mixing processes in the investigated solution systems are spontaneous. Besides, the values of $\Delta_m H$ are mostly negative, which means the mixing processes are mainly exothermic. Generally, the thermodynamic properties of mixing are affected by the properties of solvents.

Table 10. Mixing thermodynamic properties of 1,5-pentanediamine adipate dihydrate in water + methanol mixtures [a,b].

x_w	$\Delta_m H/\text{kJ·mol}^{-1}$	$\Delta_m G/\text{kJ·mol}^{-1}$	$\Delta_m S/\text{J·mol}^{-1}\cdot\text{K}^{-1}$
		$T = 278.15$ K	
0.360	−3.767	−1.873	−6.813
0.491	−4.283	−1.997	−8.223
0.600	−4.480	−2.017	−8.858
0.692	−4.879	−2.050	−10.17
0.771	−4.992	−1.983	−10.82
0.900	−5.399	−1.814	−12.89
1.000	−5.435	−1.450	−14.33
		$T = 283.15$ K	
0.360	−4.036	−1.953	−7.360
0.491	−4.676	−2.100	−9.102
0.600	−4.882	−2.119	−9.763
0.692	−4.961	−2.072	−10.20
0.771	−5.152	−2.019	−11.06
0.900	−5.435	−1.810	−12.80
1.000	−5.464	−1.434	−14.24
		$T = 288.15$ K	
0.360	−4.656	−2.114	−8.825
0.491	−5.038	−2.195	−9.872
0.600	−5.184	−2.197	−10.37
0.692	−5.388	−2.176	−11.15
0.771	−5.555	−2.113	−11.95
0.900	−5.758	−1.876	−13.48
1.000	−5.692	−1.464	−14.68

Table 10. *Cont.*

x_w	$\Delta_m H/\text{kJ}\cdot\text{mol}^{-1}$	$\Delta_m G/\text{kJ}\cdot\text{mol}^{-1}$	$\Delta_m S/\text{J}\cdot\text{mol}^{-1}\cdot\text{K}^{-1}$
		$T = 293.15$ K	
0.360	−5.315	−2.290	−10.32
0.491	−5.563	−2.330	−11.03
0.600	−5.658	−2.315	−11.41
0.692	−5.754	−2.265	−11.90
0.771	−5.789	−2.165	−12.36
0.900	−5.976	−1.911	−13.87
1.000	−5.999	−1.495	−15.37
		$T = 298.15$ K	
0.360	−5.714	−2.408	−11.09
0.491	−5.828	−2.404	−11.48
0.600	−5.862	−2.370	−11.71
0.692	−5.888	−2.297	−12.04
0.771	−5.988	−2.207	−12.68
0.900	−6.167	−1.936	−14.19
1.000	−6.105	−1.483	−15.50
		$T = 303.15$ K	
0.360	−6.579	−2.667	−12.91
0.491	−6.507	−2.589	−12.92
0.600	−6.490	−2.533	−13.05
0.692	−6.439	−2.434	−13.21
0.771	−6.413	−2.305	−13.55
0.900	−6.400	−1.970	−14.62
1.000	−6.415	−1.499	−16.22
		$T = 308.15$ K	
0.360	−6.809	−2.764	−13.13
0.491	−7.018	−2.748	−13.86
0.600	−6.859	−2.638	−13.70
0.692	−6.775	−2.520	−13.81
0.771	−6.705	−2.368	−14.08
0.900	−6.572	−1.986	−14.88
1.000	−6.710	−1.532	−16.81
		$T = 313.15$ K	
0.360	−7.245	−2.957	−13.69
0.491	−7.265	−2.847	−14.11
0.600	−7.115	−2.721	−14.03
0.692	−6.993	−2.578	−14.10
0.771	−7.378	−2.063	−16.98
0.900	−7.056	−2.077	−15.90
1.000	−7.068	−1.579	−17.53

[a] The values of $\Delta_m G$, $\Delta_m H$ and $\Delta_m S$ were calculated by Equation (23). [b] The standard uncertainty are $u(T) = 0.03$ K, $u(p) = 0.3$ kPa. The combined expanded uncertainties are $u_c(\Delta_m H) = 0.060\Delta_m H$, $u_c(\Delta_m S) = 0.065\Delta_m S$, $u_c(\Delta_m G) = 0.070\Delta_m G$ (0.95 level of confidence).

Table 11. Mixing thermodynamic properties of 1,5-pentanediamine adipate dihydrate in water + ethanol mixtures [a,b].

x_w	$\Delta_m H$/kJ·mol^{-1}	$\Delta_m G$/kJ·mol^{-1}	$\Delta_m S$/J·mol^{-1}·K^{-1}
		T = 278.15 K	
0.447	−0.3891	−0.1108	−1.000
0.581	−1.808	−0.4737	−4.801
0.683	−3.093	−0.8665	−8.010
0.764	−3.769	−1.114	−9.550
0.829	−4.291	−1.295	−10.77
0.928	−5.100	−1.499	−12.95
		T = 283.15 K	
0.447	−0.7009	−0.1980	−1.776
0.581	−2.368	−0.6156	−6.194
0.683	−3.602	−0.9993	−9.197
0.764	−4.012	−1.177	−10.01
0.829	−4.530	−1.352	−11.22
0.928	−5.242	−1.517	−13.16
		T = 288.15 K	
0.447	−1.096	−0.3010	−2.762
0.581	−2.626	−0.6872	−6.734
0.683	−3.782	−1.050	−9.488
0.764	−4.318	−1.255	−10.63
0.829	−4.900	−1.440	−12.01
0.928	−5.458	−1.552	−13.56
		T = 293.15 K	
0.447	−1.382	−0.3795	−3.423
0.581	−3.102	−0.8079	−7.830
0.683	−4.231	−1.165	−10.46
0.764	−4.634	−1.334	−11.26
0.829	−5.026	−1.466	−12.15
0.928	−5.611	−1.570	−13.79
		T = 298.15 K	
0.447	−1.807	−0.4852	−4.436
0.581	−3.654	−0.9458	−9.090
0.683	−4.496	−1.236	−10.93
0.764	−4.854	−1.389	−11.62
0.829	−5.069	−1.469	−12.08
0.928	−5.835	−1.603	−14.20
		T = 303.15 K	
0.447	−2.575	−0.6620	−6.314
0.581	−4.098	−1.061	−10.02
0.683	−4.858	−1.332	−11.63
0.764	−5.186	−1.471	−12.25
0.829	−5.493	−1.566	−12.96
0.928	−6.092	−1.639	−14.69
		T = 308.15 K	
0.447	−3.240	−0.8211	−7.854
0.581	−4.444	−1.155	−10.67
0.683	−5.092	−1.397	−11.99
0.764	−5.543	−1.561	−12.96
0.829	−5.848	−1.646	−13.64
0.928	−6.264	−1.654	−14.96
		T = 313.15 K	
0.447	−4.078	−1.029	−9.741
0.581	−5.004	−1.310	−11.80
0.683	−5.648	−1.353	−13.08
0.764	−5.932	−1.664	−13.63
0.829	−6.356	−1.764	−14.67
0.928	−6.692	−1.722	−15.88

[a] The values of $\Delta_m G$, $\Delta_m H$ and $\Delta_m S$ were calculated by Equation (23). [b] The standard uncertainty are $u(T)$ = 0.03 K, $u(p)$ = 0.3 kPa. The combined expanded uncertainties are $u_c(\Delta_m H)$ = 0.060$\Delta_m H$, $u_c(\Delta_m S)$ = 0.065$\Delta_m S$, $u_c(\Delta_m G)$ = 0.070$\Delta_m G$ (0.95 level of confidence).

Table 12. Mixing thermodynamic properties of 1,5-pentanediamine adipate dihydrate in water +DMF mixtures [a,b].

x_w	$\Delta_m H$/kJ·mol^{-1}	$\Delta_m G$/kJ·mol^{-1}	$\Delta_m S$/J·mol^{-1}·K^{-1}
		$T = 278.15$ K	
0.648	−23.05	−2.277	−1.024
0.741	−17.50	−1.803	1.927
0.811	−11.39	−1.397	9.279
0.866	−7.392	−1.064	11.70
0.945	−3.686	−5.551	6.710
		$T = 283.15$ K	
0.648	−22.87	−2.271	−0.567
0.741	−17.12	−1.797	2.989
0.811	−11.34	−1.395	9.239
0.866	−7.374	−1.064	11.56
0.945	−4.046	−5.547	5.304
		$T = 288.15$ K	
0.648	−22.73	−2.263	−0.359
0.741	−16.68	−1.789	4.219
0.811	−11.11	−1.389	9.652
0.866	−7.371	−1.058	11.16
0.945	−4.172	−5.560	4.818
		$T = 293.15$ K	
0.648	−22.52	−2.253	0.0403
0.741	−16.32	−1.782	5.112
0.811	−10.89	−1.381	9.958
0.866	−7.454	−1.055	10.58
0.945	−4.501	−5.570	3.647
		$T = 298.15$ K	
0.648	−22.35	−2.244	0.308
0.741	−15.67	−1.767	6.708
0.811	−10.82	−1.377	9.888
0.866	−7.546	−1.052	10.00
0.945	−4.552	−5.499	3.178
		$T = 303.15$ K	
0.648	−22.14	−2.239	0.819
0.741	−15.47	−1.757	6.946
0.811	−10.66	−1.364	9.832
0.866	−7.644	−1.043	9.199
0.945	−4.822	−5.145	1.065
		$T = 308.15$ K	
0.648	−21.69	−2.223	1.733
0.741	−15.06	−1.742	7.638
0.811	−10.65	−1.351	9.300
0.866	−7.894	−1.036	8.023
0.945	−5.467	−5.538	0.2324
		$T = 313.15$ K	
0.648	−21.30	−2.209	2.528
0.741	−14.67	−1.719	8.059
0.811	−10.71	−1.335	8.432
0.866	−8.115	−1.029	6.952
0.945	−5.760	−5.469	−0.9297

[a] The values of $\Delta_m G$, $\Delta_m H$ and $\Delta_m S$ were calculated by Equation (23). [b] The standard uncertainty are $u(T) = 0.03$ K, $u(p) = 0.3$ kPa. The combined expanded uncertainties are $u_c(\Delta_m H) = 0.060\Delta_m H$, $u_c(\Delta_m S) = 0.065\Delta_m S$, $u_c(\Delta_m G) = 0.070\Delta_m G$ (0.95 level of confidence).

5. Conclusions

In this paper, the solubility data of 1,5-pentanediamine adipate dihydrate in binary solvent mixtures (water + methanol, water + ethanol, water + DMF) were measured under atmospheric pressure at temperatures ranging from 278.15 K to 313.15 K by gravimetric method. The solubility of 1,5-pentanediamine adipate dihydrate increased with the rising of temperature. At fixed temperature and solvent composition, the solubility order of 1,5-pentanediamine adipate dihydrate is (water + methanol) > (water + ethanol) > (water + DMF), which is consistent with the solvent polarity of methanol, ethanol and DMF. In water + methanol binary mixtures, the trend of solubility with components changes as the temperature rises, and the cut-off point is 303.15 K. Meanwhile, the cosolvency phenomenon was observed in water + ethanol system and maximum solubility values were observed when molar content of the water is about 0.8–0.9. As for water + DMF mixed solvent, a progressive increase in the initial content of organic solvent results in a decrease in its solubility. Furthermore, the solubility data were fitted by the Apelblat model, the NRTL model, the CNIBS/R–K model, and the Jouyban–Acree model. The results show satisfied fitting consistency. Finally, the mixing thermodynamic data indicate that the mixing process is spontaneous.

Author Contributions: Data curation, H.F. and Y.M.; Methodology, D.L.; Writing—original draft, L.L. and Y.Z.; Writing—review & editing, B.H., N.W., T.W. and H.H. All authors have read and agreed to the published version of the manuscript.

Funding: This research was funded by [Tianjin Natural Science Foundation] grant number [18JCZD JC38100].

Institutional Review Board Statement: Not applicable.

Informed Consent Statement: Not applicable.

Data Availability Statement: Not applicable.

Acknowledgments: The authors are very grateful for the financial support of the Tianjin Natural Science Foundation (grant number 18JCZDJC38100). And Liang Li and Yihan Zhao are co-first authors of the article.

Conflicts of Interest: The authors declare no conflict of interest.

References

1. Yang, P.; Peng, X.; Wang, S.; Li, D.; Li, M.; Jiao, P.; Zhuang, W.; Wu, J.; Wen, Q.; Ying, H. Crystal structure, thermodynamics, and crystallization of bio-based polyamide 56 salt. *CrystEngComm* **2020**, *22*, 3234–3241. [CrossRef]
2. Wang, Y.; Kang, H.; Guo, Y.; Liu, R.; Hao, X.; Qiao, R.; Yan, J. The structures and properties of bio-based polyamide 56 fibers prepared by high-speed spinning. *J. Appl. Polym. Sci.* **2020**, *137*, 49344. [CrossRef]
3. Kricheldorf, H.R.; Zolotukhin, M.G.; Cárdenas, J. Non-Stoichiometric Polycondensations and the Synthesis of High Molar Mass Polycondensates. *Macromol. Rapid Commun.* **2012**, *33*, 1814–1832. [CrossRef] [PubMed]
4. Gao, Y.Y.; Xie, C.; Wang, J.K. Effects of low magnetic field on batch crystallisation of glycine. *Mater. Res. Innov.* **2009**, *13*, 112–115. [CrossRef]
5. Hao, H.; Yin, Q.; Zhong, J.; Wang, X. Crystallization for Pharmaceutical and Food Science. *Curr. Pharm. Des.* **2018**, *24*, 2327–2328. [CrossRef]
6. Zhang, C.; Wang, A.J.; Wang, Y. Solubility of Ceftriaxone Disodium in Acetone, Methanol, Ethanol, N,N-Dimethylformamide, and Formamide between 278 and 318 K. *J. Chem. Eng. Data* **2005**, *50*, 1757–1760. [CrossRef]
7. Mao, Y.; Li, F.; Wang, T.; Cheng, X.; Li, G.; Li, D.; Zhang, X.; Hao, H. Enhancement of lysozyme crystallization under ultrasound field. *Ultrason. Sonochem.* **2020**, *63*, 104975. [CrossRef]
8. Zhang, H.; Chen, Y.; Wang, J.; Gong, J. Investigation on the Spherical Crystallization Process of Cefotaxime Sodium. *Ind. Eng. Chem. Res.* **2010**, *49*, 1402–1411. [CrossRef]
9. Yang, J.; Hou, B.; Huang, J.; Li, X.; Tian, B.; Wang, N.; Bi, J.; Hao, H. Solution thermodynamics of tris-(2,4-ditert-butylphenyl)-phosphite in a series of pure solvents. *J. Mol. Liq.* **2019**, *283*, 713–724. [CrossRef]
10. Apelblat, A.; Manzurola, E. Solubilities of L-aspartic, DL-aspartic, DL-glutamic, p-hydroxybenzoic, o-anisic, p-anisic, and itaconic acids in water from T = 278 K to T = 345 K. *J. Chem. Thermodyn.* **1997**, *29*, 1527–1533. [CrossRef]
11. Apelblat, A.; Manzurola, E. Solubilities of l-glutamic acid, 3-nitrobenzoic acid, p-toluic acid, calcium-l-lactate, calcium gluconate, magnesium-dl-aspartate, and magnesium-l-lactate in water. *J. Chem. Thermodyn.* **2002**, *34*, 1127–1136. [CrossRef]
12. Yan, H.; Wang, Z.; Wang, J. Correlation of Solubility and Prediction of the Mixing Properties of Capsaicin in Different Pure Solvents. *Ind. Eng. Chem. Res.* **2012**, *51*, 2808–2813. [CrossRef]
13. Choi, P.B.; Mclaughlin, E. Effect of a phase transition on the solubility of a solid. *AIChE J.* **1983**, *29*, 150–153. [CrossRef]
14. Zong, S.; Wang, J.; Xiao, Y.; Wu, H.; Zhou, Y.; Guo, Y.; Huang, X.; Hao, H. Solubility and dissolution thermodynamic properties of lansoprazole in pure solvents. *J. Mol. Liq.* **2017**, *241*, 399–406. [CrossRef]

15. Renon, H.; Prausnitz, J.M. Local compositions in thermodynamic excess functions for liquid mixtures. *AIChE J.* **1968**, *14*, 135–144. [CrossRef]

16. You, Y.; Gao, T.; Qiu, F.; Wang, Y.; Chen, X.; Jia, W.; Li, R. Solubility Measurement and Modeling for 2-Benzoyl-3-chlorobenzoic Acid and 1-Chloroanthraquinone in Organic Solvents. *J. Chem. Eng. Data* **2013**, *58*, 1845–1852. [CrossRef]

17. Acree, W.E. Mathematical representation of thermodynamic properties: Part 2. Derivation of the combined nearly ideal binary solvent (NIBS)/Redlich-Kister mathematical representation from a two-body and three-body interactional mixing model. *Thermochim. Acta* **1992**, *198*, 71–79. [CrossRef]

18. Jouyban, A.; Soltani, S.; Chan, H.-K.; Acree, W.E. Modeling acid dissociation constant of analytes in binary solvents at various temperatures using Jouyban–Acree model. *Thermochim. Acta* **2005**, *428*, 119–123. [CrossRef]

19. Jouyban, A.; Fakhree, M.A.A.; Acree, J.W.E. Comment on "Measurement and Correlation of Solubilities of (Z)-2-(2-Aminothiazol-4-yl)-2-methoxyiminoacetic Acid in Different Pure Solvents and Binary Mixtures of Water + (Ethanol, Methanol, or Glycol)". *J. Chem. Eng. Data* **2012**, *57*, 1344–1346. [CrossRef]

20. Huang, Q.; Li, Y.; Yuan, F.; Xiao, L.; Hao, H.; Wang, Y. Thermodynamic properties of enantiotropic polymorphs of glycolide. *J. Chem. Thermodyn.* **2017**, *111*, 106–114. [CrossRef]

21. Wang, G.; Wang, Y.; Ma, Y.; Hao, H.; Luan, Q.; Wang, H. Determination and correlation of cefuroxime acid solubility in (acetonitrile + water) mixtures. *J. Chem. Thermodyn.* **2014**, *77*, 144–150. [CrossRef]

22. Ruidiaz, M.A.; Delgado, D.R.; Martinez, M.A.R.; Marcus, Y. Solubility and preferential solvation of indomethacin in 1,4-dioxane+water solvent mixtures. *Fluid Phase Equilibria* **2010**, *299*, 259–265. [CrossRef]

23. Zou, F.; Zhuang, W.; Wu, J.; Zhou, J.; Liu, Q.; Chen, Y.; Xie, J.; Zhu, C.; Guo, T.; Ying, H. Experimental measurement and modelling of solubility of inosine-5′-monophosphate disodium in pure and mixed solvents. *J. Chem. Thermodyn.* **2014**, *77*, 14–22. [CrossRef]

24. Jouyban, A. Review of the cosolvency models for predicting solubility of drugs in water-cosolvent mixtures. *J. Pharm. Pharm. Sci.* **2008**, *11*, 32–58. [CrossRef]

crystals

Solution-Mediated Polymorphic Transformation of L-Carnosine from Form II to Form I

Yanan Zhou [1], Shuyi Zong [2], Jie Gao [1], Chunsong Liu [3,*] and Ting Wang [2,*]

[1]   College of Chemical Engineering, North China University of Science and Technology, Tangshan 063210, China; zynzyn@tju.edu.cn (Y.Z.); jiegao@ncst.edu.cn (J.G.)
[2]   National Engineering Research Center of Industrial Crystallization Technology, School of Chemical Engineering and Technology, Tianjin University, Tianjin 300072, China; shuyizong@tju.edu.cn
[3]   Graduate School, North China University of Science and Technology, Tangshan 063210, China
*   Correspondence: liuchunsong@ncst.edu.cn (C.L.); wang_ting@tju.edu.cn (T.W.); Tel./Fax: +86-22-2737-4971 (T.W.)

Abstract: In this study, L-carnosine was chosen as the model compound to systematically study solution-mediated polymorphic transformation by online experiment and theoretical simulation. Form II, a new polymorph of L-carnosine, was developed using an antisolvent crystallization method. The properties of form I and form II L-carnosine were characterized by powder X-ray diffraction, polarizing microscope, thermal analysis, and Raman spectroscopy. In order to explore the relative stability, the solubility of L-carnosine form I and form II in a (water + DMAC) binary solvent mixture was determined by a dynamic method. During the solution-mediated polymorphic transformation process of L-carnosine in different solvents, Raman spectroscopy was employed to detect the solid-phase composition of suspension in situ, and the gravimetric method was used to measure the liquid concentration. In addition, the effect of the solvent on the transformation process was evaluated and analyzed. Finally, a mathematical model of dissolution–precipitation was established to simulate the kinetics of the polymorphic transformation process based on the experimental data. Taking the simulation results and the experimental data into consideration, the controlling step of solution-mediated polymorphic transformation was discussed.

Keywords: L-carnosine; polymorphic transformation; solvent; kinetics; mathematical model

Citation: Zhou, Y.; Zong, S.; Gao, J.; Liu, C.; Wang, T. Solution-Mediated Polymorphic Transformation of L-Carnosine from Form II to Form I. *Crystals* **2022**, *12*, 1014. https://doi.org/10.3390/cryst12071014

Academic Editors: Jingxiang Yang and Xin Huang

Received: 28 June 2022
Accepted: 19 July 2022
Published: 21 July 2022

Publisher's Note: MDPI stays neutral with regard to jurisdictional claims in published maps and institutional affiliations.

1. Introduction

Polymorphism is defined as the ability of a compound to exist in multiple crystalline forms with different molecular arrangements or molecular conformations in a crystal lattice [1]. Polymorphism is a frequent phenomenon in the pharmaceutical industry. It is found that more than half of solid drugs have polymorphic forms. Due to differences in crystal structure, different polymorphs of the same solid drug generally present various physicochemical characteristics, such as powder property, melting point, enthalpy of fusion, dissolution behavior, and stability, which may lead to a different drug bioavailability, curative effect, and half-life of the drug [2]. As a consequence, the characterization and analysis of the physicochemical properties of different polymorphs, including the thermodynamic and kinetic properties of the polymorphic system, is essential to guide the development, manufacture, and application of solid drugs [3–6].

According to the Ostwald rule, solution-mediated polymorphic transformation would happen on a metastable form due to its higher Gibbs free energy, which finally transforms it into its stable polymorph [7]. Solution-mediated polymorphic transformation can be mainly divided into three steps: (i) dissolution of the metastable form, (ii) nucleation of the stable form, and (iii) growth of the stable form [8]. Additionally, the slowest step among (i)–(iii) was the so-called rate-controlling step. Based on examining the solution and solid-phase compositions for solution-mediated polymorphic transformation, O 'Mahony et al. summarized four kinds of principal scenarios, including "dissolution-controlled", "growth-controlled", "nucleation-dissolution-controlled", and "nucleation-growth-controlled" polymorphic transformations [9].

L-carnosine ($C_9H_{14}N_4O_3$, Figure 1, CAS Registry No. 305−84−0, molar mass: 226.235 g/mol), a bioactive peptide found in the brain and muscle tissues of mammal, was chosen as the model compound [10]. Owing to its strong antioxidant effects, L-carnosine has been widely used in treating ulcers, arthritis, atherosclerosis, cataracts, diabetes, hypertension, heart disease, and cancer. Through consulting a large volume of literature, it can be observed that studies about L-carnosine concentrate on its preparation, characterization, function, and application [11–14]. However, the polymorphism of L-carnosine has not been reported before. In this work, a new polymorph of L-carnosine was developed by the antisolvent crystallization method. It was named as form II, and the existing polymorph was named as form I. Different methods were employed to characterize and analyze these two forms of L-carnosine. Furthermore, the solvent-mediated polymorph transformation from form II to form I was investigated by online Raman, in which the influence of the solvent was further discussed. The rate-determining step in the transformation process was determined using the method of offline sampling, and the kinetics of crystal dissolution, nucleation, and growth were simulated and analyzed according to a dissolution-precipitation model.

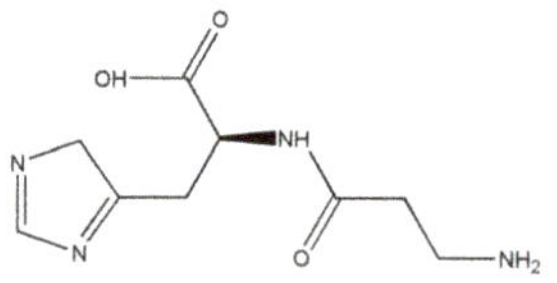

Figure 1. The molecular structure of L-carnosine.

2. Experimental Section

2.1. Materials

Form I of L-carnosine (≥0.990 mass fraction) was offered by Shanghai Yuanye Bio-Technology Co., Ltd., Shanghai, China. All the organic solvents used in the experiments, including methanol, ethanol, 2-propanol, acetone, dimethyl formamide (DMF), and dimethylacetamide (DMAC), were purchased from Tianjin Chemical Reagent No. 6 Factory, Tianjin China. The deionized water was supplied by Tianjin Yongqingyuan Co., Ltd., Tianjin, China. More details regarding the materials are listed in Table 1. All chemicals were used without further purification. Form II of L-carnosine was prepared in the laboratory by the antisolvent crystallization method, in which water acted as a solvent and DMF or DMAC served as an antisolvent.

Table 1. Sources and mass fraction purity of MAC chemicals used in this article.

Chemical Name	Source	Mass Purity	Purification Method	Analysis Method
L-carnosine (Form I)	Shanghai Yuanye Bio-Technology Co., Ltd., China	≥0.990	None	HPLC [a]
Methanol	Tianjin Chemical Reagent No.6 Factory, China	>0.995	None	GC [b]
Ethanol	Tianjin Chemical Reagent No.6 Factory, China	>0.995	None	GC [b]
2-Propanol	Tianjin Chemical Reagent No.6 Factory, China	>0.995	None	GC [b]
Acetone	Tianjin Chemical Reagent No.6 Factory, China	>0.995	None	GC [b]
DMF	Tianjin Chemical Reagent No.6 Factory, China	>0.995	None	GC [b]
DMAC	Tianjin Chemical Reagent No.6 Factory, China	>0.995	None	GC [b]
Deionized Water	Tianjin Yongqingyuan Co., Ltd., China	≥18.25 MΩ·cm	None	CT [c]

[a] High performance liquid chromatography, which was determined by Shanghai Yuanye Bio-Technology Co., Ltd., Shanghai, China. [b] Gas chromatography, which was determined by Tianjin Chemical Reagent No.6 Factory, Tianjin, China. [c] Conductivity test, which was carried out by Tianjin Yongqingyuan Co., Ltd., Tianjin, China.

2.2. Development of New Polymorph

The new polymorph of L-carnosine was developed by an antisolvent crystallization method through the following procedures. Firstly, 42 mL DMF or DMAC was gently poured into a 100 mL jacketed crystallizer, which was equipped with a mechanical stirrer. A thermostat (CF41, Julabo, Seelbach, Germany) was used to control the system at 303.15 K. Then, 2.45 g L-carnosine raw material was dissolved in water at room temperature to prepare a 0.175 g/mL L-carnosine solution. Finally, 14 mL L-carnosine solution was added into the jacketed crystallizer by a peristaltic pump (BT100-1F, Longer, Baoding, China) with a dropping rate of 333.3 μL/min. To fully mix the organic solvent and the L-carnosine solution, the mechanical stirrer was adjusted to 300 rpm during the entire experiment. Once the adding process finished, the suspension was filtered and dried in a vacuum oven at 298.15 K for further characterization.

2.3. Characterization Methods

2.3.1. Powder X-ray Diffraction

Powder X-ray diffraction, the most classic and commonly used method in qualitative and quantitative analysis for polymorphism, was used to measure the crystal form of L-carnosine. The data were collected by a D/max-2500 diffractometer (Rigaku, Tokyo, Japan) with Cu Ka radiation (0.15405 nm). Samples were determined at the diffraction angle (2θ) from 2° to 40° with a scanning rate of 8 °/min and a step size of 0.02°.

2.3.2. Polarizing Microscope

Polymorphism is one of the main factors that affect the morphology of solid drugs. A polarizing microscope (BX51, Olympus, Tokyo, Japan) was used to study the morphology of form I and form II of L-carnosine.

2.3.3. Thermal Analysis

Differential scanning calorimetry (1/500, Mettler-Toledo, Greifensee, Switzerland) and thermogravimetry (1/SF, Mettler Toledo, Greifensee, Switzerland) were carried out to obtain the melting temperature and decomposition temperature of form I and form II of L-carnosine. The measurements were conducted from 303.15 to 548.15 K at the rate of 2 K/min under the protection of a nitrogen atmosphere.

2.3.4. Raman Spectroscopy

A Raman spectrometer (RXN2, Mettler Toledo, Greifensee, Switzerland), equipped with an MR probe head and a PhAT probe head, was implemented to monitor the solid-phase composition of the suspension in situ during the solution-mediated polymorphic transformation process [15]. The data were collected in the wavenumber range from 150 to 1890 cm^{-1} at a laser wavelength of 514.5 nm.

2.4. Solubility Experiments

The solubility of L-carnosine form I and form II in binary solvent (water + DMAC) was determined using a dynamic method based on previous studies [16]. The molar ratio of water to DMAC was 9:1 in a binary solvent mixture. The solubility experiments were carried out in temperatures ranging from 278.25 K to 323.15 K under atmospheric pressure.

2.5. Solution-Mediated Polymorphic Transformation Experiments

The polymorphic transformation experiments from form II to form I of L-carnosine were performed in six kinds of binary solvent mixtures, including water + methanol, water + ethanol, water + 2-propanol, water + acetone, water + DMF, and water + DMAC, which can be divided into two parts. Water + methanol, water + ethanol, water + 2-propanol, and water + acetone have been used in antisolvent crystallization to prepare form I, and water + DMAC and water + DMF were used to produce form II in this work. The volume ratio of water to organic solvent was 1:3 in a binary solvent mixture. First, 100 mL binary solvent mixture was gently poured into a 150 mL jacketed crystallizer, which was equipped with a mechanical stirrer. A thermostat (CF41, Julabo, Seelbach, Germany) was used to control the system at 303.15 K. Then, 3.00 g of form II of L-carnosine was added into the jacketed crystallizer to prepare the initial suspension of form II. The initial concentration of L-carnosine in the liquid phase is supersaturated for form I. To fully mix the suspension, the mechanical stirrer was adjusted to 300 rpm. During the polymorphic transformation process, a Raman MR probe was inserted into the slurry to monitor the solid-phase composition of the two polymorphs in situ. Meanwhile, the solid-phase composition was also measured and analyzed using powder X-ray diffraction by intermittent sampling [17]. In addition, the liquid concentration of L-carnosine was determined by the gravimetric method at a certain time interval [18]. All the transformation experiments were performed three times in this study. The experimental setup of solution-mediated polymorphic transformation between the two polymorphs of L-carnosine is shown in Figure 2.

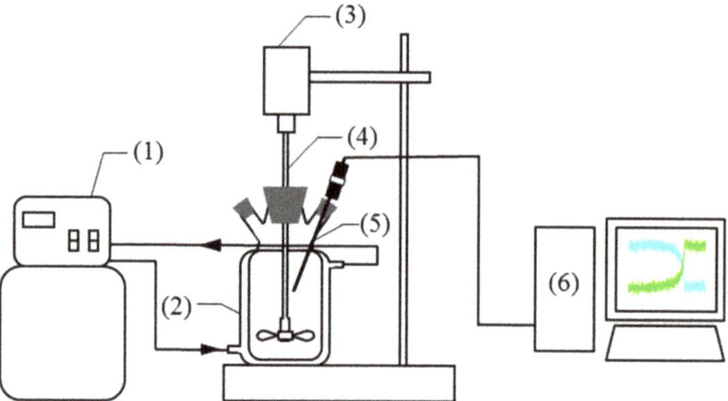

Figure 2. The experimental setup of solution-mediated polymorphic transformation between the two polymorphs of L-carnosine: (**1**) thermostat; (**2**) crystallizer; (**3**) mechanical stirrer; (**4**) stirring paddle; (**5**) Raman probe; (**6**) computer.

2.6. Theoretical Model

In this study, a theoretical model of dissolution-precipitation was established to simulate the kinetics of the polymorphic transformation process from form II to form I of L-carnosine [19]. In the dissolution-precipitation model, the transformation process contains three steps: dissolution of the metastable form, nucleation of the stable form, and growth of the stable form. Based on these steps, the change in the amount of undissolved L-carnosine solid can be expressed via Equation (1).

$$\frac{\mathrm{d}A_\mathrm{d}}{\mathrm{d}t} = -\frac{\mathrm{d}A_\mathrm{s}}{\mathrm{d}t} = D - J - G \tag{1}$$

where A_d and A_s represent the amounts of dissolved and undissolved solid L-carnosine. D, J, and G refer to the dissolution rate, nucleation rate, and growth rate, respectively.

Assuming the dissolution rate of the crystals in the suspension is size-independent, the dissolution rate can be defined as a proportional function of the amount of solid in the suspension, which is written as follows:

$$D_i = k_\mathrm{diss,i} \cdot A_\mathrm{s,i} \tag{2}$$

where subscript i stands for the i th polymorph of crystals. k_diss is the dissolution rate constant, which reflects the properties of the particles and solution [20].

The nucleation process, the first step of the crystallization process, is the spontaneous formation of clusters past the critical size [21]. At a microscopic level, the nucleation process can be simplified as the encounter of dissolved solute molecules. Therefore, the nucleation rate can be obtained as follows:

$$J_i = k_{\text{nuc},i} \cdot V \cdot C^{\alpha,i} \tag{3}$$

where k_{nuc} is the nucleation rate constant, which represents the possibility of generating aggregates. V and C are the volume and concentration of the solution. α refers to the nucleation molecularity index, denoting the average number of molecules needed to form nucleation.

During the growth process, discrete solute molecules in solution continuously aggregate onto the pre-existing crystals. Because of this, the growth rate is related to the concentration of the dissolved solute and the amount of undissolved solid in the suspension [22]. It can be shown as Equation (4).

$$G_i = k_{\text{growth},i} \cdot A_{\text{s},i} \cdot C \tag{4}$$

where k_{growth} is the growth rate constant, reflecting the reaction rate of dissolved solute molecules with pre-existing crystals.

Clearly, according to the above assumptions, the kinetics of undissolved L-carnosine solid can be calculated as in Equation (5).

$$-\frac{\mathrm{d}A_{\text{s},i}}{\mathrm{d}t} = k_{\text{diss},i} \cdot A_{\text{s},i} - k_{\text{nuc},i} \cdot V \cdot C^{\alpha,i} - k_{\text{growth},i} \cdot A_{\text{s},i} \cdot C \tag{5}$$

where i = 1, 2 represents form I and form II of L-carnosine.

The relationship between solution concentration and undissolved solid amount can be described by Equation (6), when $C = A_{\text{d}}/V$ is substituted into Equation (1).

$$V\frac{\mathrm{d}C}{\mathrm{d}t} = -\frac{\mathrm{d}A_{\text{s},1}}{\mathrm{d}t} - \frac{\mathrm{d}A_{\text{s},2}}{\mathrm{d}t} \tag{6}$$

When the system reaches equilibrium between the solid and liquid phases, the nucleation term can be ignored [19]. Thus, the dissolution rate equals the growth rate. Additionally, the solution concentration is the solubility of the solid solute, which can be described as following:

$$C_{\text{sol},i} = \frac{k_{\text{diss},i}}{k_{\text{growth},i}} \tag{7}$$

where C_{sol} is the solubility of L-carnosine in the given binary solvent mixture.

Substituting Equation (7) into Equations (5) and (6), changes in the solution concentration and undissolved L-carnosine solid amount over time can be calculated by Equations (8) and (9), respectively.

$$\frac{\mathrm{d}C}{\mathrm{d}t} = \sum_{i=1}^{2} \left(\frac{k_{\text{growth},i} \cdot A_{\text{s},i} \cdot (C_{\text{sol},i} - C)}{V} - k_{\text{nuc},i} \cdot C^{\alpha,i} \right) \tag{8}$$

$$\frac{\mathrm{d}A_{\text{s},i}}{\mathrm{d}t} = -V \cdot \left(\frac{k_{\text{growth},i} \cdot A_{\text{s},i} \cdot (C_{\text{sol},i} - C)}{V} - k_{\text{nuc},i} \cdot C^{\alpha,i} \right) \tag{9}$$

where i = 1, 2, denotes form I and form II of L-carnosine.

Based on the experimental data and dissolution-precipitation model, the solution-mediated polymorphic transformation process of L-carnosine was simulated using MATLAB (2015 version). The model parameters, including k_{diss}, k_{nuc}, k_{growth}, and α for form I and form II, were obtained by a nonlinear dynamic parameter fitting procedure based on the least squares method. In addition, the set ordinary differential equations were solved using the ode15s function with a variable integration step.

3. Results and Discussion

3.1. Characterizations

Polymorphs of L-carnosine, including form I and form II, were characterized by powder X-ray diffraction, polarizing microscope, thermal analysis, and Raman spectroscopy. The powder X-ray diffraction patterns are shown in Figure 3. It can be found that form I and form II exhibit distinct peaks, indicating that form II is a new polymorph of L-carnosine [12]. A polarizing microscope was utilized to research the crystal habit of the two polymorphs. As illustrated in Figure 4, form I exhibits needle-shaped crystals, whereas form II exhibits short rod-shaped crystals. Furthermore, the results of differential scanning calorimetry and thermogravimetry are displayed in Figure 5. It can be seen that there are obvious endothermic peaks at 517 K and 491 K for form I and form II of L-carnosine, respectively. Considering that samples of form I and form II started to decrease in weight at 517 K and 491 K, 517 K and 491 K were recognized as the decomposition temperatures of form I and form II, respectively. The results demonstrate that both polymorphs are decomposing before melting, and the chemical stability of form I is higher than that of form II. The Raman spectra of L-carnosine polymorphs are shown in Figure 6. Significantly distinguishable characteristic peaks of the two crystal forms are found in the wavelength range of 1600 cm^{-1}–1500 cm^{-1} and 1300 cm^{-1}–1200 cm^{-1}, which indicates the difference in crystal structure. In this work, characteristic peaks at 1227.3 cm^{-1} and 1524.8 cm^{-1} were chosen to represent the changes in the solid content of form I and form II during the solution-mediated polymorphic transformation.

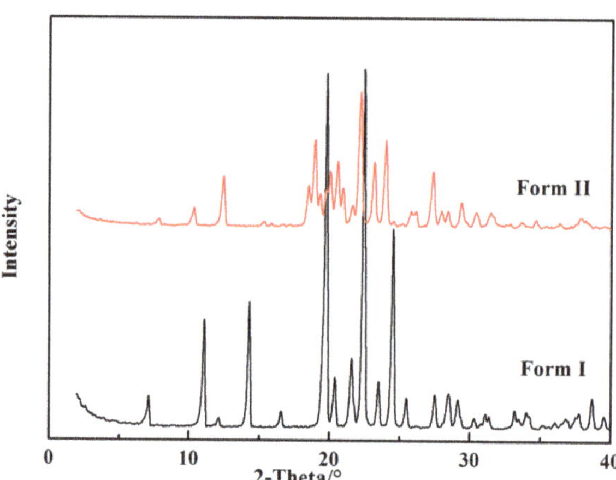

Figure 3. Powder X-ray diffraction patterns of L-carnosine form I and form II.

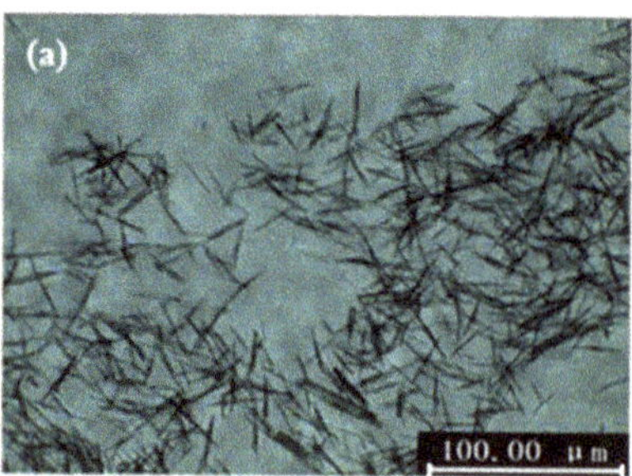

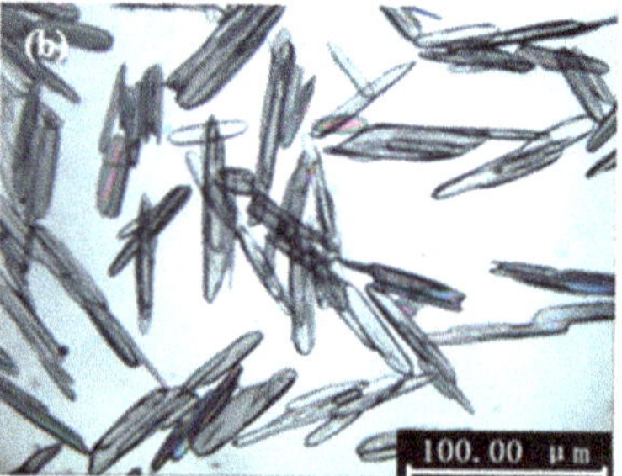

Figure 4. Polymorphic morphology of L-carnosine in a polarizing microscope: (**a**) form I; (**b**) form II.

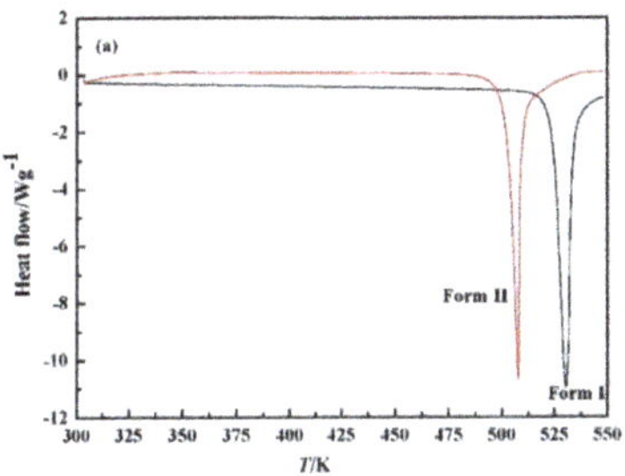
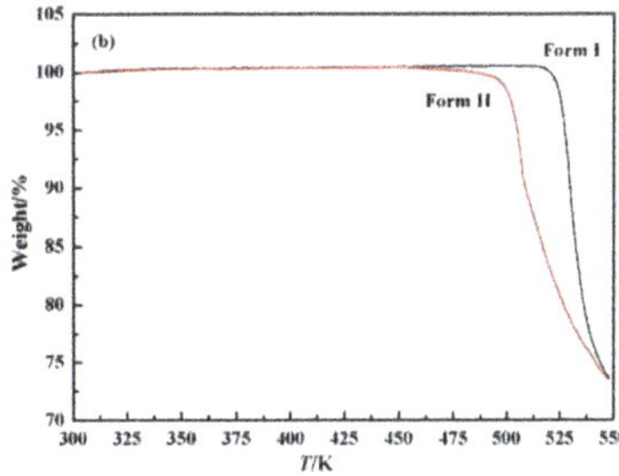

Figure 5. Thermal analysis curves of L-carnosine form I and form II: (**a**) differential scanning calorimetry; (**b**) thermogravimetry.

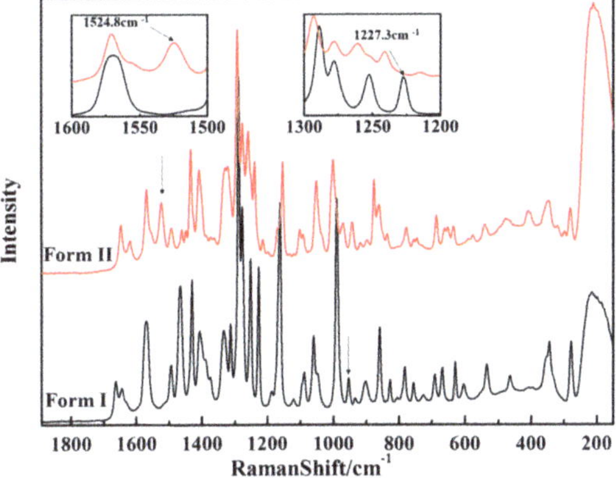

Figure 6. Raman spectra of L-carnosine form I and form II.

3.2. Solubility Data of L-carnosine Polymorphs

In this work, the solubility data of the L-carnosine polymorphs were measured to compare the stability of the two polymorphs. The mole fraction solubility of L-carnosine is graphically depicted in Figure 7. The results indicate that the solubility of the two polymorphs of L-carnosine is positively correlated with temperature. In addition, the solubility of form II is higher than that of form I throughout the whole temperature range studied. It confirms that form I is the stable form and form II is a metastable form, which is consistent with the results of thermal analysis [23].

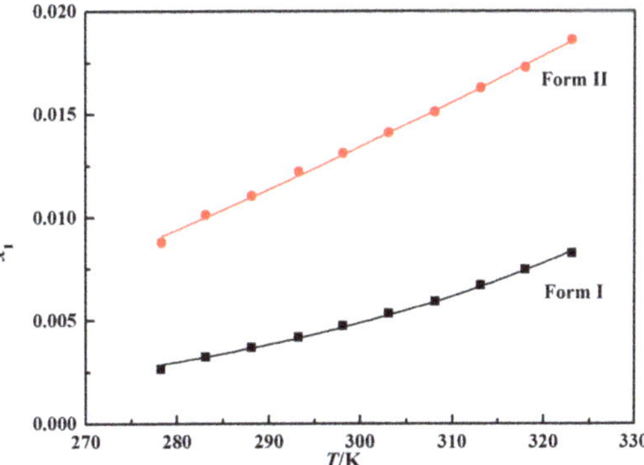

Figure 7. Solubility of L-carnosine form I and form II in a binary solvent mixture of water and DMAC.

3.3. Solution Mediated Transformation

The solution-mediated polymorphic transformation of L-carnosine in different binary solvent mixtures of water + organic solvent, including methanol, ethanol, 2-propanol, acetone, DMF, and DMAC, was detected by online Raman spectroscopy. The results demonstrate that the transformation experiments of L-carnosine in a 100 mL binary solvent mixture were repeatable. The changes in Raman relative intensity during the polymorphic transformation process from form II to form I are shown in Figure 8. It can be seen that after the induction period of form I, the characteristic peak at 1227.3 cm^{-1} of form I increases with a corresponding decrease in the characteristic peak at 1524.8 cm^{-1} of form II. Finally, the characteristic peak of form II disappeared, and the characteristic peak intensity of form I remained basically stable, which means form II had completely transformed into form I at the end of this transformation process. In addition, the time needed for performing the crystal transformation of L-carnosine varies from tens of minutes to tens of hours in different binary solvent mixtures. The duration time of the transformation process for L-carnosine in the tested solvent systems was (water + acetone) < (water + 2-propanol) < (water + ethanol) < (water + methanol) < (water + DMAC) < (water + DMF), which represents the combined effects of solute conformation and solute−solvent interactions [24]. It indicates that the transformation rate from form II to form I of L-carnosine can be effectively regulated by changing the solvent.

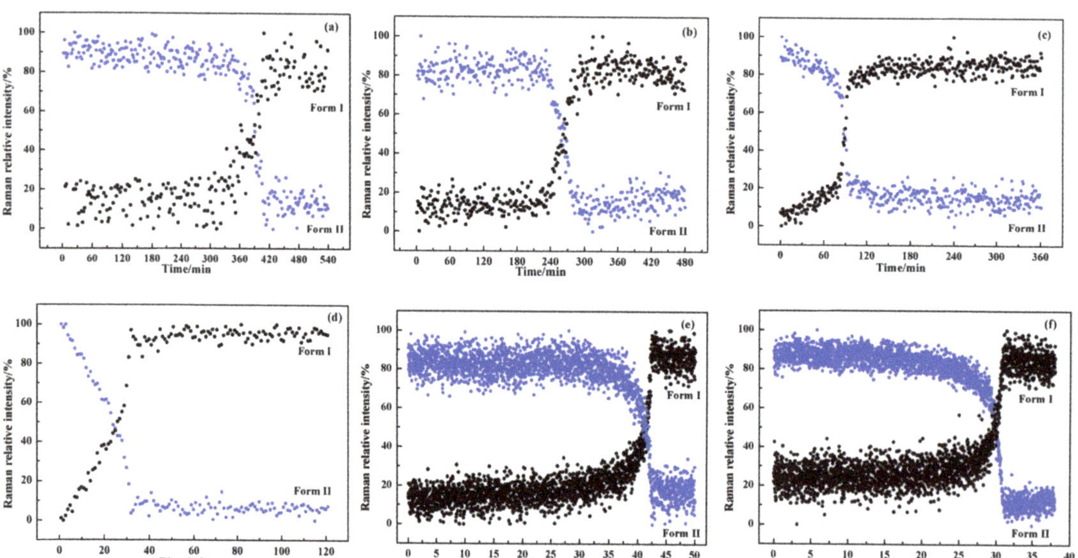

Figure 8. Changes of Raman relative intensity during the solution-mediated polymorphic transformation of L-carnosine in different binary solvent mixtures: (**a**) water + methanol; (**b**) water + ethanol; (**c**) water + 2-propanol; (**d**) water + acetone; (**e**) water + DMF; (**f**) water + DMAC.

Powder X-ray diffraction was carried out at certain time intervals to verify the polymorphic transformation results from form II to form I of L-carnosine. The powder X-ray diffraction patterns of solid samples withdrawn from suspension in a water + ethanol binary solvent mixture are shown in Figure 9. It can be seen that the characteristic peak intensity of form I gradually increased, while the characteristic peak intensity of form II gradually decreased after 180 min. At this point, form I and form II of L-carnosine coexisted in the system. Finally, the characteristic peak of form II disappeared, while the characteristic peak intensity of form I reached its maximum. Accordingly, form I was the only polymorph detected in the solid phase after 300 min, which was consistent with the results of Raman spectroscopy.

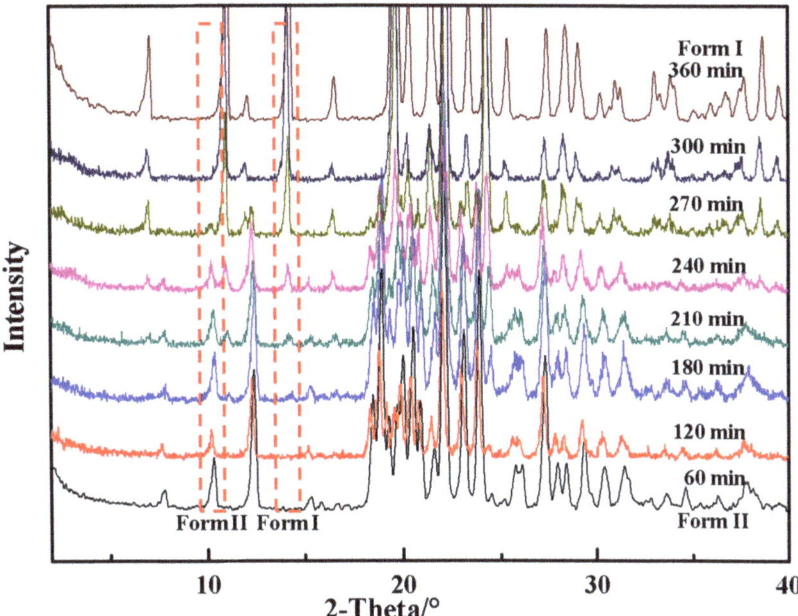

Figure 9. Powder X-ray diffraction patterns at different times during the solution-mediated polymorphic transformation of L-carnosine in a water + ethanol binary solvent mixture.

In combination with the changes of Raman relative intensity during the solution-mediated polymorphic transformation, the transformation process in the first kind of solvent systems can be divided into those with and without induction time. Therefore, three typical binary solvent mixtures, water + acetone, water + ethanol, and water + DMF, were chosen for further study, which are able to fully reflect the transformation mechanism. To explore the transformation mechanism, the solution concentration was measured at specific time intervals by the gravimetric method. The results in typical binary solvent mixtures, including water + acetone, water + ethanol, and water + DMF, are plotted in Figure 10. It can be seen that the driving force for transformation, namely the difference in solubility between the two polymorphs, decreased with the increase in solvent polarity. As a result, the transformation time for L-carnosine in the tested solvent systems increased with the increase in solvent polarity [25]. The transformation started immediately after form II of L-carnosine was added into a water + acetone binary solvent mixture. However, the induction time of the polymorphic transformation process in the water + ethanol and water + DMF binary solvent mixtures was 180 min and 30 h respectively, which indicates that the nucleation of form I is the controlling step of polymorphic transformation in water + ethanol and water + DMF binary solvent mixtures. In addition, the solution concentration in water + acetone and water + ethanol binary solvent mixtures held steady at the solubility of the metastable form II during the increase in form I, and started to decrease when form II had dissolved completely. It indicates that the dissolution rate of form II is higher than the growth rate of form I, and the growth of form I is the controlling step of polymorphic transformation in water + acetone and water + ethanol binary solvent mixtures. Nevertheless, the solution concentration in water + DMF decreased to the solubility of stable form I when the amount of form I started to increase, suggesting that the dissolution rate of form II is smaller than the growth rate of form I, and the dissolution of form I is the controlling step of polymorphic transformation in a water + DMF system. In conclusion, the polymorphic transformation of L-carnosine is "growth-controlled" in water + acetone solution, "nucleation-growth-controlled" in water + ethanol solution, and "nucleation-dissolution-controlled" in water + DMF solution [9].

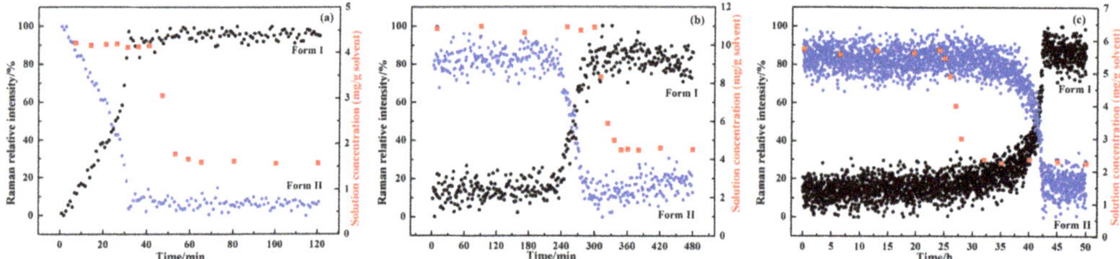

Figure 10. Changes of solution concentration during the solution-mediated polymorphic transformation of L-carnosine in different solvent mixtures: (**a**) water + acetone; (**b**) water + ethanol; (**c**) water + DMF.

3.4. Transformation Process Simulation

The dissolution-precipitation kinetics model was carried out to simulate and analyze the polymorphic transformation of L-carnosine based on the experimental data. The values of the model parameters are listed in Table 2. Figure 11 illustrate the calculated results of the liquid concentration and solid-phase composition of L-carnosine in three binary solvent systems. It was found that the simulated solution concentration curves can fit the experimental values well, and the simulated solid-phase compositions of the two polymorphs are also close to the measured online Raman data. The delicate discrepancies of the solid-phase composition in water + ethanol and water + DMF binary solvent mixtures are acceptable, considering the assumptions proposed to simplify the calculation process. The consistency between the model results and experimental data supposes that the model established in our work is reliable to simulate the solution-mediated polymorphic transformation process of L-carnosine.

Table 2. Model parameters for the solution-mediated polymorphic transformation of L-carnosine in different solvent mixtures.

Solvent	Solid Form	k_{diss}/min^{-1}	C_{sol}	k_{nuc}/min^{-1}(g/mg)$^{\alpha-1}$	k_{growth}/min^{-1}	α
Water + Acetone	form I	0.176	1.57×10^{-3}	5.89×10^{-3}	112	1.46
	form II	0.157	4.21×10^{-3}	5.53×10^{-2}	37.3	2.71
Water + Ethanol	form I	0.758	4.49×10^{-3}	4.81×10^{-5}	169	1.74
	form II	0.267	1.10×10^{-2}	2.22×10^{-2}	24.3	2.19
Water + DMF	form I	0.318	2.24×10^{-3}	4.65×10^{-6}	142	1.87
	form II	0.389	5.74×10^{-3}	2.34×10^{-2}	67.8	2.79

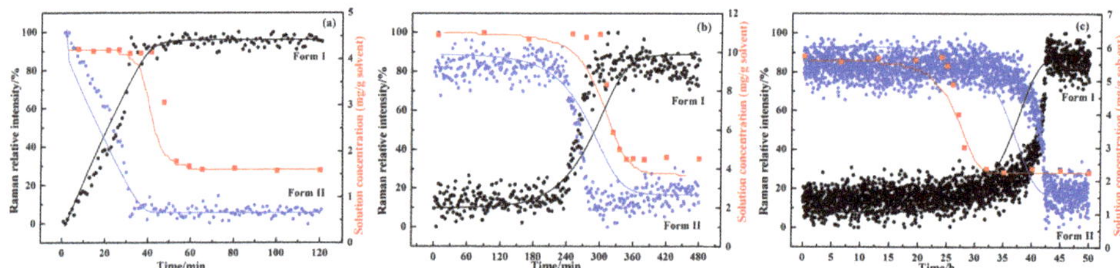

Figure 11. Calculated solid composition profiles and solution concentration profile during the solution-mediated polymorphic transformation from form II to form I of L-carnosine in different solvent mixtures: (**a**) water + acetone; (**b**) water + ethanol; (**c**) water + DMF.

4. Conclusions

In this paper, form II—a new polymorph of L-carnosine—was successfully developed by an antisolvent crystallization method. The properties of the two polymorphs were characterized by powder X-ray diffraction, polarizing microscope, thermal analysis, and Raman spectroscopy. The solubility of L-carnosine form I and form II in a (water + DMAC) binary solvent mixture was determined by a dynamic method, based on which the relative stability of the two forms was analyzed. The results suggest that form I is the thermodynamically stable form compared with form II. In addition, the solution-mediated polymorphic transformation of the two polymorphs in different binary solvent systems was detected by online and offline analytical techniques. It was found that the transformation process became longer with increasing solvent polarity. The polymorphic transformation of L-carnosine was "growth-controlled" in a water + acetone binary solvent, while "nucleation−growth" and "nucleation-dissolution" were the rate-controlling steps in water + ethanol and water + DMF binary solvents, respectively. A theoretical model of dissolution-precipitation was built to simulate the kinetics of the polymorphic transformation. It was verified that the simulated curves are basically consistent with the experimental results, which means that the model established in this work is reliable to simulate the solution-mediated polymorphic transformation process of L-carnosine.

Author Contributions: Investigation and writing—original draft preparation, Y.Z.; software, S.Z.; data curation, J.G.; visualization and project administration, C.L.; supervision and writing—review and editing, T.W. All authors have read and agreed to the published version of the manuscript.

Funding: This research was funded by the Fundamental Research Funds for the Universities of Hebei Province, grant number JQN2020030.

Institutional Review Board Statement: Not applicable.

Informed Consent Statement: Not applicable.

Data Availability Statement: Not applicable.

Acknowledgments: The authors are very grateful for the financial support of the Fundamental Research Funds for the Universities of Hebei Province (grant number JQN2020030).

Conflicts of Interest: The authors declare no conflict of interest.

References

1. Brog, J.P.; Chanez, C.L.; Crochet, A.; Fromm, K.M. Polymorphism, what it is and how to identify it: A systematic review. *Rsc Adv.* **2013**, *3*, 16905–16931. [CrossRef]
2. Zhou, Y.; Wang, J.; Xiao, Y.; Wang, T.; Huang, X. The Effects of Polymorphism on Physicochemical Properties and Pharmacodynamics of Solid Drugs. *Curr. Pharm. Des.* **2018**, *24*, 2375–2382. [CrossRef] [PubMed]
3. Fandaruff, C.; Rauber, G.S.; Araya-Sibaja, A.M.; Pereira, R.N.; de Campos, C.E.M.; Rocha, H.V.A.; Monti, G.A.; Malaspina, T.; Silva, M.A.S.; Cuffini, S.L. Polymorphism of Anti-HIV Drug Efavirenz: Investigations on Thermodynamic and Dissolution Properties. *Cryst. Growth Des.* **2014**, *14*, 4968–4975. [CrossRef]
4. Singhal, D.; Curatolo, W. Drug polymorphism and dosage form design: A practical perspective. *Adv. Drug Deliv. Rev.* **2004**, *56*, 335–347. [CrossRef] [PubMed]
5. Yang, C.; Zhang, Z.; Zeng, Y.; Wang, J.; Wang, Y.; Ma, B. Structures and characterization of m-nisoldipine polymorphs. *Cryst. Eng. Comm.* **2012**, *14*, 2589–2594. [CrossRef]
6. Kitamura, M. Polymorphism in the crystallization of L-glutamic acid. *J. Cryst. Growth* **1989**, *96*, 541–546. [CrossRef]
7. Nývlt, J. The Ostwald Rule of Stages. *Cryst. Res. Technol.* **1995**, *30*, 443–449. [CrossRef]
8. Guo, N.; Hou, B.; Wang, N.; Xiao, Y.; Huang, J.; Guo, Y.; Zong, S.; Hao, H. In Situ Monitoring and Modeling of the Solution-Mediated Polymorphic Transformation of Rifampicin: From Form II to Form I. *J. Pharm. Sci.* **2017**, *107*, 344–352. [CrossRef]
9. O'Mahony, M.A.; Maher, A.; Croker, D.M.; Rasmuson, Å.C.; Hodnett, B.K. Examining Solution and Solid State Composition for the Solution-Mediated Polymorphic Transformation of Carbamazepine and Piracetam. *Cryst. Growth Des.* **2012**, *12*, 1925–1932. [CrossRef]
10. Gulewitsch, W.; Amiradžibi, S. Ueber das Carnosin, eine neue organische Base des Fleischextractes. *Eur. J. Inorg. Chem.* **1900**, *33*, 1902–1903. [CrossRef]
11. Sharif, S.; Schagen, D.; Toney, M.D.; Limbach, H.-H. Coupling of Functional Hydrogen Bonds in Pyridoxal-5′-phosphate−Enzyme Model Systems Observed by Solid-State NMR Spectroscopy. *J. Am. Chem. Soc.* **2007**, *129*, 4440–4455. [CrossRef]

12. Zhou, Y.; Wang, J.; Wang, T.; Wang, N.; Xiao, Y.; Zong, S.; Huang, X.; Hao, H. Self-Assembly of Monodispersed Carnosine Spherical Crystals in a Reverse Antisolvent Crystallization Process. *Cryst. Growth Des.* **2019**, *19*, 2695–2705. [CrossRef]
13. Wu, J.W.; Liu, K.-N.; How, S.-C.; Chen, W.-A.; Lai, C.-M.; Liu, H.-S.; Hu, C.-J.; Wang, S.S.S. Carnosine's Effect on Amyloid Fibril Formation and Induced Cytotoxicity of Lysozyme. *PLoS ONE* **2013**, *8*, e81982. [CrossRef]
14. Banerjee, S.; Mukherjee, B.; Poddar, M.K.; Dunbar, G.L. Carnosine improves aging-induced cognitive impairment and brain regional neurodegeneration in relation to the neuropathological alterations in the secondary structure of amyloid beta (Aβ). *J. Neurochem.* **2021**, *158*, 710–723. [CrossRef]
15. Jiang, C.; Wang, Y.; Yan, J.; Yang, J.; Xiao, L.; Hao, H. Formation Mechanism and Phase Transformation Behaviors of Pantoprazole Sodium Heterosolvate. *Org. Process Res. Dev.* **2015**, *19*, 1752–1759. [CrossRef]
16. He, F.; Wang, Y.; Yin, Q.; Tao, L.; Lv, J.; Xu, Z.; Wang, J.; Hao, H. Effect of polymorphism on thermodynamic properties of cefamandole nafate. *Fluid Phase Equilibria* **2016**, *422*, 56–65. [CrossRef]
17. Wu, S.; Du, S.; Chen, M.; Li, K.; Jia, L.; Zhang, D.; Macaringue, E.G.J.; Hou, B.; Gong, J. Crystal Structures and Phase Behavior of Sulfadiazine and a Method for the Preparation of Aggregates with Good Performance. *Chem. Eng. Technol.* **2017**, *41*, 532–540. [CrossRef]
18. Zhou, Y.; Wang, J.; Wang, T.; Gao, J.; Huang, X.; Hao, H. Solubility and dissolution thermodynamic properties of L-carnosine in binary solvent mixtures. *J. Chem. Thermodyn.* **2020**, *149*, 106167. [CrossRef]
19. Jakubiak, P.; Schuler, F.; Alvarez-Sánchez, R. Extension of the dissolution-precipitation model for kinetic elucidation of solvent-mediated polymorphic transformations. *Eur. J. Pharm. Biopharm.* **2016**, *109*, 43–48. [CrossRef]
20. Zhu, M.; Wang, Y.; Li, F.; Bao, Y.; Huang, X.; Shi, H.; Hao, H. Theoretical Model and Experimental Investigations on Solution-Mediated Polymorphic Transformation of Theophylline: From Polymorph I to Polymorph II. *Crystals* **2019**, *9*, 260. [CrossRef]
21. Xiao, Y.; Wang, J.; Huang, X.; Shi, H.; Zhou, Y.; Zong, S.; Hao, H.; Bao, Y.; Yin, Q. Determination Methods for Crystal Nucleation Kinetics in Solutions. *Cryst. Growth Des.* **2017**, *18*, 540–551. [CrossRef]
22. Jakubiak, P.; Wagner, B.; Grimm, H.P.; Petrig-Schaffland, J.; Schuler, F.; Alvarez-Sánchez, R. Development of a Unified Dissolution and Precipitation Model and Its Use for the Prediction of Oral Drug Absorption. *Mol. Pharm.* **2016**, *13*, 586–598. [CrossRef]
23. Grunenberg, A.; Henck, J.-O.; Siesler, H. Theoretical derivation and practical application of energy/temperature diagrams as an instrument in preformulation studies of polymorphic drug substances. *Int. J. Pharm.* **1996**, *129*, 147–158. [CrossRef]
24. Kitamura, M.; Umeda, E.; Miki, K. Mechanism of Solvent Effect in Polymorphic Crystallization of BPT. *Ind. Eng. Chem. Res.* **2012**, *51*, 12814–12820. [CrossRef]
25. Marcus, Y. The properties of organic liquids that are relevant to their use as solvating solvents. *Chem. Soc. Rev.* **1993**, *22*, 409–416. [CrossRef]

crystals

Article

Croconic Acid Doped Glycine Single Crystals: Growth, Crystal Structure, UV-Vis, FTIR, Raman and Photoluminescence Spectroscopy

Elena Balashova [1,*], Aleksandr A. Levin [1] Valery Davydov [1], Alexander Smirnov [1], Anatoly Starukhin [1], Sergey Pavlov [1], Boris Krichevtsov [1], Andrey Zolotarev [2], Hongjun Zhang [3], Fangzhe Li [4] and Hua Ke [4]

1   Ioffe Institute, Politechnicheskaya 26, 194021 Saint Petersburg, Russia
2   Institute of Earth Sciences, Saint-Petersburg State University, Universitetskaya Nab. 7/9, 199034 Saint-Petersburg, Russia
3   Functional Materials and Acoustooptic Instruments Institute, School of Instrument Science and Engineering, Harbin Institute of Technology, Harbin 150080, China
4   Scool of Materials Sciences and Engineering, Harbin Institute of Technology, Harbin 150080, China
*   Correspondence: balashova@mail.ioffe.ru

Abstract: Glycine (Gly) single crystals doped with croconic acid (CA) were grown by evaporation from aqueous solutions. Depending on the weight ratio of Gly and CA in solutions, the crystals take on a plate or pyramidal shape. Both powder and single crystal XRD analyses indicate that the crystal lattices of plates (α-Gly:CA) and pyramides (γ-Gly:CA) correspond to the lattices of pure α-Gly and γ-Gly polymorphs, respectively. Raman and FTIR spectra of Gly:CA crystals are very close to the spectra of undoped crystals, but include bands associated with CA impurity. Analysis of UV-Vis absorption spectra indicates that doping does not remarkably change bandgap value Eg~5.2 eV but results in appearance of strong absorption bands in the transparency region of pure glycine crystals, which result from local electronic transitions. Incorporation of CA molecules in Gly creates strong green photoluminescence in a wide spectral range 1.6–3.6 eV. Comparison of the optical spectra of Gly:CA and previously studied TGS:CA crystals indicates that in both cases, the modifications of the optical spectra induced by CA doping are practically identical and are related to the interaction between CA molecules located in the pores of the host Gly crystals and neighboring Gly molecules.

Keywords: glycine; polymorphism; croconic acid; crystal structure; XRD; FTIR and Raman spectroscopy; UV-Vis absorption; photoluminescence

Citation: Balashova, E.; Levin, A.A.; Davydov, V.; Smirnov, A.; Starukhin, A.; Pavlov, S.; Krichevtsov, B.; Zolotarev, A.; Zhang, H.; Li, F.; et al. Croconic Acid Doped Glycine Single Crystals: Growth, Crystal Structure, UV-Vis, FTIR, Raman and Photoluminescence Spectroscopy. *Crystals* **2022**, *12*, 1342. https://doi.org/10.3390/cryst12101342

Academic Editor: Jingxiang Yang

Received: 19 August 2022
Accepted: 17 September 2022
Published: 22 September 2022

Publisher's Note: MDPI stays neutral with regard to jurisdictional claims in published maps and institutional affiliations.

1. Introduction

Protein amino acid glycine (Gly) NH_2-CH_2-COOH was discovered in 1820 by the chemist, botanist, and pharmacist Henri Braconnot. It was the first case wherein a pure amino acid was obtained from a protein [1]. Since glycine is produced in the human body during metabolism, it is safe and is widely used in medicine as a metabolic regulator that normalizes and activates protective inhibition processes in the central nervous system and reduces psycho-emotional stress.

Piezoelectricity is in increasing demand as it provides diverse entries into electronic, electromechanical, optical, and optoelectronic applications. A very high polarization response, which is promising for capacitors, memories, and piezoelectric applications, has been discovered mostly in inorganic oxides containing toxic lead or rare elements. Purely organic ferroelectrics, which are expected to be used as key materials in organic, printable, and bendable electronic device applications, are being pursued as possible alternatives. Recently, it was demonstrated that a basic component of biological structure, γ-glycine, presents nanoscale ferroelectricity, with an exhibited technologically significant piezoelectric response. Glycine is one of the simplest and smallest biological molecules.

It therefore presents a starting point for the design of functional materials as it can be synthetically modified to optimize properties.

Crystallization of glycine is characterized by polymorphism. Currently, α-, β-, γ-, δ-, and ε-Gly modifications of glycine crystal structure are known. The most stable among them are α- and γ-modifications. It is believed that the α modification is the most stable one because γ- and β-glycine are transformed to α-glycine at certain conditions. However, there exists an opposite perspective in the literature, which relates the γ-modification to the most stable one [2]. Γ-Gly transforms in α-Gly at $T \approx 460$ K [3,4]. B-Gly transforms in α- Gly or γ-Gly at ambient conditions [2]. Similarly, under high pressure, phase transitions of β-Gly to δ-Gly and γ-Gly to ε-Gly occur [5].

α-Gly crystallizes in centrosymmetric crystal structure [space group $P2_1/n$ (14)]. In contrast, β- and γ-Gly crystallize in non-centrosymmetric structure of monoclinic [sp.gr. $P2_1$ (4)], and trigonal [sp.gr. $P3_2$ (145) or $P3_1$ (144)] symmetry, respectively [2–7]. Nominally pure glycine single crystals are transparent in a wide spectral range. Strong increase of absorption due to band-to-band electronic transitions begins in ultraviolet (UV) region for the light wavelength λ shorter than cut-off wavelength λ_{gap}. In both α-Gly [8] and γ-Gly [9–12] $\lambda_{gap} \approx 240$ nm. Estimations of the bandgap E_g are based on use of Tauc plot and provide the values of $E_g = 5.11 \pm 0.02$ eV in α-Gly [8] and $E_g = 5.09$–5.3 eV in γ-Gly [11–13]. A much higher value of $E_g = 6.2$ eV has been reported in Ref. [14].

Non-centrosymmetric γ-Gly attracts much attention due to remarkable piezoelectric, pyroelectric, and non-linear optical (NLO) properties. The longitudinal piezoelectric coefficient d_{33} value for the γ-Gly crystals was discovered to be $d_{33} = 7.37$ pC/N [13], piezoelectric strain constant ~9.93 pm·V^{-1} [15], pyroelectric coefficient p~13–21.4 μC/m^{2}K [16,17]. The large value of piezoelectric coefficient makes γ-Gly an attractive material for design of flexible amino acid-based energy harvesting [18]. The glycine polycrystalline sensors indicate the effective piezoelectric constants of $d_{33} = 0.9$ pC/N and of $g_{33} = 60$ mV·m/N, respectively. Note that the latter one exceeds that of commercial piezoelectric lead zirconium titanate (PZT) [18].

Since glycine crystals grow from aqueous solutions, the addition of various substances to solutions makes it possible to significantly influence the process of glycine crystallization and obtain doped glycine crystals with modified physical properties. In particular, the type of additive and its amount can determine what polymorph of glycine will grow from solution. It was demonstrated in [19] that the addition to the solution of a small amount of zinc sulphate $ZnSO_4$ with a molar concentration (0.2–0.4 M) is accompanied by growth of α-Gly, and an increase in concentration (0.6–1 M) leads to the appearance of the γ-polymorph. In [20], the effects of seven common salts on primary nucleation of the metastable α-Gly and the stable γ-Gly were studied. It was demonstrated that addition of $(NH_4)_2SO_4$, NaCl and KNO_3, in general, results in primary nucleation of γ-Gly simultaneously inhibiting α-Gly primary nucleation. Addition of Ca $(NO_3)_2$ and $MgSO_4$ also promotes γ-Gly and inhibits α-Gly primary nucleation but not sufficiently to induce γ-Gly. Na_2SO_4 and K_2SO_4 promote not only γ-Gly but also α-Gly primary nucleation.

Polymorphic control of glycine by using the microdroplet technique for NaCl additive was investigated in [21]. Analysis of thermodynamics and nucleation kinetics of glycine polymorphs in microdroplets indicates that NaCl reduces the nucleation energy barrier of the γ-form of glycine, and for the same form of glycine, crystallization by the conventional method has a higher interfacial tension than in microdroplets.

Comparison of the crystallization process of glycine polimorphs and its salts/co-crystals in the presence of organic carboxylic acids through implementation of several different preparative techniques was studied in [6]. It was discovered that for the outcome of crystallization in "glycine– carboxylic acid" systems upon slow evaporation of aqueous solutions, both the choice of the acid and its concentration are of importance.

Additives can also influence a morphology of glycine crystals. As indicated in [22], the γ-Gly crystals grown from the solution containing glycine and sodium fluoride, sodium hydroxide, or sodium acetate had only three well developed faces (101), (0–11), and (−111) with the remaining faces developed in the form of a pyramid. In contrast, adding to the solution of sodium nitrate resulted in a different morphology with four well-developed faces.

It is important that additions can strongly affect a nonlinear properties of glycine crystals. It was determined that the addition of alkali metal nitrides, bromides, chlorides, or acetates to a solution can significantly change the second-harmonic generation (SHG) efficiency in γ-Gly [11,23–25], reaching a maximum value for Cs chloride 5.06 times higher than that in a potassium dihydrogen phosphate (KDP) crystal [26].

From aqueous solutions of glycine with various inorganic and organic acids and salts, many new crystals were obtained [6,7], particularly the well-known ferroelectric triglycine sulfate (TGS), which has record values of the pyroelectric coefficient [27], and other ferroelectric crystals based on glycine and methylated glycine [7,28,29].

The effect of organic croconic acid (CA) additives on the glycine crystallization process has not yet been studied. Croconic acid crystals are a promising material in which CA molecules bonded by O-H$\cdots$O hydrogen bonds provide excellent ferroelectricity at room temperature (RT) [30]. However, CA is corrosive and requires another acid in the solution to crystallize. In this regard, it is of considerable interest to study the properties of mixed Gly-CA crystals. The CA molecule, $C_5H_2O_5$, belongs to a series of planar monocyclic $C_nH_2O_n$ oxocarbon acids. It has highly symmetric (D_{5h}) plane topology and large value of molecular dipole momentum $d\sim$ (9–10) D [30]. CA molecule is small and may be incorporated into the host lattice as an impurity in a wide concentration range.

It was recently discovered that doping above RT ferroelectric TGS crystals with CA molecules results in appearance of a strong green luminescence band in region of photon energy E_{ph} = 1.6–3.5 eV with the band maximum at E_{ph} = 2.55 eV. The doping also leads to the appearance of absorption bands related to local optical transitions in the transparency region of nominally pure TGS crystals and change of dielectric hysteresis loops [31]. CA doping does not change significantly TGS crystal structure, does not result in appearance of new phases, and does not influence frequencies of molecular vibrations measured by Raman and Fourier-transform infrared spectroscopy (FTIR).

Since, unlike TGS (which includes ordinary glycinium ions, zwitterions, and sulfuric acid molecules), glycine crystals are formed only by glycine zwitterions; the crystal structure of glycine can be either centrosymmetric (α-Gly) or non-centrosymmetric (γ-Gly). It is interesting to compare the changes in the crystal structure, luminescence, and optical absorption caused by doping TGS and Gly crystals with croconic acid. For these reasons, in this work Gly single crystals were grown from aqueous solutions of glycine and croconic acid at the same relative ratios (80:20, 90:10, 98:2) as in previous studies of TGS:CA crystals.

Particularly, the primary questions of interest for the present study were the following: how does CA doping affect the crystal structure, crystal morphology, unit cell parameters of various glycine isomorphs, the amplitudes and frequencies of glycine molecular vibrations, the electronic band structure, in particular, the value of the band gap E_g, optical properties including absorption and luminescence spectroscopy of doped crystals. Generally, the aim of this research was the synthesis of CA–doped α- and γ-Gly single crystals and the study of their structural and optical properties. Preliminary results on Raman and photoluminescence spectroscopy were presented in our conference paper [32].

2. Materials and Methods

2.1. Crystal Synthesis

Figure 1 presents images of crystals used in this research. Nominally pure single crystals of α-Gly were grown by evaporation from saturated aqueous solutions of aminoacetic acid (α-Gly) (Figure 1a). Colorless "pure" γ-Gly crystals were grown from aqueous solution of α-Gly crystals with adding 5% of citric acid $C_6H_8O_7$ and may contain a small admixture of the acid (Figure 1b). Croconic acid crystals (Figure 1c) were prepared from the CA reagent (Alfa Aesar B21809, 98% purity) according to the procedure described in [33].

Croconic acid doped Gly crystals were grown from aqueous solutions with different weight rations of Gly and CA—80:20, 90:10, 98:2. Gly:CA crystals grown from 80:20 and

98:2 solutions have a pyramidal shape with a hexagonal base (Figure 1d,f) analogous to nominally pure γ-Gly (Figure 1b). Crystals grown from 90:10 solution have plate shape like pure α-Gly (Figure 1a).

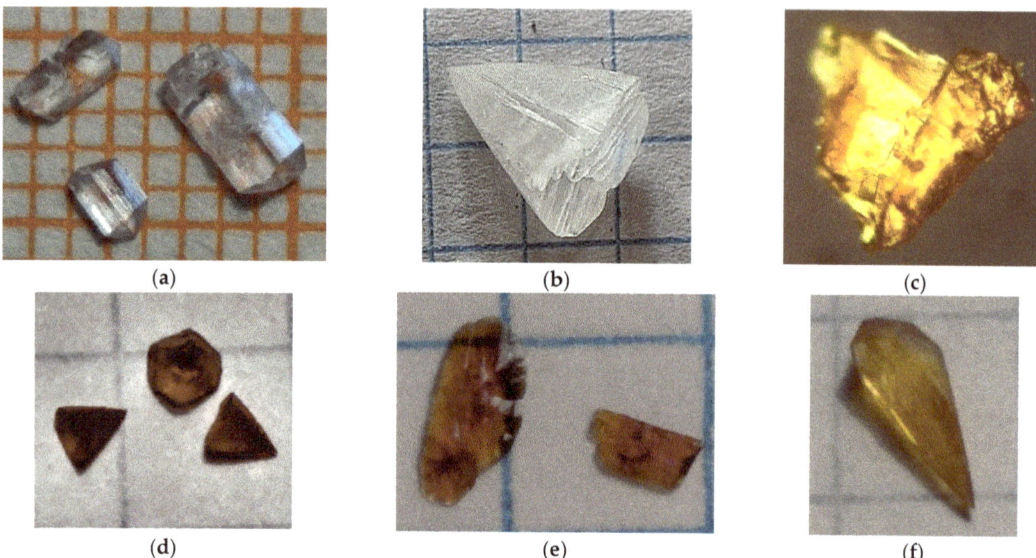

Figure 1. Images of crystals used in this research. α-Gly crystals grown from aqueous solution (**a**). γ-Gly crystal grown from aqueous solution with adding 5% of citric acid C6H8O7 (**b**). Croconic acid crystal (**c**). Gly:CA crystals grown from solutions with relative weight ratios of Gly and CA 80:20 (**d**), 90:10 (**e**), 98:2 (**f**).

2.2. Experimental Details of Single Crystal XRD

Single crystal X-ray diffraction (XRD) studies of Gly:CA (90:10) and Gly:CA (80:20) were performed in the Harbin Institute of Technology, China, by a Bruker D8 Venture setup (50 kV, 30 mA) at RT using Mo-K_α radiation, and Gly:CA (80:20) by Rigaku XtaLAB Synergy diffractometer at 100 K using Cu-K_α radiation (50 kV, 1 mA) in the Resource Center "X-ray Diffraction Methods" of St. Petersburg State University. The structures have been solved by the direct methods by means of the SHELX program [34] incorporated in the OLEX2 program package [35]. The carbon and nitrogen-bound H atoms were placed in calculated positions and were included in the refinement in the 'riding' model approximation. Empirical absorption correction was applied in CrysAlisPro program complex [CrysAlisPro, Version 1.171.36.32, Agilent Technologies Inc., Santa Clara, CA, USA] using spherical harmonics, implemented in SCALE3 ABSPACK scaling algorithm [for Gly:CA (80:20)] and using the Bruker software SADABS-2016/2 [Bruker, 2016/2] [for Gly:CA (90:10)]. Supplementary crystallographic data have been deposited in the Cambridge Crystallographic Data Centre (CCDC) (2194293-2194294) and can be obtained free of charge via www.ccdc.cam.ac.uk/data_request/cif (accessed on 8 April 2022). The drawings of the Gly:CA (90:10) and Gly:CA (80:20) crystal structures were made using the Vesta program [36].

2.3. Experimental Details of Powder XRD

To prepare the samples for powder XRD investigations, the grown crystals were grinded in a corundum mortar. Low-background holders in the form of polished single-crystal Si (119) plate were used to prepare powder samples for XRD measurements.

Powder XRD measurements were performed utilizing a D2 Phaser X-ray powder diffractometer (Bruker AXS, Karlsruhe, Germany) equipped with an X-ray tube with a copper anode and a Ni K_β filter (Cu-K_α radiation, λ = 1.54184 Å). The XRD patterns were recorded in a symmetric scanning θ-2θ mode using a semiconductor position-sensitive linear X-ray detector LYNXEYE (Bruker AXS, Karlsruhe, Germany). During the XRD measurements, temperature in sample chamber was 313 ± 1 K. Additional XRD measurements of the powder samples mixed with internal XRD powder standard Si640f (NIST, Gaithersburg, MD, USA) were performed to win the angular corrections of the observed Bragg angles of the XRD reflections to zero shift and displacement.

After introducing the angular corrections to the Bragg angles $2\theta_B$ of XRD reflections with attributed Miller indices *hkl*, at first stage of analysis, unit cell parameters of the crystalline phases were calculated using the program Celsiz [37], utilizing the least-square technique. Microstructural parameters (average size D of crystallites and absolute values of mean microstrain ε_s in them) for crystalline phases were estimated from parameters of XRD reflections [Bragg angles $2\theta_B$, full width at half maximum $FWHM_{obs}$, maximum (Imax) and integral (Iint) intensities] by graphical methods of Williamson–Hall plot (WHP) [38] and size-strain plot (SSP) [39] adapted for pseudo-Voigt (pV) type of XRD reflections observed in the XRD patterns. All microstructure calculations [including the correction of the $FWHM_{obs}$ to instrumental broadening for pV reflections (resulting in $FWHM_{corr}$)] were performed using the program SizeCr [40]. Detailed description of the formalism of microstructural calculations can be found elsewhere [40,41].

To confirm the reliability of the crystal structure model and the results of single-crystal XRD analysis, to obtain more accurate values of the unit cell and microstructure parameters, and to verify the single-phase nature of powder samples, the powder XRD patterns were studied by the Le Bail (LB) and Rietveld methods. The TOPAS program [42] was used for LB and Rietveld fittings. For LB fitting the XRD patterns, a structure model is not required, but only an approximate value of the unit cell parameters and symmetry space group knowledge [43]. Rietveld fitting is used for structure model refinement, including atomic coordinates and temperature parameters [44].

The XRD patterns of both samples are characterized by a noticeable influence of preferential orientation effects. Unlike the Rietveld method in the LB fitting, the influence of preferred orientation effects to XRD pattern is corrected automatically. When fitted by the Rietveld method, considering the effect of preferential orientation within the March–Dollase model [45] led to a decrease in the weighted profile factor R_{wp}, characterizing the quality of the fit by ~2.5 and ~4 times for α-Gly:CA (90:10) and γ-Gly:CA (80:20), respectively. Additional consideration of the remaining effects of preferential orientation in the framework of the 8th-order spherical harmonics model [46] led to a further decrease in R_{wp} by ~1% for both samples.

Understated due to serial correlations [47], estimated standard deviations (e.s.d.s) of the refined parameters obtained using LB or Rietveld refinement were corrected by multiplying by the coefficient $m_{e.s.d.}$ calculated using the RietESD program [48], based on the formalism developed in [47,49].

Other parameters of powder XRD measurements, WHP and SSP analysis, LB, and Rietveld fittings are the same as detailed in [31,41,50] and briefly discussed in Section S2 of the Supplementary Materials.

2.4. Raman Scattering, FTIR, UV Vis Absorption, and Photoluminescence

Raman and micro-photoluminescence (μ-PL) measurements in the Gly:CA single crystals were carried out using LabRAM HREvo UV-VIS-NIR-Open spectrometer (Horiba Jobin–Yvon, Lille, France) equipped with a confocal microscope and a silicon CCD cooled by liquid nitrogen. Polarized micro-Raman measurements were performed at RT, in the spectral range 5–4000 cm^{-1} at different backscattering geometries: −Z (XX)Z, −Z (YY)Z, and −Z (XY)Z. Here, the Z-axis is oriented normally to the crystal surface (100), and X and Y are along c^*- and *b*-crystal axes, respectively. The line at λ = 532 nm (2.33 eV) of Nd:YAG laser (Torus, Laser Quantum, Inc., Edinburg, UK) was used as the excitation source. The laser power on the samples was as low as ~25–80 μW with a spot size of ~1 μm in diameter, to avoid sample heating. We used 1800 lines/mm grating and 100× ($NA = 0.90$) objective lens to measure Raman spectra. In the low frequency spectral region, the Rayleigh line was suppressed using three BragGrate notch filters (OptiGrate Corp., Oviedo, FL, USA) with an $OD = 4$ and a spectral bandwidth <0.3 nm.

The μ-PL measurements were performed in microscope stage Linkam THMS600 (Linkam Sci. Inst. Ltd., Salfords, Surrey, UK). The line at $\lambda = 325$ nm (2.81 eV) of HeCd laser (Plasma JSC, Ryazan, Russia) was used for continuous wave (CW) excitation. We used 600 lines/mm grating and a large working distance lens [Mitutoyo 50× UV ($NA = 0.40$)] with a spot size of ~2 μm and power density of 6 kWt/cm^{-2} on a sample was used to measure μ-PL.

Measurements of the infrared (IR) absorption spectra (FTIR spectra) were performed using an IR-Fourier spectrophotometer IRPrestige-21 with an IR microscope AIM-8000 (Shimadzu Corp., Kyoto, Japan), both in the specular reflection mode and in the transmission mode, followed by the Kramers–Kronig transformation. The results were then converted to absorbance. The measured spectral range was from 650 to 5000 cm^{-1}.

Optical absorption spectra of Gly:CA in the UV-visible (UV-Vis) spectral range were obtained using a UV-3600i Plus UV-Vis-NIR spectrophotometer (Shimadzu Corp., Kyoto, Japan) operating at RT in the wavelength range of 200–2000 nm. The test was conducted in reflection mode using an integrating sphere. $BaSO_4$ was used as a reference sample.

3. Results and Discussion

3.1. Single Crystal XRD Analysis

Single crystal XRD analysis indicates that the pyramid-like samples of Gly:CA (80:20) crystallize in a lattice typical to γ-Gly (space group $P3_1$ (144)) (γ-Gly:CA), and the plate-like crystals of Gly:CA (90:10) crystallize in α-Gly-like structure (space group $P2_1/n$ (14)) (α-Gly:CA) (Table 1). Structures of α-Gly:CA (90:10) and γ-Gly:CA (80:20) are drawn in Figure 2 using the data of Table 1 and refined atomic coordinates (CCDC 2194293-2194294). Relative coordinates of atoms (x/a, y/b, z/c), their isotropic temperature factors U, and atomic displacement parameters U^{ij} in α-Gly:CA (90:10) and γ-Gly:CA (80:20) are presented in Supplementary Materials (Tables S1–S4). Figure S1 portrays the structure of glycine molecule in (a) α-Gly:CA (90:10) and (b) γ-Gly:CA (80:20) with atoms represented by thermal ellipsoids by means of program ORTEP [51].

Table 1. Details of crystal data and structure refinement of α-Gly:CA (90:10) and γ-Gly:CA (80:20) single crystals.

	α-Gly:CA (90:10)	γ-Gly:CA (80:20)
Chemical formula	$C_2H_5NO_2$	$C_2H_5NO_2$
Formula weight, D_a	75.07	75.07
Space group	$P2_1/n$ (14)	$P3_1$ (144)
a, Å	5.1004 (5)	6.9848 (3)
b, Å	11.9664 (9)	6.9848 (3)
c, Å	5.4570 (5)	5.4744 (2)
β, °	111.714 (3)	90
V_{cell}, Å^3	309.43 (5)	231.30 (2)
Z	4	3
D_{calc}, g·cm^{-3}	1.611	1.617
F (000)	160.0	120.0
μ, mm^{-1}	0.143	1.25

Table 1. *Cont.*

	α-Gly:CA (90:10)	γ-Gly:CA (80:20)
Radiation (λ, Å)	Mo-K_α (0.71073)	Cu-K_α (1.54184)
Θ max, °	26.35	77.18
h, k, l max	6, 14, 6	8, 8, 6
Reflections collected	3240	3120
Independent reflections	629	645
Data/restraints/parameters	629/0/47	645/1/48
GOOF	1.158	1.081
Final R indexes [Reflections $I \geq 2\sigma$ (I)]	$R_1 = 0.0345, wR_2 = 0.1109$ [629]	$R_1 = 0.0299, wR_2 = 0.0735$ [645]
Final R indexes [Reflections all]	$R_1 = 0.0416, wR_2 = 0.1173$ [538]	$R_1 = 0.0299, wR_2 = 0.0735$ [645]
Largest difference peak/hole, e·Å^{-3}	0.19/−0.19	0.16/−0.20
Temperature of measurements, K	298	100

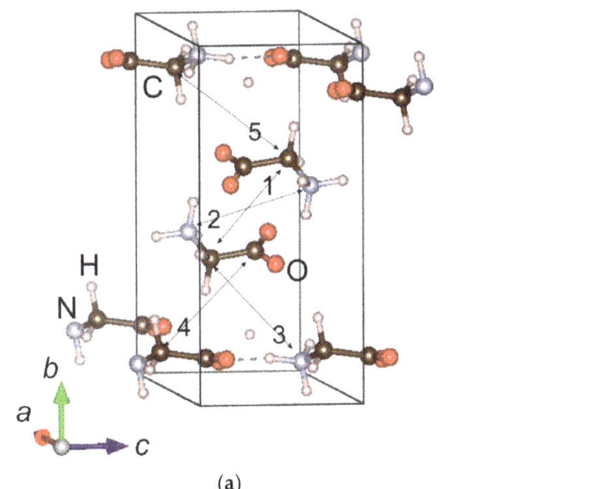
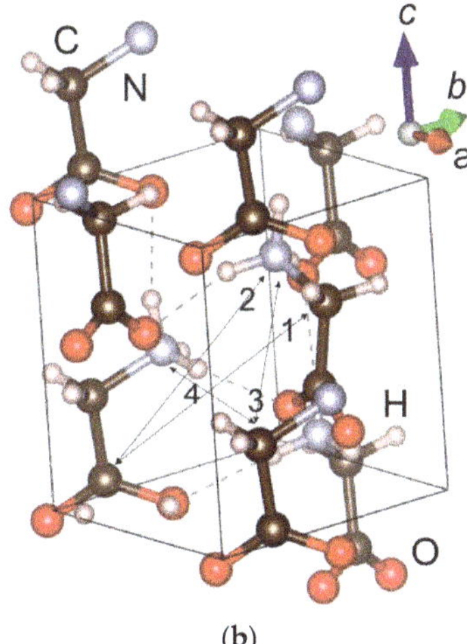

(a)   (b)

Figure 2. Crystal structures of (**a**) α-Gly:CA (90:10) and (**b**) γ-Gly:CA (80:20). Selected interatomic distances are marked in (**a**) for α-Gly:CA (90:10) (1: 0.44 nm, 2: 0.53 nm, 3: 0.45 nm, 4: 0.49 nm, 5: 0.50 nm) and in (**b**) for γ-Gly:CA (80:20) (1: 0.60 nm, 2: 0.55 nm; 3: 0.54 nm; 4: 0.69 nm). Maximum size of flat CA molecule is ~0.5 nm.

The crystal structures of all glycine phases are built from chains of glycine molecules linked by N1-H3···O2 hydrogen bonds and directed along the crystal *c* axis of all polymorphic modifications. In α-Gly, somewhat longer hydrogen bonds involving N1-H4···O1 atoms form the second set of hydrogen bond chains running along the *a* crystal axis. These chains form a layer parallel to the (010) plane. The layers are arranged perpendicular to the *b* crystal axis. Due to the strong hydrogen bonds N1-H5···O1, a two-layer structure is formed, in which each pair of layers is connected to each other by weak hydrogen bonds of the C1-H1···O1 and C1-H1···O2.

γ-Glycine is trigonal, crystallizing in the chiral space groups $P3_1$ (144). Its crystal structure does not consist of layers. Molecules form helices around the 3_1 screw-axis coinciding with the *c* crystal axis, successive molecules being linked by N1-H5···O2 type hydrogen bonds. Hydrogen chains N1-H3···O2 type along the *c* axis, like alpha-glycine, are formed between each molecule and symmetry-equivalent molecules away along the helix, i.e., connected by one repetition of the lattice along the c crystal axis.

Analysis of single crystal XRD data indicates that the best refinement of the structure of γ-Gly:CA crystals can be achieved by considering twinning (twin components are related by screw axis 3_1). The crystals of α-Gly:CA do not demonstrate twinning. Comparison with the literature data indicates (Table 2) that in centrosymmetric α-Gly:CA (90:10), the unit cell volume V_{cell} = 309.43 (5) Å3 is little more than in pure α-Gly [V_{cell} = 309.00 (3) Å^3] observed in [52], and smaller than V_{cell} = 310.10 (4) Å3 in [53]. In the magnitude of lattice parameters, CA doping manifests itself most clearly in decrease of c parameter and β angle in α-Gly:CA and a and c parameters in γ-Gly:CA (Tables 2 and 3).

Table 2. Comparison of unit cell parameters a, b, c, angle β, and unit cell volume V_{cell} in α-Gly:CA (90:10) and nominally pure α-Gly crystals from [52,53].

	α-Gly:CA (90:10) 298 K	α-Gly [53] 294 K	α-Gly [52] 301 K
a, Å	5.1004 (5)	5.1047 (3)	5.0999 (3)
b, Å	11.9664 (9)	11.9720 (14)	11.9516 (6)
c, Å	5.4570 (5)	5.4631 (3)	5.4594 (3)
β, °	111.714 (3)	111.740 (5)	111.781 (2)
V_{cell}, Å^3	309.43 (5)	310.10 (4)	309.00 (3)

Table 3. Comparison of unit cell parameters a, c, and unit cell volume V_{cell} in γ-Gly:CA (80:20) and nominally pure γ-Gly crystals from [53,54].

	γ-Gly:CA (80:20) 100 K	γ-Gly [53] 150 K	γ-Gly [54] 100 K	γ-Gly [53] 294 K	γ-Gly [54] 300 K
a, Å	6.9848 (3)	6.998 (16)	6.9869 (1)	7.0383 (7)	7.0402 (1)
c, Å	5.4744 (2)	5.4784 (18)	5.4768 (1)	5.4813 (8)	5.4813 (1)
V_{cell}, Å^3	231.30 (2)	232.5 (1)	231.540 (6)	235.15 (3)	235.280 (6)

Comparison of bond lengths and angles of glycine molecules in CA-doped and pure α- and γ-glycine crystals indicates that glycine zwitterions ($^+NH_3CH_2CO_2{}^-$) in these crystals have close magnitudes of corresponding bond angles and bond lengths (Tables 4 and 5). Nevertheless, some small differences, which can be attributed to CA doping, may be indicated. CA doping of α- and γ-Gly causes small increase of angle O1-C1-O2 and decrease of C2–C1–O1 that means the change of O1 atom position. However, doping does not influence other angles C2–C1–O2 and C1–C2–N1 in glycine molecules (Tables 4 and 5).

Thus, CA doping of α- and γ-modifications of glycine single crystals does not lead to a change in symmetry and noticeable changes in the lattice parameters of the corresponding nominally pure glycine isomorphs. Like TGS:CA crystals [31], the incorporation of CA molecules into glycine crystals leads only to a slight modification of the unit cell parameters. In contrast to TGS:CA, the presence of twins is observed in noncentrosymmetric γ-Gly:CA crystals.

Note that both in TGS:CA and Gly:CA crystals, changes in bond angles and bond lengths were observed in glycine zwitterions. In TGS:CA, all bond angles in glycine zwitterions ($^+NH_3CH_2CO_2{}^-$) are changed to bring it closer to the $N^+H_3CH_2COOH$ [31] glycinium ion. Conversely, only a slight change in the position of O1 atoms in the carboxyl group and N atoms in the amino group of zwitterions is observed in Gly:CA, which indicates the formation of O … H–O and O … H–N types of hydrogen bonds between glycine zwitterions and croconic acid molecules.

Table 4. Bond angles and bond lengths C1–O1, C1–O2, C2–N1, and C1–C2 of glycine molecules in CA-doped and nominally pure α-Gly crystals [52,53].

	α-Gly:CA (90:10)	α-Gly [53]	α-Gly [52]
	bond angles, °		
O1–C1–O2	125.86 (15)	125.62 (7)	125.45 (7)
C2–C1–O1	116.69 (15)	116.93 (7)	117.16 (7)
C2–C1–O2	117.44 (14)	117.44 (6)	117.38 (6)
C1–C2–N1	111.78 (13)	111.70 (6)	111.83 (5)
	bond lengths, Å		
C1–O1	1.251 (2)	1.2549 (9)	1.2512 (10)
C1–O2	1.247 (2)	1.2517 (9)	1.2480 (10)
C2–N1	1.475 (2)	1.4777 (10)	1.4743 (8)
C1–C2	1.526 (2)	1.5268 (9)	1.5252 (9)

Table 5. Bond angles and bond lengths C1–O1, C1–O2, C2–N1, and C1–C2 of glycine molecule in CA-doped and nominally pure γ-Gly crystals [53,54].

	γ-Gly:CA (80:20) 100K	γ-Gly [53] 150 K	γ-Gly [54] 100 K
	bond angles, °		
O1–C1–O2	126.0 (3)	125.3 (2)	125.70 (7)
C2–C1–O1	116.4 (2)	116.8 (2)	116.61 (7)
C2–C1–O2	117.6 (3)	117.8 (2)	117.63 (6)
C1–C2–N1	111.6 (2)	111.2 (2)	111.55 (6)
	bond lengths, Å		
C1–O1	1.260 (4)	1.263 (3)	1.2572 (9)
C1–O2	1.248 (4)	1.251 (3)	1.2487 (9)
C2–N1	1.479 (4)	1.480 (3)	1.4719 (10)
C1–C2	1.524 (4)	1.529 (3)	1.5249 (9)

3.2. Powder XRD Analysis

Phase analysis has led to a conclusion that both the measured XRD patterns of α-Gly:CA (90:10) and γ-Gly:CA (80:20) contain the XRD reflections attributed to one phase expected according to preparation, namely, α-Gly-like or γ-Gly like phases (Table 1). However, a thorough inspection revealed that the reflections of the α-Gly:CA (90:10) and γ-Gly:CA (80:20) has an asymmetric shape, especially clearly seen for reflections non-overlapped with neighboring ones (see examples in small insets of Figure 3).

Recently, a similar asymmetric shape of powder XRD reflections of organic ferroelectric 2-methylbenzimidazole (MBI) has been successfully described in the framework of models of two or three crystalline phases with the same MBI structure but with slightly different unit cell parameters and differing microstructure parameters. The formation of the MBI modifications has been associated with penetration of solvent molecules (ethanol, acetone, deuterated acetone, or water) into the voids of the MBI crystal structure.

The structures of α-Gly:CA (90:10) and γ-Gly:CA (80:20) are also characterized by rather large pores with sizes of ~0.4 nm–0.6 nm (Figure 2). The CA molecules used during the synthesis of samples have a flat shape and a maximum size of ~0.5 nm. Based on these geometric observations, it can be assumed that CA molecules occupy the pores of structures during the sample synthesis. If CA molecules do not fill the lattice voids homogeneously, but only some regions, then the filled lattice regions will stretch, forming modifications of the same structure with slightly larger lattice parameters. The average size of such regions with approximately the same content of CA molecules will determine the size of the crystallites of these different modifications. Simultaneously, due to the entry of CA molecules into the pores and the neighborhood of crystallites with slightly different lattice parameters, the presence of local microstrains in the crystallites of different modifications is possible.

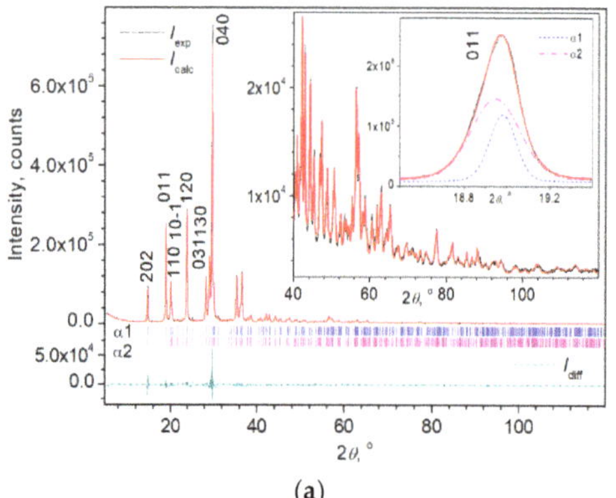

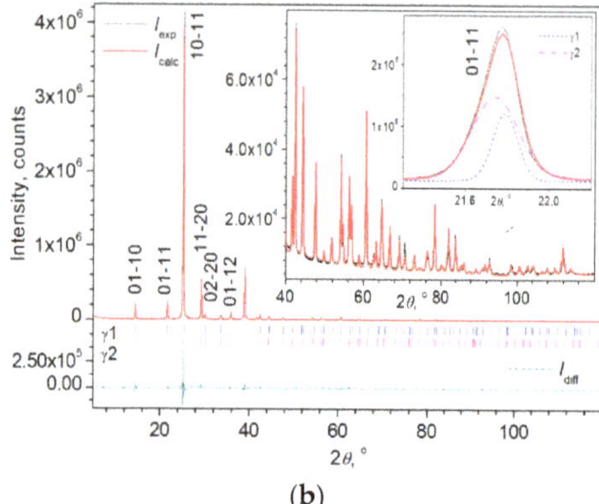

(a)   (b)

Figure 3. Rietveld fit results for (**a**) α-Gly:CA (90:10) and (**b**) γ-Gly:CA (80:20). During the Rietveld fit the atomic coordinates of the phases were fixed to values determined in single-crystal investigation of this work and not refined. The Miller indices *hkl* and Miller–Bravais indices *hkil* of some selected reflections are indicated in (**a**) and (**b**), respectively. Large insets in (**a**,**b**) portray the quality of fitting the high 2θ angle regions of the XRD patterns in a larger scale. Small insets present examples of Rietveld fitting the asymmetric non-overlapping reflections.

To verify these assumptions, which appear reasonable, the WHP and SSP methods at first stage were used to perform a profile analysis of XRD reflections (XRD Line profile analysis, LPA) observed in XRD patterns, assuming two phases with close parameters of unit cells (correspondingly, α1 and α2 modifications or γ1 and γ1 ones for α-Gly:CA (90:10) and γ- Gly:CA (80:20) XRD patterns). To do this, each observed reflection was separated into two reflections with the same known Miller indices hkl [for α-Gly:CA (90:10)] or Miller–Bravais indices hkil [for γ-Gly:CA (80:20)], where possible (some reflections, especially in the high-angle region of $2\theta > \sim45°$, are highly overlapped, and it is difficult to extract them). The resulting WHP and SSP graphs are presented in the Figures S2 and S3 of Supplementary Materials. Table 6 presents the results of Celsiz and LPA calculations.

Table 6. Results of LPA (WHP and SSP) and LB and Rietveld fitting of the α-Gly:CA (90:10) and γ-Gly:CA (80:20) XRD powder patterns measured at 313 ± 1 K.

a, Å b, Å	c, Å β, °	V_{cell}, Å	D (nm) ε_s, %
α-Gly:CA (90:10), LPA			
phase α1			
5.1050 (16) [a] 11.9771 (40) [a]	5.4614 (17) [a] 111.74 (2) [a]	310.18 (18)	52 (9)/78 (14) [b] 0.07 (6)/0.16 (4) [b]
phase α2			
5.1216 (20) [a] 12.0064 (64) [a]	5.4715 (22) [a] 111.70 (3) [a]	312.61 (25)	44 (3)/54 (3) [b] 0.06 (4)/0.14 (2) [b]
α-Gly:CA (90:10), LB fitting			

Table 6. *Cont.*

a, Å b, Å	c, Å β, °	V_{cell}, Å	D (nm) ε_s, %
		phase α1	
5.1002 (10) [c] 11.9709 (6) [c]	5.4613 (10) [c] 111.68 (2) [c]	309.85 (9)	96 (4) [c,d] 0.050 (6) [c,d]
		phase α2	
5.1107 (5) [c] 12.0102 (8) [d]	5.4694 (6) [c] 111.70 (1) [c]	311.31 (5)	37.4 (2) [c,d] 0.11 (2) [c,d]
α-Gly:CA (90:10), Rietveld fitting (without refinement of atomic coordinates)			
		phase α1	
5.1000 (16) [c] 11.9704 (8) [c]	5.4613 (9) [c] 111.71 (2) [c]	399.76 (12)	98 (5) [c,d] 0.050 (6) [c,d]
		phase α2	
5.1107 (5) [c] 12.0102 (8) [d]	5.4694 (6) [c] 111.70 (1) [c]	311.31 (5)	38.4 (6) [c,d] 0.11 (1) [c,d]
α-Gly:CA (90:10), Rietveld fitting (with refinement of atomic coordinates)			
		phase α1	
5.0999 (10) [c] 11.9701 (7) [c]	5.4613 (10) [c] 111.71 (2) [c]	309.74 (10)	98 (5) [c,d] 0.050 (6) [c,d]
		phase α2	
5.1109 (7) [c] 12.0102 (10) [d]	5.4684 (7) [c] 111.69 (2) [c]	311.90 (8)	38.9 (5) [c,d] 0.12 (1) [c,d]
γ-Gly:CA (80:20), LPA			
		phase γ1	
7.0388 (6) [a] a	5.4797 (9) [a] 90	235.12 (4)	70 (12)/82 (7) [b] 0/0.057 (16) [b]
		phase γ2	
7.0541 (14) [a] a	5.4890 (18) [a] 90	236.54 (9)	48 (8)/48 (8) [b] 0/0[b]
γ-Gly:CA (80:20), LB fitting			
		phase γ1	
7.0404 (4) [c] a	5.4838 (28) [c] 90	235.40 (12)	92 (3) [c,d] 0.044 (16) [c,d]
		phase γ2	
7.0600 (7) [c] a	5.4903 (21) [c] 90	236.99 (10)	35.7 (7) [c,d] 0[c]
γ-Gly:CA (80:20), Rietveld fitting (without refinement of atomic coordinates)			
		phase γ1	
7.0402 (5) [c] a	5.4817 (35) [c] 90	235.30 (15)	92 (4) [c,d] 0.051 (17) [c,d]
		phase γ2	
7.0602 (9) [c] a	5.4903 (9) [c] 90	237.01 (5)	35.7 (9) [c,d] 0 [c]

[a] calculated by program Celsiz. [b] calculated by WHP/SSP techniques using program SizeCr. [c] Parameters portrayed are corrected by multiplication of the e.s.d. values obtained in LB/Rietveld fitting refinement on $m_{e.s.d.}$ factor (Table 7). [d] $D = Lvol\_FWHM$ and ε_s (%) = $2 \cdot e_0 \cdot 100$%, where $Lvol\_FWHM$ and e_0 are, respectively, crystallite size and microstrain parameters refined by *TOPAS* (see thorough explanation with examples in Refs. [40,41]).

Table 7. Agreement factors obtained in Rietveld and LB fitting (refinement), weight content *Wt* of crystalline phases p1 (α1 or γ1) and p2 (α2 or γ2) according to results of Rietveld fitting and the factor $m_{\text{e.s.d.}}$ for correction of e.s.d.s of refined parameters calculated by RietESD.

R_{wp}, % R_{p}, %	cR_{wp}, % [a] cR_{p}, % [a]	R_B, % p1 R_B, % p2	*Wt*, wt.% p1 *Wt*, wt.% p2	$m_{\text{e.s.d.}}$ [a]
α-Gly:CA (90:10), LB fitting				
4.82 3.09	7.04 5.24	- -	- -	5.05
α-Gly:CA (90:10), Rietveld fitting (without refinement of atomic coordinates)				
6.35 4.17	9.42 7.36	0.98 1.13	24.4 (1) [b] 75.6 (1) [b]	5.24
α-Gly:CA (90:10), Rietveld fitting (with refinement of atomic coordinates)				
5.77 3.87	8.54 6.82	0.71 0.58	23.4 (1) [b] 76.6 (1) [b]	5.33
γ- Gly:CA (80:20), LB fitting				
9.96 6.21	13.14 9.58	- -	- -	6.97
γ-Gly:CA (80:20), Rietveld fitting (without refinement of atomic coordinates)				
11.36 7.40	14.97 11.31	1.94 2.92	33.2 (5) [b] 66.8 (5) [b]	7.09

[a] Parameters calculated by program *RietESD* (see Supplementary Materials). [b] E.s.d. of *Wt* presented is corrected by multiplication of the e.s.d. value obtained in Rietveld fitting refinement on $m_{\text{e.s.d.}}$ factor.

Based on the results of LPA, to perform the tasks set out in Section 2.3, the LB and Rietveld fittings of XRD patterns were performed in the next steps assuming the existence of two close-phase modifications in the samples (α1 and α2 in α-Gly:CA (90:10) and γ1 and γ2 in γ-Gly:CA (80:20)).

Final graphic results of Rietveld fitting of the simulated XRD patterns to experimental ones (with atomic coordinates of the both phase modifications fixed to the vales obtained in single-crystal refinement) for investigated powder samples are presented in Figure 3. The graphic results of LB fitting are provided in Figure S4 of Supplementary Materials.

Final values of agreement factors characterizing the quality of fitting simulated XRD patterns to experimental ones [weighted profile (R_{wp}) and profile (R_p) factors and their analogues after subtracting the background contribution (cR_{wp} and cR_p)] and the degree of correspondence of the structure model to the observed integral intensities of reflections attributed to crystalline phases (Bragg factors R_B for crystalline phases) for LB and Rietveld fitting are summarized in Table 7. Definitions of the agreement factors can be found, for example, in Ref. [55].

When only one phase is considered to exist, and all other things during fittings being equal, the weight profile factors Rwp increase by ~1.7 times, and the profile factors R_p by more than 2 times compared to the factors provided in Table 7. In turn, the Bragg factor R_B increases by ~3 times. Thus, the fitting of XRD patterns by LB and Rietveld methods unambiguously demonstrated that the assumption of only one phase does not allow us to describe XRD patterns and individual XRD reflections well enough. The powder samples contain at least two phases with a small difference in the unit cell parameters [α1 and α2 in α-Gly:CA (90:10) and γ1 and γ2 in γ-Gly:CA (80:20)] at least.

As noted above, the coordinates of the atoms in the Rietveld fitting were not refined. In the case of γ-Gly:CA (80:20), in addition, a structural model was obtained by measuring a single crystal at 100 K, while the powder XRD pattern of this sample was taken at 313 K. Nevertheless, as can be observed from Figure 3 and Figure S4 (Supplementary Materials), the quality of powder XRD pattern fittings is quite high not only by the LB method, but also by the Rietveld method. The high quality of the fittings of powder XRD patterns indicates the high quality of the results of the determination (refinement) of the structures of the samples performed on single crystals (which is expected), in addition to the absence of crystalline phases in the powder with the structures other than α-Gly-like and γ-Gly-like in α-Gly:CA (90:10) and γ-Gly:CA (80:20), respectively.

An attempt to refine the atomic coordinates for modifications of the α-Gly:CA (90:10) does not reveal a noticeable improvement in fitting quality (see Table 7 for agreement factors and Figure S5 of the Supplementary Materials). The values of the parameters of the unit cells of the modifications α-Gly:CA (90:10) and γ-Gly:CA (80:20) phases and the values of their microstructural parameters obtained in LB and Rietveld refinements practically coincide within one e.s.d. and are close (within two to three e.s.d.s) with values obtained independently by LPA methods (Table 6). Thus, for comparison with the results of single-crystal experiments (Tables 2 and 3), it is possible to consider the results with more precise values, for example, ones obtained in LB fitting (Table 6).

The similarity of the values of the parameters and volume of the unit cells of $\alpha1$ phase in powder and α-Gly:CA (90:10) single-crystal phase with the values for pure α-Gly according to the literature data [52,53] (cf. Tables 2 and 6) indicates that in the phase $\alpha1$ of powder and in the single-crystal α-Gly:CA (90:10), likely only a small number of CA molecules have penetrated into the voids of the structures. The same conclusion can be drawn from the comparison with the literature data [53,54] for pure γ-Gly, apparently about the powder phase $\gamma1$ and the single-crystal γ-Gly:CA (80:20) (cf. RT and low-temperature data in Tables 3 and 6 for $\gamma1$ powder, γ-Gly:CA (80:20) single crystal and pure γ-Gly). Simultaneously, likely as a result of a larger number of CA molecules embedded in the pores of $\alpha2$ and $\gamma2$ modifications, their parameters and volumes of unit cells are noticeably larger than, respectively, in $\alpha1$ and $\gamma1$ (cf. V_{cell} = 311.31 (5) Å^3 in $\alpha2$ and 236.99 (10) Å^3 in $\gamma2$, in comparison to V_{cell} = 309.85 (9) Å^3 in $\alpha1$ and 235.40 (12) Å^3 in $\gamma1$ (LB fit)).

According to the refinement by the Rietveld method, the weight content of phases $\alpha1$ and $\gamma1$ with a small number of embedded CA molecules in powders is significantly less than phases $\alpha2$ and $\gamma2$ with an increased volume due to the introduction of a higher number of CA molecules [correspondingly, ~24 wt.% and ~33 wt.% in $\alpha1$ and $\gamma1$ in comparison to ~76 wt.% and ~67 wt.% in $\alpha2$ and $\gamma2$ (Table 7)]. Nevertheless, as noted above, studies of single crystals have revealed precisely phases that are close in terms of unit cell parameters $\alpha1$ and $\gamma1$ (cf. Tables 1–3 and 6). Apparently, this fact can be explained by the results of the determination of microstructural parameters (Table 6). According to the results of LPA and refinements by LB and Rietveld methods, modifications $\alpha1$ and $\gamma1$ are characterized by large values of crystallite sizes in comparison with $\alpha2$ and $\gamma2$ (cf. D~95 nm in $\alpha1$ and $\gamma1$ and D~40 nm in $\alpha2$ and $\gamma2$). As a result, the reflections of $\alpha1$ and $\gamma1$ modifications are much narrower, and those of $\alpha2$ and $\gamma2$ phases are more blurred (see small insets in Figure 3a,b). It is likely that single crystals with unit cell parameters close to $\alpha2$ and $\gamma2$ were also characterized by more diffuse reflections and were rejected by quality when selected for single crystal study.

Likely due to the embedding of CA molecules into structural pores and the neighborhood of crystallites of different modifications, microstrains are observed in the crystallites of $\alpha1$ and $\gamma1$ phases, which are approximately the same size (ε_s~0.05 % (Table 6)). Simultaneously, in $\alpha2$ crystallites with a more expanded crystal lattice, the absolute value of the average microstrain increases to ε_s~0.10%. However, microstrains were not detected in the crystallites of modifications $\gamma2$ also characterized by an increased volume of the unit cell, which requires further investigation.

Thus, XRD studies of α-Gly:CA (90:10) and γ-Gly:CA (80:20) powders revealed the presence in them of at least two crystalline modifications with the same α-Gly or γ-Gly type structure, respectively, but with slightly different unit cell parameters and noticeably different sizes of crystallites. The formation of two-phase modifications is associated with the embedding of different amounts of CA molecules into the pores of the α-Gly and γ-Gly structures.

3.3. Raman and FTIR Spectroscopies

Polarized Raman spectra measured in Gly:CA crystals of both pyramidal (Figure 4a) and plate-like shapes (Figure 4b) are like those of nominally pure α- and γ-Gly [56–58]. They consist of many lines, originated predominantly from intramolecular vibrations of different types. Analysis of the Raman and FTIR spectra makes it possible to ascertain the chemical structure and polymorphous modification of the molecular crystals studied. Tables 8 and 9 portray the position of the lines in the polarized Raman and FTIR absorption spectra in pyramidal [γ-Gly:CA (80:20)] and plate-like [α-Gly:CA (90:10)] Gly:CA crystals, correspondingly. The positions of the lines in pyramidal Gly:CA crystals coincide with the positions of most lines in both nominally pure γ-Gly, and in plate-like α-Gly:CA crystals [56–58]. The shifts in the positions of high-frequency spectral lines at high frequency in each pair of crystals (α-Gly, α-Gly:CA) and (γ-Gly, γ-Gly:CA) do not exceed several cm^{-1}. This indicates that the types and frequencies of molecular modes active in in Raman spectra of α- and γ-Gly doped with CA are close to those in pure Gly crystals. This is in agreement with the results of XRD analysis of Gly:CA single crystals and powders, which demonstrate very small changes in the crystal structure and position of atoms caused by doping with CA. Weak lines observed in α-Gly:CA spectrum portrayed in Figure 4c by red arrows are absent in pure Gly crystals and can be attributed to CA ions [59] (see Table 9).

Table 8. Positions of lines (ν, cm^{-1}) observed in polarized Raman and FTIR spectra of γ-Gly:CA (80:20) crystals and assignments of analogous lines in nominally pure γ-Gly from Ref. [58].

Raman (ν, cm^{-1})			FTIR (ν, cm^{-1})	Assignments [57]
X(zz)X	X(yy)X	X(yz)X		
-	3095.4 m	-	3100 m	νNH (3)$\cdots$O (1)
-	2999.1 s	2999.1 s	-	ν_aCH$_2$
2963.7 s	2962.6 vs	2964.8 m	2963 w	ν_sCH$_2$
2849.6 w	-	-	-	-
2744.5 vw	-	-	-	νNH (1)$\cdots$O (1)
2695.7 vw	-	-	-	νNH (1)$\cdots$O (1)
2611.6 w	-	-	-	νNH (1)$\cdots$O (1)
-	-	-	1736 w	-
-	1676.1 w	-	1684 m	-
1659.5 w	-	-	1665 w	δ_aNH$_3{}^+$
-	-	-	1630 s	δ_aNH$_3{}^+$
-	-	-	1601 vs	ν_aCOO$^-$
-	-	1589.8 w	-	-
-	1576.5 m	-	-	ν_aCOO$^-$ (?)
-	-	-	1557 m	-
-	1506.7 m	-	1506 m	-
-	-	1502.3 vw	1499 s	δ_sNH$_3{}^+$
1483.5 vw	-	-	-	δ_sNH$_3{}^+$
1438.1 w	1437 s	1440.3 m	1439 w	δCH$_2$
1389.4 vw	1393.8 s	-	1398 m	ν_sCOO$^-$
-	1335.2 s	-	1344 m	ωCH$_2$
-	-	1321.9 m	1321 w	τCH$_2$
-	1154.7 w	1156.9 vw	1153 m	ρNH$_3{}^+$
-	1135.9 m	1130.3 w	-	ρNH$_3{}^+$

Table 8. *Cont.*

Raman (ν, cm^{-1})			FTIR (ν, cm^{-1})	Assignments [57]
X(zz)X	X(yy)X	X(yz)X		
-	-	-	1126 s	ρNH$_3^+$
-	1047.3 w	1048.4 m	1049 vw	ν_aCCN
	924.47 sh	931.11 w	931.6 s	ρCH$_2$
892.37 s	892.37 s	892.37 w	889.2 m	νCC
749.55 vw	-	-	-	-
683.13 m	688.67 vw	687.56 w	692.4 m	σCOO$^-$
-	-	-	667.37 m	-
607.85 w	-	607.85 w	-	-
560.24 vw	-	558.03 w	-	-
-	-	517.07 w	-	ωCOO$^-$
502.68 s	-	-	-	τCOO$^-$
-	-	500.46 m	-	ρCOO$^-$
354.33 m	354.33 w	359.86 w	-	ρCCN
-	213.,73 m	215.94 w	-	-
	-	171.66 sh	-	ω or τ CCN
174.98 w	-	-	-	Lattice modes
-	165.02 sh	-	-	-"-
-	149.52 s	-	-	-"-
-	-	138.45 s	-	-"-
-	127.38 sh	-	-	-"-
108.56 vw	101.92 m	103.02 m	-	-"-
90.9 vw	86.42 vs	87.528 m	-	-"-

Table 9. Positions of lines (ν, cm^{-1}) observed in polarized Raman and FTIR spectra of α-Gly:CA (90:10) crystals and assignments of analogous lines in nominally pure α-Gly from Ref. [58].

X(ZZ)X	X(YY)X	X(YZ)X	FTIR	Assignments [58]
			3171 s	νNH$\cdots$O
-	3143 m	-	-	νNH$\cdots$O
3008 vs	3007 s	3009 s	3005 m	ν_aCH$_2$
2972 vs	2973 vs	2973 s	2974 m	ν_sCH$_2$
-	2903 w	-	2918 w	νNH$\cdots$O
2883 vw	2880 w	-	-	-"-
-	2818 vw	-	-	-"-
2787 vw	2746 vw	-	-	-"-
-	-	-	2718 s	-"-
2608 vw	2633 vw	-	2627 s	-"-
-	-	-	2525 m	-"-
-	-	-	2363 m	-
1723 vw	1722 vw	-	-	CA [59]: ν (CO)
-	-	1673 m	1647 vs	δ_aNH$_3^+$

Table 9. *Cont.*

X(ZZ)X	X(YY)X	X(YZ)X	FTIR	Assignments [58]
-	-	-	1611 s	$\delta_a NH_3^+$
-	-	-	1555 m	$\nu_a COO^-$
-	-	-	1541 m	-
1502 m	1506 vw	-	1508 m	$\delta_s NH_3^+$
1483 w	-	-	1474 w	$\delta_s NH_3^+$
-	-	1455 s	1458	δCH_2
-	1439 m	-	1445 m	δCH_2
1411 s	1410 m	1416 w	1421 s	$\nu_s COO^-$
-	-	-	1341 m	ω or τCH_2
1327 s	1325 s	1325 m	1314 m	τCH_2
1138 m	1137 w	-	1134 w	ρNH_3^+
-	1106 w	-	1115 w	ρNH_3^+
1037 w	-	-	1034 w	$\nu_a CCN$
-	-	-	914 s	ρCH_2
-	-	-	897 m	-
892.8 s	892.4 s	892.4 m	-	νCC
802 vw	-	-	-	CA [59]: δ (CO)
695.9 m	-	-	-	σCOO^-
643 vw	639 vw	-	-	CA [59]: Ring breath.
-	600. vw	602.3 m	-	-
496.9 w	-	483.9 m	-	ρCOO^-
357.2 m	-	-	-	ρCCN
196.3 m	197.1 m	-	-	Lattice modes
-	-	178.3 m	-	-"-
-	161.7 w	163.9 sh	-	-"-
154 s	-	-	-	-"-
139.2m	-	-	-	-"-
-	-	107.5 s	-	-"-
98.95 s	99.71 s	-	-	-"-
73.54 s	78.67 sh	-	-	-"-
-	-	70.92 s	-	-"-
50.26 s	51 s	53.21 w	-	-"-

The difference in polarized Raman spectra of γ- and α-Gly is most evident in the position of strong lines in the region of low (0–300 cm^{-1}) and high (2800–3100 cm^{-1}) frequencies of the spectrum. This difference is clearly observed in the Raman spectra of pyramidal and plate-like Gly:CA crystals (Figure 4a,b). At high frequencies (2800–3100 cm^{-1}) the stretching symmetric and antisymmetric vibrations of CH$_2$ group have been observed in α-Gly:CA at 2973 cm^{-1} and 3007 cm^{-1} and in γ-Gly:CA at 2963 cm^{-1} and 2999 cm^{-1} (Figure 5b), respectively, which corresponds to literature data related to pure α- and γ-Gly [58]. At low frequencies (0–300 cm^{-1}), positions of lines in α- and γ-Gly:CA spectra are slightly shifted in compared to pure Gly crystals [56,57] (Tables S5 and S7 in Supplementary Materials), which may be attributed to CA doping.

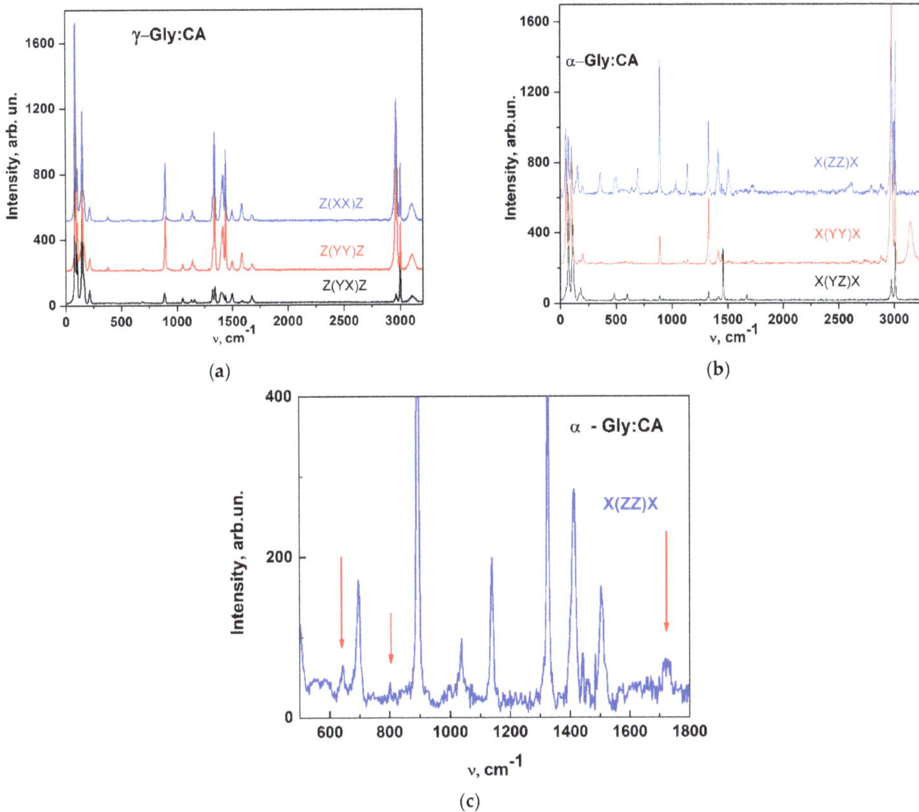

Figure 4. Polarized Raman spectra of pyramidal (**a**) and plate-like (**b**) Gly:CA crystals after subtracting the photoluminescence background. Fragment of the plate-like Gly:CA spectrum on an enlarged scale (**c**). For pyramidal crystals, the Z-axis is parallel to the polar axis (the height of the pyramid); for plate-like crystals, the Y-axis is parallel to the C_2 axis. The red arrows in (**c**) indicate positions of lines observed in CA crystals. λ_{exc} = 532 nm.

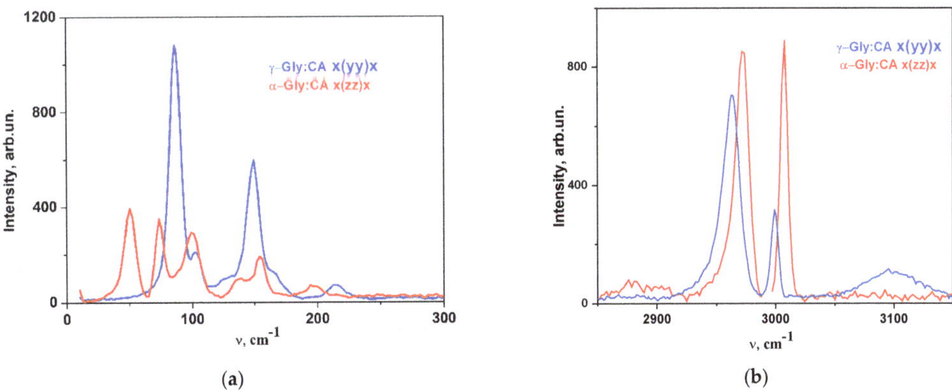

Figure 5. Comparison of Raman spectra of Gly:CA crystals of pyramidal (blue line) and plate-like (red line) shapes in low-energy (**a**) and high-energy (**b**) spectral regions.

As in the case of Raman spectra, a comparison of the FTIR absorption spectra of doped CA and nominally pure Gly indicates their close similarity (Tables 8 and 9). However, in the case of α-Gly:CA (90:10) plates, additional lines in the photon energy range E_{ph} = 0.5–0.6 eV (4000–5000 cm^{-1}) were observed in the FTIR absorption spectra measured in transmission geometry. Note that similar lines have been observed in the FTIR spectra of TGS crystals doped with CA [31] (Figure 6). No additional lines were observed in the reflection geometry, most likely due to their low intensity. Note that we did not measure FTIR absorption spectra of γ-Gly crystals in transmission geometry because of their pyramidal shape.

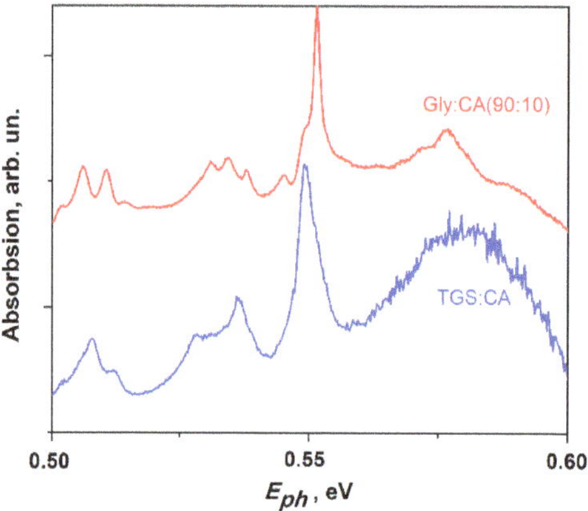

Figure 6. FTIR absorption spectra in spectral region E_{ph} = 0.5–0.6 eV in Gly:CA (90:10) measured in this work and in previously studied TGS:CA (90:10) (see text).

So, polarized Raman and FTIR absorption spectra of Gly:CA single crystals indicates that addition of croconic acid to the aqueous solution which was used to grow Gly:CA crystals does not significantly affect the polymorphic modification of the crystals grown. It has been established that γ-Gly forms pyramidal crystals, while α-Gly forms plate-like crystals. The incorporation of CA molecules in a glycine crystal does not noticeably change the spectral positions of Raman lines in the high-energy region of the spectrum (2900–3100 cm^{-1}), but leads to modest shift of several low-energy lines, which characterizes the effect of CA doping on the lattice vibration spectrum [32].

3.4. UV-Vis Absorption

RT absorption spectra of α-Gly:CA (90:10) and γ-Gly:CA (80:20), are portrayed in Figure 7. Colorless α- and γ-Gly crystals are transparent in wide spectral range. The energy band gaps of different polymorphic modifications of Gly crystals practically coincide and are about 5.1 eV [8,11,12]. Significant optical absorption is observed only for photons with energies E_{ph} > 5 eV, which manifests itself in a sharp increase of absorbance at λ < 240 nm. The doping of Gly with CA is accompanied by the appearance of absorption in the transparent region of pure Gly. This absorption manifests itself in the form of a structured long-wavelength tail adjacent to the edge of the fundamental absorption of a Gly crystal. The traits of the absorption are clearly observed when considering the absorption in Tauc coordinates $(\alpha E_{ph})^2$, E_{ph} (Figure 7c,d) (where α is the absorption coefficient of the crystal). The shortest wavelength part of the absorption spectrum in region E_{ph} = (5.8–6.2) eV in Figure 7c,d is well approximated by the linear dependence (solid black line), which can be regarded as the Tauc plot related to direct allowed interband transitions in crystals. The Tauc plot yields an estimation of the value of the optical bandgap as ~5.3 eV, which is close to the E_g value in pure Gly (~5.1 eV) (Figure 7c,d). Note that in Tauc coordinates, the shape of the absorption edge of undoped Gly is indeed well approximated by a straight line [8]. The presence of the additional absorption in the Gly:CA spectra (as compared with undoped Gly) can be associated with electronic transitions involving defects in the Gly lattice induced by the introduction of CA molecules. The contribution of these transitions to the absorption spectrum can be estimated as the difference between the absorption spectrum of Gly:CA (line 1 in the insets in Figure 7e,f) and the fragment of its fundamental absorption spectrum modeled by the Tauc plot (line 2 in the insets in Figure 7e,f). The absorption spectra thus obtained are portrayed in Figure 7e,f, which includes at least two absorption bands with maxima at ~5.2 and ~3.9 eV. It should be noted that similar structure is observed in the absorption spectrum of CA solution in ethanol (olive line in Figure 7e,f) or in water [60]. In Gly:CA crystals the absorption bands can presumably be associated with the creation of excitons localized at the lattice defects, including impurity CA molecules or neighboring Gly ones.

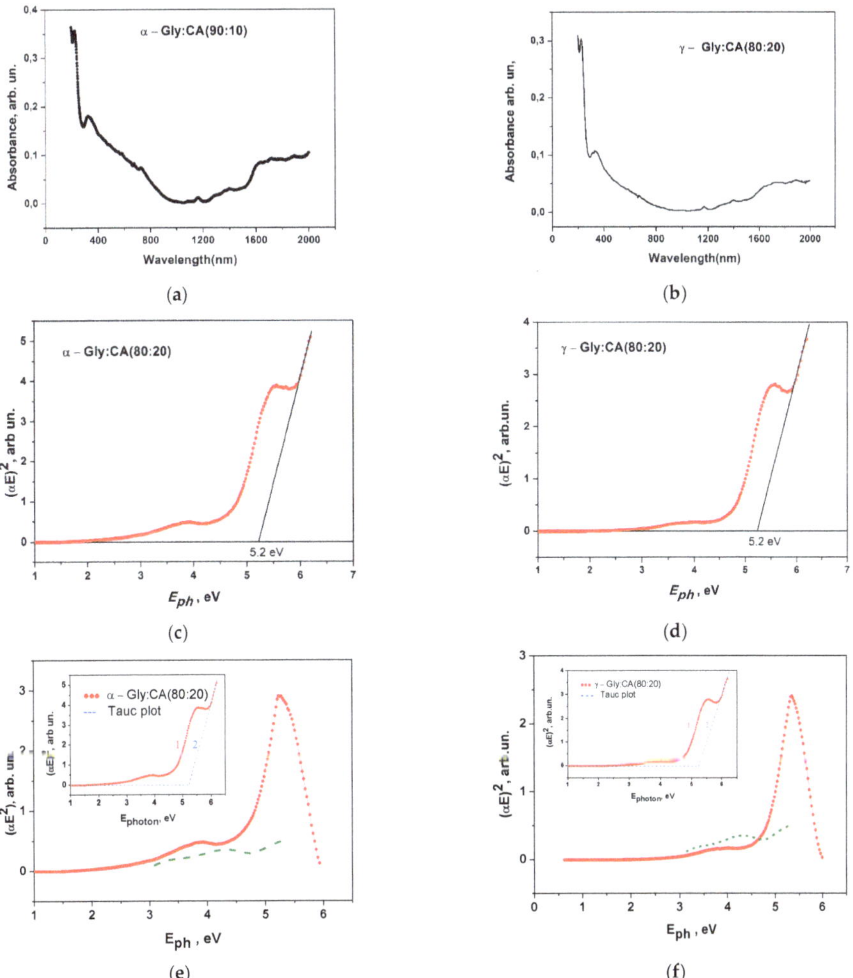

Figure 7. Absorbance spectra of (**a**) α-Gly:CA (90:10), (**b**) γ-Gly:CA (80:20), (**c**) α-Gly:CA (90:10), and (**d**) γ-Gly:CA (80:20) in Tauc coordinates. (**e**,**f**) Change in the Gly crystal absorption caused by incorporating CA molecules into the Gly crystals. The olive lines represent the absorption spectrum of CA solution in ethanol [60]. The inset portrays the absorption spectra of doped and pure Gly crystals (see text).

Table 10 presents a comparison of the E_g values, and high and low frequency local bands in α-Gly:CA (90:10), γ-Gly:CA (80:20), and TGS:CA (80:20) [31]. It is observed that CA doping results in practically the same changes of absorption spectra in these crystals.

Table 10. Energy gap E_g, eV, photon energy E_{ph} of high- and low-energy local bands and normalized amplitude of low energy band in absorption spectra of α-Gly:CA (90:10), γ-Gly:CA (80:20) and TGS:CA (80:20) [31].

	E_g, eV	High Energy Local band E_{ph}, eV	Low Energy Local Band E_{ph}, eV	Normalized Amplitude of Low Energy Band
α-Gly:CA (90:10)	5.2	5.25	3.9	0.16
γ-Gly:CA (80:20)	5.4	5.33	3.7	0.07
TGS:CA (80:20)	5.2	5.39	3.9	0.2

Thus, doping of crystalline Gly with CA significantly affects the shape of the absorption edge of Gly, which manifests itself in smearing of the sharp absorption edge and the appearance of noticeable absorption at frequencies $h\nu < E_g$. Note that the same changes of absorption spectrum have been observed earlier in CA-doped triglycine sulfate crystals [31].

3.5. Photoluminescence

The photoluminescence (PL) of "pure" glycine crystals is very weak. As a result, in the emission spectrum of "pure" γ-Gly narrow Raman lines are clearly observed, whose intensity is comparable to that of the luminescence band (Figure 8a). Like pure glycine, solid croconic acid is also practically non-luminescent [61]. However, doping glycine crystals with croconic acid leads to a dramatic increase in the emission intensity, while the intensity of Raman scattering does not change (Figure 8b). Thus, the relatively bright luminescence of γ-Gly:CA is a distinctive characteristic of the doped crystals. A similar situation also occurs in the case of α- Gly:CA.

The RT PL spectra of α- and γ-Gly:CA crystals excited by light with λ_{exc} = 325 nm are portrayed in Figure 8b. The spectra of both γ- and α-modifications consist of broad structured emission bands with their maxima at ~2.50 eV (γ-Gly:CA) and ~2.56 eV (α-Gly:CA), respectively. Comparison of the PL spectra of pure and doped Gly crystals (Figure 8) portrays that the shapes of the PL bands are practically identical, but doping Gly with croconic acid leads mainly to a dramatic increase in the PL intensity. Considering that the band gap of γ-glycine crystals is about 5 eV [8,11–13] the optical electron transitions responsible for the emission involve localized electron states whose energy levels are located within the crystal band gap.

Figure 8c,d portrays the decomposition of the emission band profile into four Gaussian components both in the case of α-Gly:CA and γ-Gly:CA. The positions of the Gaussian components and their relative intensity in the PL spectra of α-Gly:CA (90:10), γ-Gly:CA (80:20), and TGS:CA (90:10) [31] are presented in Table 11. The observed emission-band structure implies the presence of at least four different emitting centers that contribute to the emission of Gly:CA crystals. Note that the expansion of the TGS:CA (90:10) PL spectrum in terms of the Gaussian component presented in our previous paper [31] contains only three components. Here, we present the results of more accurate calculations that include four Gaussian components which provide a better fit. Both in Gly:CA crystals and in TGS:CA crystals, the positions and normalized intensities of the Gaussian components in both Gly:CA and TGS:CA crystals are close, which demonstrates that in all the crystals compared the PL originates from the same emission centres. The similar PL behavior in CA-doped Gly and TGS crystals clearly indicates that the appearance of PL is not associated with the SO_4 groups, which are present in TGS but absent in Gly crystals. The PL is probably the result of the interaction of CA with glycine ions and, most likely, with glycine zwitterions, since, as XRD phase analysis demonstrates, the structural parameters of these ions are affected by CA doping.

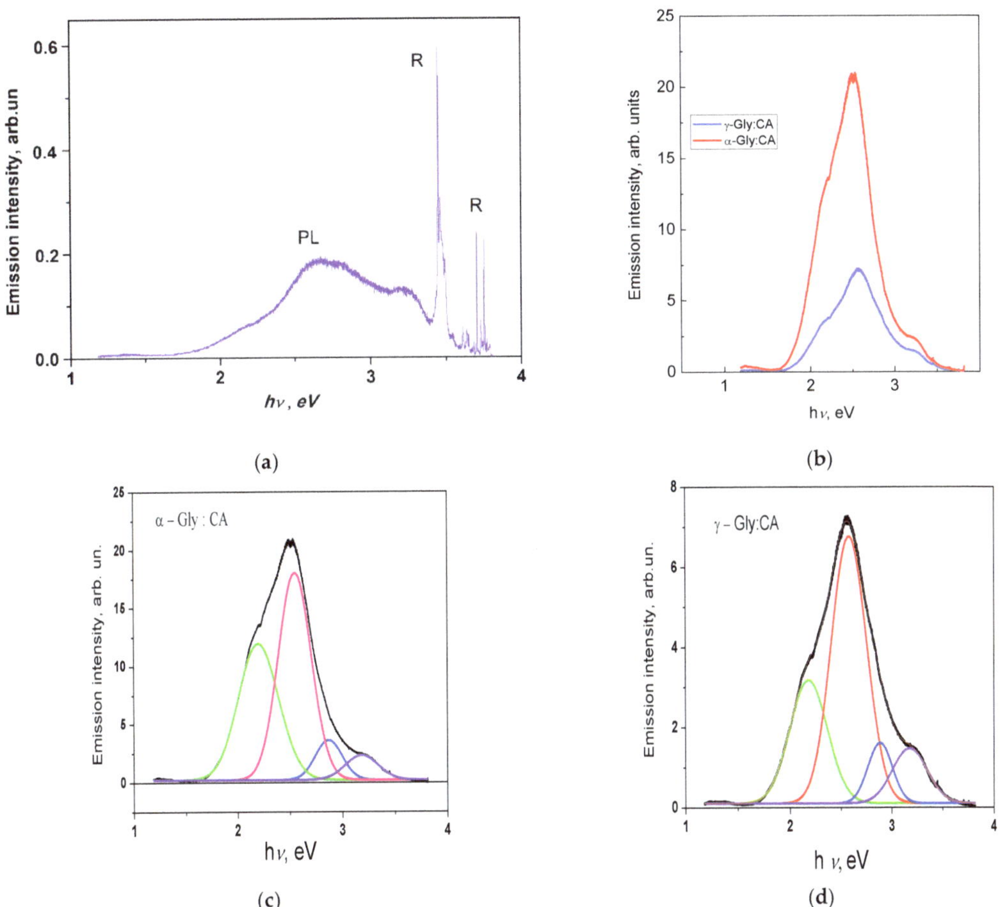

Figure 8. (**a**) Emission spectrum of "pure" γ-glycine wih a small admixture of citric acid. Here, R labels the Raman lines. (**b**) Emission spectra of γ- and α-glycine crystals doped with croconic acid. (**c**) Decomposition of the α-Gly:CA emission band profile into four Gaussian components $h\nu_{exc} = 3.8$ eV. $T = 300$ K. (**d**) Decomposition of the γ-Gly:CA emission band profile into four Gaussian components. The Gaussian components in (**c**,**d**) are shown by different colors.

Table 11. Resonance photon energies E_{ph}, eV, normalized intensity Ai/A_2 obtained after decomposition of PL spectra in α-Gly:CA, γ-Gly:CA and TGS:CA in four Gauss components.

	α-**Gly:CA**		γ-**Gly:CA**		**TGS:CA**	
Band, i	E_{ph}, eV	A_i/A_2	E_{ph}, eV	A_i/A_2	E_{ph}, eV	A_i/A_2
1	2.20	0.66	2.18	0.47	2.20	0.59
2	2.54	1	2.57	1	2.57	1
3	2.86	0.20	2.88	0.24	2.86	0.33
4	3.18	0.13	3.17	0.22	3.19	0.22

From Table 11 we observe that the normalized intensities of Gauss components may depend on crystal symmetry because bands 3 and 4 in non-centrosymmetric γ-Gly, and TGS:CA are more intensive than in centrosymmetric α-Gly:CA and band 1 has higher amplitude in α-Gly:CA.

Although the specific structure of these centers is currently not clear, it can be assumed that they include molecules of both glycine and croconic acid (glycine and CA crystals are not luminescent by themselves). The presence of an analogous emission band in the luminescence spectrum of "pure" γ-Gly (Figure 8a) with a small admixture of citric acid demonstrates that not only CA but also other organic acids can promote the formation of emitting centers (glycine lattice defects) with a similar structure of radiative transitions.

4. Conclusions

Crystals of α- and γ-glycine doped with croconic acid were grown from aqueous solutions of glycine and croconic acid. α-Gly crystals grown from a solution with a relative weight content Gly and CA of 90:10 have a plate shape. Pyramid-shaped γ-Gly crystals were obtained from a solution with a relative weight content of 98:2 and 80:20. A study of the structural and optical properties of the grown crystals has been performed. The study indicates that the crystal symmetry and morphology of CA-doped-doped glycine single crystals grown from aqueous solutions with different components weight ratios correspond well to well-known α- and γ-isomorphs of nominally pure glycine. Incorporation of CA molecules in the host glycine crystal structures does not change remarkably the parameters of the unit cells, but results in small changes of unit cell volumes, and bond angles and lengths of glycine zwitterions.

In many aspects, this situation is quite similar to that observed in CA-doped TGS single crystals [31]. CA doping both Gly and TGS crystals does not significantly change the frequencies of molecular vibrations in Raman and FTIR spectra. Also, it does not remarkably influence the magnitude of forbidden gap E_g, caused by direct allowed electronic transitions.

The most significant changes caused by CA doping glycine (Gly:CA), as in the case TGS crystals, are associated with absorption and photoluminescence spectra. In the UV Vis region CA doping results in appearance of a strong absorption band caused by local electronic transitions near the band gap ($E_{ph} \approx 5.2$–5.4 eV), and a broad band of lower intensity with a maximum in the region of $E_{ph} \approx 3.7$–3.9 eV. Weak absorption bands were detected by FTIR in TGS:CA and Gly:CA in the region of $E_{ph} \approx 0.5$–0.7 eV.

Strong PL spectra in α-, γ-Gly:CA, and TGS:CA crystals demonstrate the contributions of four emission centers emitting photons of close energies (~2.2 eV, 2.57 eV, 2.86 eV, 3.18 eV) and intensity ratios. It can be assumed that the appearance of these centers is due to the incorporation of small CA ions into the pores of the host crystal structure and the formation of hydrogen bonds between the CA molecule and glycine zwitterions.

In α- and γ-Gly:CA, in contrast to TGS:CA, different phases (regions) with a crystal structure of the same symmetry but slightly different unit cell parameters arising from the inhomogeneous doping distribution of the CA molecule in the host crystal were revealed by powder X-ray diffraction methods. The presence of twins in non-centrosymmetric γ-Gly:CA was detected using single-crystal XRD. These crystal structure details do not appear in optical spectroscopy experiments due to the integral nature of these methods.

Supplementary Materials: The following supporting information can be downloaded at: https://www.mdpi.com/article/10.3390/cryst12101342/s1. Table S1: Results of refinement of the α-Gly:CA (90:10) structure (space group $P2_1/n$ (14)) using single crystal XRD data. Table S2: Atomic displacement parameters for the α-Gly:CA (90:10) (Å^2). Table S3: Results of refinement of the γ-Gly:CA (80:20) structure (space group $P3_1$ (N 144)) using single crystal XRD data. Table S4: Atomic displacement parameters for the γ-Gly:CA (80:20) (Å^2). Figure S1: Structures of (a) α-Gly:CA (90:10) and (b) γ-Gly:CA (80:20) with atoms represented by thermal ellipsoids by means of program ORTEP using, correspondingly, the data presented in Tables S1–S4. Figure S2: (a), (c) WHP and (b), (d), (f) SSP of α1 and α2 modifications of α-Gly. Figure S3: (a), (c) WHP and (b), (d), (f) SSP of γ1 and γ2 modifications of γ-Gly. Figure S4: LB fit results for (a) α-Gly:CA (90:10 and (b) γ-Gly:CA (80:20). Figure S5: Rietveld fit results for α-Gly:CA (90:10). Table S5: Low frequency intermolecular modes and symmetry modes in Gly:CA plate in different experimental geometries, and in nominally pure α-Gly [5]. Table S6: High frequency intramolecular modes in Gly:CA plate in different experimental geometries, and in nominally pure α-Gly [6]. Table S7: Low frequency intermolecular modes in Gly:CA pyramid in different experimental geometries, and in nominally pure γ- Gly [7]. Table S8. High frequency intramolecular modes in Gly:CA pyramid.

Author Contributions: Formal analysis, E.B., A.Z., A.A.L., B.K., V.D., A.S. (Alexander Smirnov), A.S. (Anatoly Starukhin), S.P., H.Z., F.L. and H.K.; Investigation, E.B., A.Z., A.A.L., V.D., A.S. (Alexander Smirnov), A.S. (Anatoly Starukhin), S.P., H.Z., F.L. and H.K.; Writing—original draft, E.B., A.Z., A.A.L., A.S. (Anatoly Starukhin), B.K., H.Z., F.L. and H.K.; Writing—review and editing, E.B., B.K. and H.K.; Funding Crystals 2022, 12, 679 23 of 25 acquisition, E.B. and H.K.; Conceptualization, E.B. All authors have read and agreed to the published version of the manuscript.

Funding: This research was funded by RFBR (project number 21-52-53015) and NSFC (project number 52111530040).

Institutional Review Board Statement: Not applicable.

Informed Consent Statement: Not applicable.

Data Availability Statement: The data presented in this study are available on request from the corresponding author.

Acknowledgments: The X-ray experiment for TGS:CA (90:10) was carried out using facilities of the X-ray diffraction Resource Centers of St. Petersburg University. The XRD powder characterizations were performed using equipment and software of the Joint Research Center "Material science and characterization in advanced technology" (Ioffe Institute, St.-Petersburg, Russia).

Conflicts of Interest: The authors declare no conflict of interest.

References

1. Vickery, H.B.; Schmid, C.A. The history of the discovery of the amino acids. *Chem. Rev.* **1931**, *9*, 169–318. [CrossRef]
2. Towler, C.S.; Davey, R.J.; Lancaster, R.W.; Price, C.J. Impact of Molecular Speciation on Crystal Nucleation in Polymorphic Systems: The Conundrum of γ Glycine and Molecular 'Self Poisoning'. *J. Am. Chem. Soc.* **2004**, *126*, 13347–13353. [CrossRef]
3. Sakai, H.; Hosogai, H.; Kawakita, T. Transformation of α-glycine to γ-glycine. *J. Cryst. Growth* **1992**, *116*, 421–426. [CrossRef]
4. Tylczyński, Z.; Busz, P. Transformation from γ to α modification in glycine crystal. *Ph. Transit.* **2014**, *87*, 1157–1164. [CrossRef]
5. Dawson, A.; Allan, D.R.; Belmonte, S.A.; Clark, S.J.; David, W.I.F.; McGregor, P.A.; Parsons, S.; Pulham, C.R.; Sawyer, L. Effect of High Pressure on the Crystal Structures of Polymorphs of Glycine. *Cryst. Growth Des.* **2005**, *5*, 1415–1427. [CrossRef]
6. Losev, E.A.; Mikhailenko, M.A.; Achkasov, A.F.; Boldyreva, E.V. The effect of carboxylic acids on glycine polymorphism, salt and co-crystal formation. A comparison of different crystallisation techniques. *New J. Chem.* **2013**, *37*, 1973–1981. [CrossRef]
7. Fleck, M.; Petrosyan, A.M. *Salts of Amino Acids. Crystallization, Structure and Properties*; Springer International Publication: Cham, Switzerland, 2014.
8. Flores, M.Z.S.; Freire, V.N.; dos Santos, R.P.; Farias, G.A. Optical absorption and electronic band structure first-principles calculations of α-glycine crystals. *Phys. Rev. B* **2008**, *77*, 115104. [CrossRef]
9. Peter, M.E.; Ramasamy, P. Growth of gamma glycine crystal and its characterization. *Spectrochim. Acta A* **2010**, *75*, 1417–1421. [CrossRef]
10. Ahamed, Z.A.S.; Dillip, G.R.; Raghavaiah, P.; Mallikarjuna, K.; Raju, B.D.P. Spectroscopic and thermal studies of γ-glycine crystal grown from potassium bromide for optoelectronic applications. *Arab. J. Chem.* **2013**, *6*, 429–433. [CrossRef]

11. Dillip, G.R.; Raghavaiah, P.; Mallikarjuna, K.; Reddy, C.M.; Bhagavannarayana, G.; Kumar, V.R.; Raju, B.D.P. Crystal growth and characterization of γ-glycine grown from potassium fluoride for photonic applications. *Spectrochim. Acta A* **2011**, *79*, 1123–1127. [CrossRef]

12. Rodríguez, J.S.; Costa, G.; da Silva, M.B.; Silva, B.P.; Honório, L.J.; de Lima-Neto, P.; Santos, R.C.R.; Caetano, E.W.S.; Alves, H.W.L.; Freire, V.N. Structural and Optoelectronic Properties of the α-, β-, and γ-Glycine Polymorphs and the Glycine Dihydrate Crystal: A DFT Study. *Cryst. Growth Des.* **2019**, *19*, 5204. [CrossRef]

13. Kumar, R.A.; Vizhi, R.E.; Vijayan, N.; Babu, D.R. Structural, dielectric and piezoelectric properties of nonlinear optical γ-glycine single crystals. *Phys. B* **2011**, *406*, 2594–2600. [CrossRef]

14. Justin, P.; Anitha, K. Influence of formic acid on optical and electrical properties of glycine crystal. *Mater. Res. Express* **2017**, *4*, 115101. [CrossRef]

15. Tylczyński, Z.; Busz, P. Low-temperature phase transition in γ-glycine single crystal. Pyroelectric, piezoelectric, dielectric and elastic properties. *Mater. Chem. Phys.* **2016**, *183*, 254–262. [CrossRef]

16. Guerin, S.; Stapleton, A.; Chovan, D.; Mouras, R.; Gleeson, M.; McKeown, C.; Noor, M.R.; Silien, C.; Rhen, F.M.F.; Kholkin, A.L.; et al. Control of piezoelectricity in amino acids by supramolecular packing. *Nat. Mater.* **2018**, *17*, 180–186. [CrossRef] [PubMed]

17. Lemanov, V.V. Piezoelectric and pyroelectric properties of protein amino acids as basic materials of Soft State Physics. *Ferroelectrics* **2000**, *238*, 211–218. [CrossRef]

18. Okosun, F.; Guerin, S.; Celikin, M.; Pakrashi, V. Flexible amino acid-based energy harvesting for structural health monitoring of water pipes. *Cell Rep. Phys. Sci.* **2021**, *2*, 100434. [CrossRef]

19. Vijayalakshmia, V.; Dhanasekaran, P. Growth and characterization study of γ-glycine crystal grown using different mole concentrations of zinc sulphate as structure-directing agents. *J. Cryst. Growth* **2018**, *498*, 372–376. [CrossRef]

20. Han, G.; Chow, P.S.; Tan, R.B.H. Understanding the Salt-Dependent Outcome of Glycine Polymorphic Nucleation. *Pharmaceutics* **2021**, *13*, 262. [CrossRef]

21. Li, D.; Zong, S.; Dang, L.; Wang, Z.; Wei, H. Effects of inorganic additives on polymorphs of glycine in microdroplets. *CrystEngComm* **2018**, *20*, 164–172. [CrossRef]

22. Bhat, M.N.; Dharmaprakash, S.M. Effect of solvents on the growth morphology and physical characteristics of nonlinear optical g-glycine crystals. *J. Cryst. Growth* **2002**, *242*, 245–252. [CrossRef]

23. Balakrishnan, T.; Babu, R.R.; Ramamurthi, K. Growth, structural, optical and thermal properties of γ-glycine crystal. *Spectrochim. Acta Part A Mol. Biomol. Spectrosc.* **2008**, *69*, 1114–1118. [CrossRef] [PubMed]

24. Dhanaraj, P.V.; Rajesh, N.P. Growth and characterization of nonlinear optical γ-glycine single crystal from lithium acetate as solvent. *Mater. Chem. Phys.* **2009**, *115*, 413–417. [CrossRef]

25. Azhagan, S.A.C.; Ganesan, S. Crystal growth, structural, optical, thermal and NLO studies of γ-glycine single crystals. *Optik* **2013**, *124*, 6456–6460. [CrossRef]

26. Anis, M.; Ramteke, S.P.; Shirsat, M.D.; Muley, G.G.; Baig, M.I. Novel report on g-glycine crystal yielding high second harmonic generation efficiency. *Opt. Mater.* **2017**, *72*, 590–595. [CrossRef]

27. Whatmore, R.W. Pyroelectric devices and materials. *Rep. Prog. Phys.* **1986**, *49*, 1335–1386. [CrossRef]

28. Szafran, M.; Tylczyński, Z.; Wiesner, M.; Czarnecki, P.; Ghazaryan, V.V.; Petrosyan, A.M. Above-room-temperature ferroelectricity and piezoelectric activity of dimethylglycinium-dimethylglycine chloride. *Mater. Des.* **2022**, *220*, 110893. [CrossRef]

29. Balashova, E.V.; Lemanov, V.V. Dielectric and Acoustic Properties of Some Betaine and Glycine Compounds. *Ferroelectrics* **2003**, *285*, 179–205. [CrossRef]

30. Horiuchi, S.; Tokunaga, Y.; Giovannetti, G.; Picozzi, S.; Itoh, H.; Shimano, R.; Kumai, R.; Tokura, Y. Above-room-temperature ferroelectricity in a single-component molecular crystal. *Nature* **2010**, *463*, 789–792. [CrossRef]

31. Balashova, E.; Levin, A.A.; Davydov, V.; Smirnov, A.; Starukhin, A.; Pavlov, S.; Krichevtsov, B.; Zolotarev, A.; Zhang, H.; Li, F.; et al. Croconic Acid Doped Triglycine Sulfate: Crystal Structure, UV-Vis, FTIR, Raman, Photoluminescence Spectroscopy, and Dielectric Properties. *Crystals* **2022**, *12*, 679. [CrossRef]

32. Balashova, E.V.; Smirnov, A.N.; Davydov, V.Y.; Krichevtsov, B.B.; Starukhin, A.N. Raman scattering and luminescence in single crystals of the amino acid glycine $C_2H_5NO_2$ with an admixture of croconic acid $C_5H_2O_5$. *J. Phys. Conf. Ser.* **2021**, *2103*, 012070. [CrossRef]

33. Braga, D.; Maini, L.; Grepioni, F. Crystallization from hydrochloric acid affords the solid-state structure of croconic acid (175 years after its discovery) and a novel hydrogen-bonded network. *CrystEngComm* **2001**, *3*, 27. [CrossRef]

34. Sheldrick, G.M. A short history of *SHELX*. *Acta Crystallogr. A* **2008**, *64*, 112–122. [CrossRef]

35. Dolomanov, O.V.; Bourhis, L.J.; Gildea, R.J.; Howard, J.A.K.; Puschmann, H. *OLEX2*: A complete structure solution, refinement and analysis program. *J. Appl. Crystallogr.* **2009**, *42*, 339–341. [CrossRef]

36. Momma, K.; Izumi, F. *VESTA 3* for three-dimensional visualization of crystal, volumetric and morphology data. *J. Appl. Crystallogr.* **2011**, *44*, 1272–1276. [CrossRef]

37. Maunders, C.; Etheridge, J.; Wright, N.; Whitfield, H.J. Structure and microstructure of hexagonal $Ba_3Ti_2RuO_9$ by electron diffraction and microscopy. *Acta Crystallogr. B* **2005**, *61*, 154–159. [CrossRef]

38. Terlan, B.; Levin, A.A.; Börrnert, F.; Simon, F.; Oschatz, M.; Schmidt, M.; Cardoso-Gil, R.; Lorenz, T.; Baburin, I.A.; Joswig, J.-O.; et al. Effect of Surface Properties on the Microstructure, Thermal, and Colloidal Stability of VB_2 Nanoparticles. *Chem. Mater.* **2015**, *27*, 5106–5115. [CrossRef]

39. Terlan, B.; Levin, A.A.; Börrnert, F.; Zeisner, J.; Kataev, V.; Schmidt, M.; Eychmüller, A. A Size-Dependent Analysis of the Structural, Surface, Colloidal, and Thermal Properties of $Ti_{1-x}B_2$ (x = 0.03–0.08) Nanoparticles. *Eur. J. Inorg. Chem.* **2016**, *6*, 3460–3468. [CrossRef]
40. Levin, A.A. Program SizeCr for Calculation of the Microstructure parameters from X-ray Diffraction Data. Preprint. 2022. Available online: https://www.researchgate.net/profile/Alexander-Levin-6/publication/359402043 (accessed on 5 June 2022). [CrossRef]
41. Balashova, E.; Levin, A.A.; Fokin, A.; Redkov, A.; Krichevtsov, B. Structural Properties and Dielectric Hysteresis of Molecular Organic Ferroelectric Grown from Different Solvents. *Crystals* **2021**, *11*, 1278. [CrossRef]
42. Brucker AXS GmbH. *TOPAS*; Version 5, Technical reference; Bruker AXS GmbH: Karlsruhe, Germany, 2014.
43. le Bail, A.; Duroy, H.; Fourquet, J.L. Ab-initio structure determination of $LiSbWO_6$ by X-ray powder diffraction. *Mater. Res. Bull.* **1988**, *23*, 447–452. [CrossRef]
44. Rietveld, H.M. Line profiles of neutron powder-diffraction peaks for structure Refinement. *Acta Crystallogr.* **1967**, *22*, 151–152. [CrossRef]
45. Dollase, W.A. Correction of Intensities for Preferred Orientation in Powder Diffractometry: Application of the March Model. *J. Appl. Crystallogr.* **1986**, *19*, 267–272. [CrossRef]
46. Järvinen, M. Application of symmetrized harmonics expansion to correction of the preferred orientation effect. *J. Appl. Cryst.* **1993**, *26*, 525–531. [CrossRef]
47. Bérar, J.-F.; Lelann, P.J. E.s.d.'s and Estimated Probable Error Obtained in Rietveld Refinements with Local Correlations. *J. Appl. Crystallogr.* **1991**, *24*, 1–5. [CrossRef]
48. Levin, A.A. Program RietESD for Correction of Estimated Standard Deviations Obtained in Rietveld-Refinement Program. Preprint. 2022. Available online: https://www.researchgate.net/profile/Alexander-Levin-6/publication/359342753 (accessed on 5 June 2022). [CrossRef]
49. Andreev, Y.G. The Use of the Serial-Correlations Concept in the Figure-of-Merit Function for Powder Diffraction Profile Fitting. *J. Appl. Crystallogr.* **1994**, *27*, 288–297. [CrossRef]
50. Balashova, E.V.; Svinarev, F.B.; Zolotarev, A.A.; Levin, A.A.; Brunkov, P.N.; Davydov, V.Y.; Smirnov, A.N.; Redkov, A.V.; Pankova, G.A.; Krichevtsov, B.B. Crystal structure, Raman spectroscopy and dielectric properties of new semi-organic crystals based on 2-methylbenzimidazole. *Crystals* **2019**, *9*, 573. [CrossRef]
51. Farrugia, L.J. ORTEP-3 for Windows - a version of ORTEP-III with a Graphical User Interface (GUI). *J. Appl. Crystallogr.* **1997**, *30*, 565. [CrossRef]
52. Langan, P.; Mason, S.A.; Myles, D.; Schoenborn, B.P. Structural characterization of crystals of α-glycine during anomalous electrical behaviour. *Acta Crystallogr. B* **2002**, *B58*, 728–733. [CrossRef]
53. Boldyreva, E.V.; Drebushchak, T.N.; Shutova, E.S. Structural distortion of the a, b, and g polymorphs of glycine on cooling. *Z. Kristallogr. Cryst. Mater.* **2003**, *218*, 366–376. [CrossRef]
54. Aree, T.; Bürgi, H.-B.; Chernyshov, D.; Törnroos, K.W. Dynamics and Thermodynamics of Crystalline Polymorphs. 3. γ-Glycine, Analysis of Variable Temperature Atomic Displacement Parameters, and Comparison of Polymorph Stabilities. *J. Phys. Chem. A* **2014**, *118*, 9951–9959. [CrossRef]
55. Hill, R.J.; Fischer, R.X. Profile agreement indices in Rietveld and pattern-fitting analysis. *J. Appl. Crystallogr.* **1990**, *23*, 462–468. [CrossRef]
56. Surovtsev, N.V.; Malinovsky, V.K.; Boldyreva, E.V. Raman study of low-frequency modes in three glycine polymorphs. *J. Chem. Phys.* **2011**, *134*, 045102. [CrossRef] [PubMed]
57. Andrews, B.; Torrie, B.H.; Powell, B.M. Intermolecular potentials for alpha-glycine from Raman and infrared scattering measurements. *Biophys. J.* **1983**, *41*, 293–298. [CrossRef]
58. Baran, J.; Ratajczak, H. Polarised IR and Raman spectra of the γ-glycine single crystal. *Spectrochim. Acta A* **2005**, *61*, 1611–1626. [CrossRef]
59. Colmenero, F.; Escribano, R. Thermodynamic, Raman Spectroscopic, and UV–Visible Optical Characterization of the Deltic, Squaric, and Croconic Cyclic Oxocarbon Acids. *J. Phys. Chem. A* **2019**, *123*, 4241–4261. [CrossRef] [PubMed]
60. Yamada, K.; Mizuno, N.; Hirata, Y. Structure of Croconic Acid. *Bull. Chem. Soc. Jap.* **1958**, *31*, 543–549. [CrossRef]
61. Feng, S.; Yang, H.; Jiang, X.; Wang, Y.; Zhu, M. A New 3D Silver(I) Coordination Polymer with Croconate Ligand Displaying Green Luminescent. *J. Mol. Struct.* **2015**, *1081*, 1–5. [CrossRef]

Article

Study on Precipitation Processes and Phase Transformation Kinetics of Iron Phosphate Dihydrate

Huiqi Wang [1], Mingxia Guo [2], Yue Niu [1], Jiayu Dai [1], Qiuxiang Yin [1,3] and Ling Zhou [1,*

[1] State Key Laboratory of Chemical Engineering, School of Chemical Engineering and Technology, Tianjin University, Tianjin 300072, China
[2] Department of Chemical Engineering, South Kensington Campus, Imperial College London, London SW7 2AZ, UK
[3] The Co-Innovation Center of Chemistry and Chemical Engineering of Tianjin, Tianjin University, Tianjin 300072, China
* Correspondence: zhouling@tju.edu.cn

Abstract: The process of the phase transformation from amorphous to crystalline $FePO_4 \cdot 2H_2O$ was studied in this research. It was found that Fe and P are predominantly present as $FeHPO_4^+$ and $FeH_2PO_4^{2+}$ and an induction period exists during the transition from amorphous to monoclinic form. The induction period and the time required for phase transformation were shortened with the increased temperature. Phase transformation could be kinetically described by the Johnson–Mehl–Avrami (JMA) dynamics model. The dissolution rate of amorphous $FePO_4 \cdot 2H_2O$ is the rate-limiting step of this process. the activation energy of phase transformation is calculated to be 9.619 kJ/mol. The results in this study provided more guidelines for the regulation of $FePO_4 \cdot 2H_2O$ precursors by precipitation method.

Keywords: precipitation process; iron phosphate dihydrate; Johnson–Mehl–Avrami (JMA) dynamics model

Citation: Wang, H.; Guo, M.; Niu, Y.; Dai, J.; Yin, Q.; Zhou, L. Study on Precipitation Processes and Phase Transformation Kinetics of Iron Phosphate Dihydrate. *Crystals* **2022**, *12*, 1369. https://doi.org/10.3390/cryst12101369

Academic Editor: Yanfei Wang

Received: 5 September 2022
Accepted: 22 September 2022
Published: 27 September 2022

Publisher's Note: MDPI stays neutral with regard to jurisdictional claims in published maps and institutional affiliations.

1. Introduction

At present, the demand for rechargeable lithium batteries for electric vehicles is growing with increased energy efficiency requirements. Creating composite electrode materials with the required properties is a promising direction in battery technology and engineering [1–4]. Researchers have proposed numerous techniques for fabricating the positive electrode $LiFePO_4$ of reversible lithium batteries [5]. So far, the precursor of $LiFePO_4$—$FePO_4$ was mainly prepared by the: (1) solvent gel method [6], (2) hydrothermal method [7], (3) coprecipitation method [8], and others. Zhang et al. [9] synthesized nano-sized $FePO_4$ by a modified sol–gel method, which is convenient for controlling the carbon content and size of $LiFePO_4$. Chen et al. [10] produced $FePO_4$ microspheres with carbon nanotube embedded (FePO4/CNT), which enhanced their electronic conductivity by a hydrothermal process. Tong et al. [11] prepared iron phosphate with various morphologies and crystalline structures by coupling precipitation and aging. Coprecipitation is the most advantageous method for synthesizing iron phosphate owing to its simple operation, economic efficiency, and environmental friendliness among these methods [5].

$FePO_4 \cdot 2H_2O$ acts as a transient precursor in the coprecipitation method to obtain $FePO_4$. Many researchers have reported the preparation of $FePO_4 \cdot 2H_2O$ with various structures and morphologies, which were achieved by altering the sources of Fe and P as well as reaction conditions [11,12]. For example, Guo et al. [13] found that iron phosphate precursor with different morphologies can be obtained at low temperature using $Fe_2(SO_4)_3$, H_3PO_4 and different additives. Jiang et al. [14] synthesized irregular polygon shapes of amorphous $FePO_4 \cdot 2H_2O$, and elongated sticks of crystalline $FePO_4 \cdot 2H_2O$ by refluxing amorphous $FePO_4 \cdot 2H_2O$ at 100 °C for 2 h using $Fe(NH_4)_2(SO_4)_2 \cdot 6H_2O$, $NH_4H_2PO_4$, and

H_2O_2 as raw materials. As the important precursor of $LiFePO_4$ cathode material, the properties of $FePO_4$ have to be controlled in the preparation process, such as the specific amount of iron–phosphorus ratio, high specific surface area, and tap density [15]. By understanding the mechanism and kinetics of the preparation process, the properties of $FePO_4 \cdot 2H_2O$ can be better controlled and the quality of $FePO_4$ can be further improved. Many studies have demonstrated that the crystal structure of $FePO_4$ significantly impacts the morphology of $FePO_4$ [16–18]. However, insufficient literature exists regarding the phase transformation on this process.

This paper studied the precipitating process of $FePO_4 \cdot 2H_2O$ using ferrous sulfate heptahydrate, phosphoric acid, and hydrogen peroxide as reactants, and the morphology and phase transformation kinetics of $FePO_4 \cdot 2H_2O$ from amorphous to monoclinic in acidic solution (pH < 1) were investigated. The changes in the morphology, structure, and composition of the precipitate during the phase transformation were also tracked. These studies provide ideas and a basis for further improvement in the properties of iron phosphate as a potential electrode material precursor.

2. Materials and Methods

2.1. Chemicals

All the reagents used in experiments containing ferric sulfate septihydrate ($FeSO_4 \cdot 7H_2O$), hydrogen peroxide (H_2O_2), monoclinic iron phosphate dihydrate ($FePO_4 \cdot 2H_2O$), and phosphoric acid (H_3PO_4) were of analytical reagent grade. Deionized water was used throughout the process. All these chemicals were purchased from Kaimart Tianjin Chemical Technology Co., Ltd (Tianjin, China). and were used directly without further purification. Detailed information could be seen in Table 1.

Table 1. Detail of the materials specification.

Chemicals	CAS Registry No.	Molecular Weight (g/mol)	Mass Fraction Purity
$FeSO_4 \cdot 7H_2O$	7782-63-0	278.01	≥ 0.99
H_2O_2	7722-84-1	34.01	≥ 0.30
$FePO_4 \cdot 2H_2O$	13463-10-0	186.82	≥ 0.99
H_3PO_4	7664-38-2	98.00	≥ 0.85

2.2. Experimental Procedure

Ferrous sulfate septihydrate and phosphoric acid at a molar ratio of 1:1.05 were dissolved with deionized water in a 500 mL jacketed glass reactor coupled with a thermostatic bath. Excess phosphorus was used to make sure the iron precipitated completely during the reaction. The mixture was then stirred by a stirrer module until ferrous sulfate septihydrate dissolved completely. The peristaltic pump added a certain amount of hydrogen peroxide, the mole of which was round 0.6 time the ferrous, to make Fe^{2+} iron oxidized completely. When the solution temperature was raised to 60 °C the precipitation was generated from the solution. The solution temperature was kept at 90 °C for 8 h and a phase transformation occurred during this process. The precipitation was filtered out at the end of the experiment. The experimental setup is shown in Figure 1 and detailed information of the solution can be seen in Table 2.

Table 2. Compositions of the solution.

	Concentration %
$FeSO_4 \cdot 7H_2O$	8.3
H_3PO_4	3.7
H_2O_2	2.4
H_2O	85.6

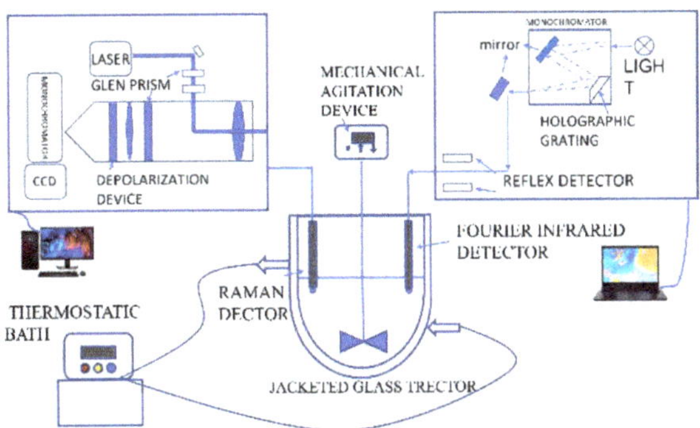

Figure 1. Experimental apparatus utilized in the experiments.

2.3. Preparation of Standard Solution for RAMAN Analysis

The difference of the two polymorphs in solution can be shown by Raman spectra through the characteristic peaks. Constructing a standard line using Raman spectra of the prepared polymorphic mixtures to quantitatively analyze changes of monoclinic $FePO_4 \cdot 2H_2O$ during the phase transformation process: the standard solution was a mixture of amorphous and monoclinic $FePO_4 \cdot 2H_2O$ at a predetermined concentration of H_2SO_4. The mole fraction of monoclinic $FePO_4 \cdot 2H_2O$ ranged from 25% to 75%. The temperature of solution was raised to 90 °C. Each measurement was carried out in three times. There was no polymorphic transformation that occurred for 1–3 min during detection, which can be indicated by Raman spectra. The function that connected the corresponding peak intensity to the mass fraction of monoclinic $FePO_4 \cdot 2H_2O$ could then be determined.

2.4. Characterization

2.4.1. Powder X-ray Diffraction (PXRD) Analysis

The samples collected during the process were analyzed by X-ray diffraction (Rigaku D/MAX-2500) with Cu Kα radiation (λ = 1.5405 Å) by Ni filter. The scanning angle ranged from 2° to 40° at a rate of 8°/min. The characterization was carried out at a voltage of 45 kV and a current of 40 mA. The melting properties of $FePO_4 \cdot 2H_2O$ are measured by thermal gravimetric analyzer (TG, MettlerToledo, Zurich, Switzerland). The sample (5−10 mg) was heated at a rate of 10 K/min from 303.15 K to 1073.15 K under a nitrogen atmosphere.

2.4.2. Morphology Characterization

The morphologies of products were observed by scanning electron microscopy (SEM; JSM-7401F, 3 kV) after being sputter-coated with Au/Pd and transmission electron microscopy (TEM; JEM-2010, 120 kV) using an accelerating voltage of 100 kV.

2.4.3. Spectra Characterization

The changes of functional groups in precipitation during the reaction were detected by FTIR (Bio-rad FTS 6000), which scanned with wavenumber from 400 to 4000 cm^{-1}; Raman analysis (RFS 100/S) used a 1064 nm Nd-YAG laser and the scanning step was 2 cm^{-1} with a 50 kHz scanning frequency, utilizing fiber-coupled probe optic technology for in situ monitoring. This system was equipped with a probe head for direct insertion or non-contact sampling.

3. Results and Discussion

3.1. The Involved Precipitation Reactions Analysis

The complexation of Fe^{3+} with phosphate is complicated in solution [19,20]. In this work, complexation ions such as $Fe(PO_4)_2^{3-}$, $Fe(OH)PO_4^-$, $FeH_3(PO_4)_3^{3-}$, $Fe(PO_4)_2^{3-}$, $Fe_2PO_4^{3+}$, etc., were not taken into account owing to small equilibrium constants ($<10^{-8}$).

The possible reactions in solution and corresponding equilibrium constants are listed in Table 3 [21–23]. On the basis of mass conservation and reactions in Table 3, the mass balance of Fe and P can be expressed, as given in formulas (1) and (2). The total concentration of Fe and P was measured by redox titration of potassium dichromate and the gravimetric method. According to pH and the equilibrium constants of possible reactions, the corresponding concentration of complexation ions can be represented by free Fe^{3+} ion and H_3PO_4 molecules.

$$[Fe]_T = [Fe^{3+}] + [FeHPO_4^+] + [FeH_2PO_4^{2+}] + [FeH_2(PO_4)^{2-})] + [FeH_4(PO_4)^{2+}] \\ + [FeH_8(PO_4)_4^-] + 2 \times [Fe_2HPO_4^{4+}] + 2 \times [Fe_2H_3(PO_4)_2^{3+}] + 3 \times [Fe_3H_6(PO_4)_4^{3+}] \tag{1}$$

$$[P]_T = [H_3PO_4] + [H_2PO_4^-] + [HPO_4^{2-}] + [PO_4^{3-}] + [FeHPO_4^+] + [FeH_2PO_4^{2+}] \\ + 2 \times [FeH_2(PO_4)_2^{2-})] + 2 \times [FeH_4 (PO_4)^{2+})] + 4 \times [FeH_8 (PO_4)_4^-] + [Fe_2H(PO_4)^{4+})] \\ + 2 \times [Fe_2H_3 (PO_4)_2^{3+}] + 4 \times [Fe_3H_6(PO_4)_4^{3+}] \tag{2}$$

where $[Fe]_T$ and $[P]_T$ is the total concentration of Fe and P in solution; $[x]$ is the concentration of x; x is the species of complexation.

Table 3. Possible complexation reactions for Fe^{3+} and phosphate.

No.	Reactions	LogK
1	$H_3PO_4 = H^+ + H_2PO_4^-$	−2.12
2	$H_2PO_4^- = H^+ + HPO_4^{2-}$	−7.20
3	$HPO_4^{2-} = H^+ + PO_4^{3-}$	−12.36
4	$Fe^{3+} + H_3PO_4 = FeHPO_4^+ + 2H^+$	1.55
5	$Fe^{3+} + H_3PO_4 = FeH_2PO_4^{2+} + H^+$	1.60
6	$Fe^{3+} + 2H_3PO_4 = FeH_2(PO_4)_2^- + 4H^+$	−4.34
7	$Fe^{3+} + 2H_3PO_4 = FeH_4(PO_4)_2^+ + 2H^+$	1.19
8	$Fe^{3+} + 4H_3PO_4 = FeH_8(PO_4)_4^- + 4H^+$	1.73
9	$2Fe^{3+} + H_3PO_4 = Fe_2HPO_4^{4+} + 2H^+$	1.92
10	$2Fe^{3+} + 2H_3PO_4 = Fe_2H_3(PO_4)_2^{3+} + 3H^+$	0.18
11	$3Fe^{3+} + 4H_3PO_4 = Fe_3H_6(PO_4)_2^{3+} + 3H^+$	0.23

The unit of equilibrium constants K is $(mol \cdot L^{-1})^{\Delta v}$, Δv is sum of stoichiometric numbers in reaction.

And the concentration of free Fe^{3+} ion and H_3PO_4 molecules, under pH ranging from 0.2 to 1.3, can be calculated by the total concentration and mass balance of Fe and P. The equations for the concentration of Fe^{3+} and H_3PO_4 were presented in Formulas (3) and (4). In addition, the concentration of various complexation species with phosphate groups under different pH environments is listed in Tables S1–S4 (Supporting Information). The distribution of the various complexation species is presented in Figure 2.

$$x + (A + B)xy + (C + D)xy^2 + Exy^4 + 2Fx^2y + 2Gx^2y^2 + 3Hx^3y^4 = 0.293 \tag{3}$$

$$(1 + a + b)y + (A + B)xy + 2(C + D)xy^2 + 4Exy^4 + Fx^2y + 2Gx^2y^2 \\ + 4Hx^3y^4 = 0.2978 \tag{4}$$

where x is the concentration of Fe^{3+} and y is the concentration of H_3PO_4; $a, b, A, B, C, D, E, F, G, H$ are the constants in different pH environment based on the equilibrium constants of possible reactions respectively.

It can be found that Fe and P are predominantly presented as $FeHPO_4^+$ and $FeH_2PO_4^{2+}$ in the pH range. The pH of the mixture at room temperature in this work is around 0.85 and the proportion of main complexation ions are calculated to be 53.82% $FeHPO_4^+$, and 9.79% $FeH_2PO_4^{2+}$. Therefore, the main precipitation reactions, even during the process of raising the temperature, can be written as the following reactions in Table 4 (the pH decreased with the increase of temperature).

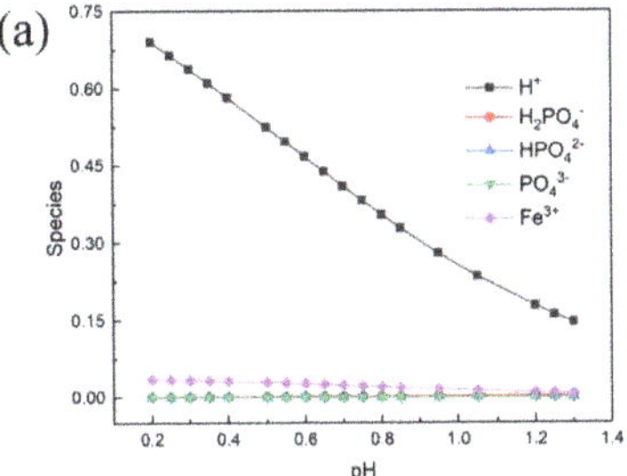
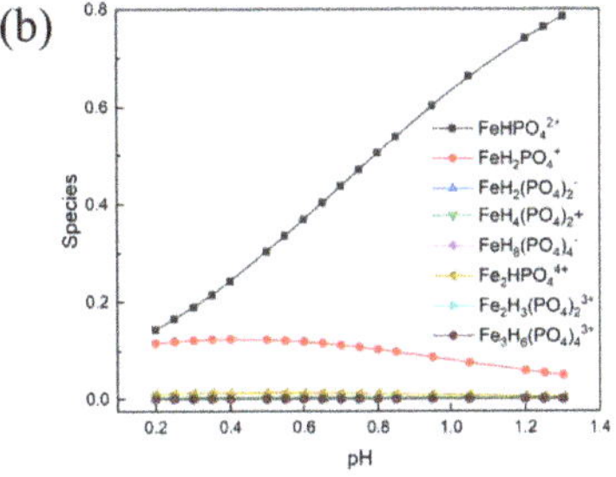

Figure 2. Distribution of the various complexation species under different pH environments: (**a**) shows distribution of H^+, $H_2PO_4^-$, HPO_4^{2-}, PO_4^{3-} and Fe^{3+}; (**b**) shows distribution of the rest of the ions.

Table 4. Possible reactions in solution with an initial pH value of 0.85.

No.	Reactions
1	$FeHPO_4^+ + H_2O = FePO_4 \cdot 2H_2O\downarrow + H^+$
2	$FeH_2PO_4^{2+} + 2H_2O = FePO_4 \cdot 2H_2O\downarrow + 2H^+$

3.2. Insight into the Evolution of the Precipitate

3.2.1. The Change of Structure

The crystal structure and phase purity of the samples were analyzed by PXRD. The powder X-ray diffraction pattern was shown in Figure 3. There were not any intensive diffraction peaks detected before 235 min, indicating that the precipitation initially produced by the reaction was amorphous. After 235 min, new characteristic diffraction peaks appeared (17.18°, 18.88°, and 19.94°, respectively), and the intensity of these peaks increased with the evolution of time. The results demonstrated that there was a phase transition during the reaction and the crystallinity gradually increased over time. Since the PXRD pattern of samples after 235 min is consistent with standard data of JCPDS file No. 15-0390, it is found that the transformed product possessed monoclinic structure and a space group of P21/n [15,24–26]. The crystal structure of monoclinic $FePO_4 \cdot 2H_2O$ was shown in Figure 4.

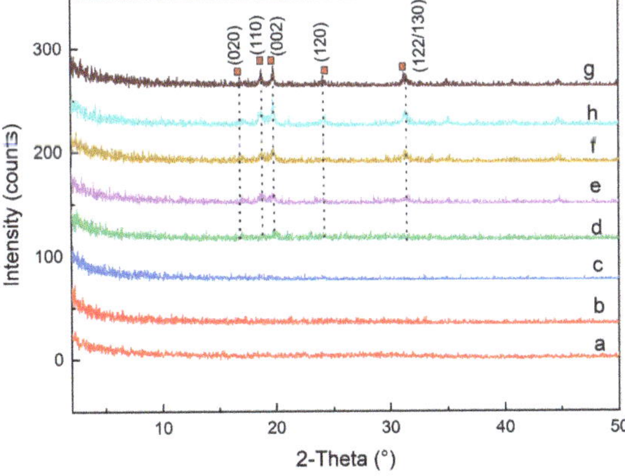

Figure 3. The Powder X-ray diffraction pattern of samples taken at 0 min (**a**), 180 min (**b**), 205 min (**c**), 225 min (**d**); 235 min (**e**), 245 min (**f**); 255 min (**h**), 450 min (**g**), respectively.

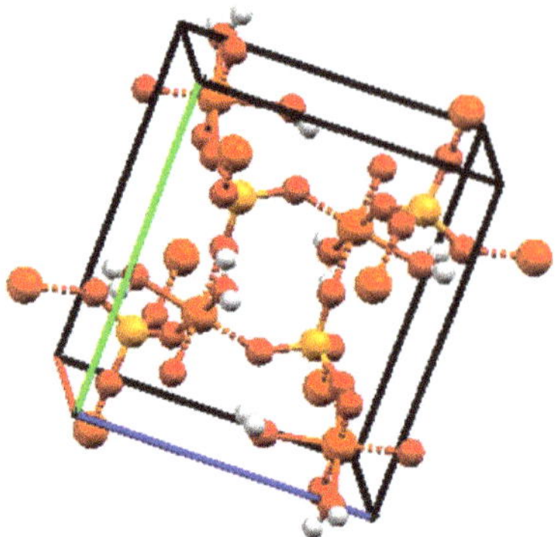

Figure 4. Crystal structure of monoclinic $FePO_4 \cdot 2H_2O$.

3.2.2. Change of Morphology

The morphology during the reaction was studied by SEM and TEM as shown in Figures 5 and 6. At the initial stage of the reaction, the sample mainly exists in the form of amorphous microsheet agglomeration and the average size is about 2 µm to 4 µm, as shown in Figures 5a and 6a. The size of microsheets gradually decreased and the extent of agglomeration increased slightly. They grew and agglomerated together by continuous dissolution–recrystallization. Figure 5c shows that plentiful nanoparticles formed on the surface of agglomerates at 150 min, which were found to be a mixture of mostly amorphous and little crystalline $FePO_4 \cdot 2H_2O$ by selected area electron diffraction (SAED)—which shows an inconspicuous diffraction pattern, as shown in Figure 6b. But there is still no obvious diffraction peak of powder X-ray diffraction pattern at this time, at which the content of crystalline $FePO_4 \cdot 2H_2O$ is lower than the minimum content standard of powder X-ray diffraction analysis. The monoclinic $FePO_4 \cdot 2H_2O$ nuclei on the active surface of amorphous agglomerates gradually grew and agglomerated accompanying the dissolution of amorphous $FePO_4 \cdot 2H_2O$ into ions. Ultimately, dense spheroid-like monoclinic $FePO_4 \cdot 2H_2O$ particles were produced continuously by dissolution–recrystallization of agglomerates. It was proven that the spheroid-like particles are comprised of long flakes of monoclinic $FePO_4 \cdot 2H_2O$ as illustrated in Figure 6c.

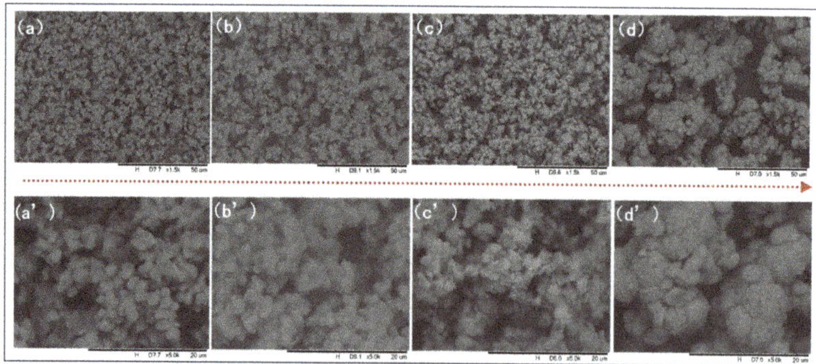

Figure 5. SEM images of morphology evolvement of $FePO_4 \cdot 2H_2O$ during the experiment: (**a**) 0 min; (**b**) 30 min; (**c**) 150 min; (**d**) 450 min; (**a'**–**d'**) are images with the higher magnification at the corresponding times (0 min, 30 min, 150 min, 450 min).

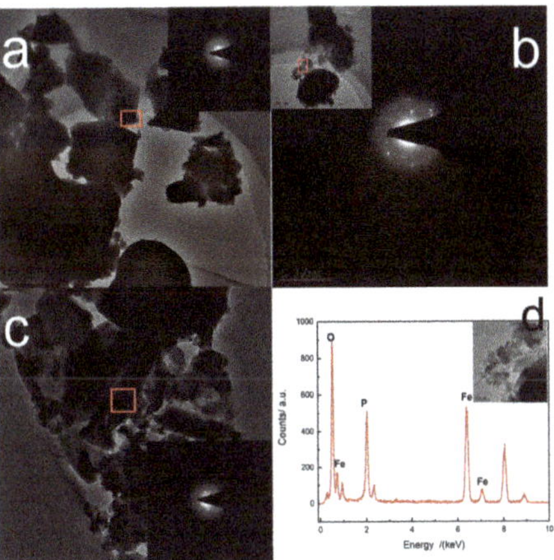

Figure 6. TEM images of $FePO_4 \cdot 2H_2O$ at (**a**) initial, (**b**) 150 min and (**c**) end of reaction, respectively; (**d**) typical EDS spectrum of initially produced precipitate.

3.2.3. Changes in the Process of Reaction

The samples in the different reaction stages were also analyzed by TG, as shown in Figure 7. The heating process was conducted in an N_2 environment and the weight loss was on account of the dehydration [27,28], which was around 19.30%. With the increase of crystallinity, the endothermic peak for dehydration is shifting in the higher temperature direction gradually [29] (115.417 °C→135.063 °C→161.769 °C→176.833 °C), which proved that water molecules gradually enter into the lattice monoclinic $FePO_4 \cdot 2H_2O$ as the phase transformation progresses.

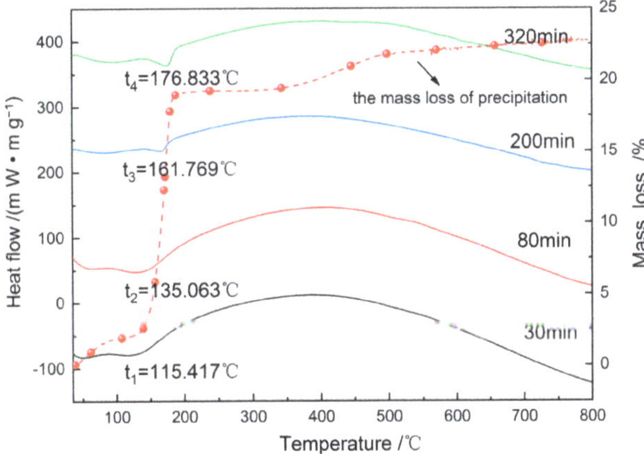

Figure 7. Heat-flow and mass change of precipitation at different reaction time during heating.

The reaction process was characterized by off-line FTIR, as shown in Figure 8. The characteristic peaks of the O-H stretching and bending vibrational mode are at 3500–3000 cm^{-1} and about 1600 cm^{-1} respectively [30,31]. The peak band at 513 cm^{-1} is attributed to the bending vibration mode and stretching band of O-P-O bonds. The P-O-P bonds are also reflected at 980 cm^{-1} to1100 cm^{-1}, which corresponds to the tetrahedral PO_4^{3-} anions.

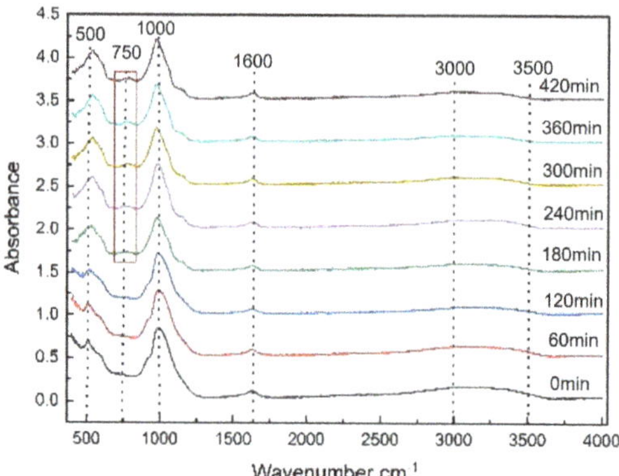

Figure 8. FTIR spectra of the precipitates at different reaction times.

After 180 min, a new peak at 750 cm^{-1} appeared, corresponding to the P-O vibration caused by the coupling effect between PO_4^{3-} polyanion and Fe-O within the structure [32–35]. The symmetrical vibration characteristic peaks of PO_4^{3-} anions at 997 cm^{-1} disappeared, and asymmetrical vibration peak at 1165 cm^{-1} appeared over time, as shown in Figure 8. These changes indicate that the $FePO_4 \cdot 2H_2O$ orderly rearranged and monoclinic $FePO_4 \cdot 2H_2O$ formed.

The result shown in Figure 9 demonstrated that the proportion of Fe in the precipitate increased at the early stage of phase transformation and then decreased with the increase of crystallization, which can also be proven by the energy dispersive spectrometer (EDS) information in Table 5. At the highest proportion of iron, the corresponding precipitation is amorphous, and the characteristic vibration peaks of iron binding with other functional groups are not found in the FTIR spectrum, indicating that there is not any intermediate during transformation.

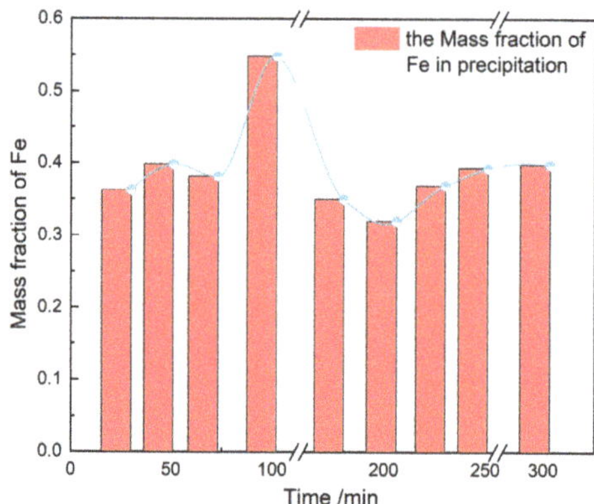

Figure 9. The mass fraction trend of Fe in precipitation at different reaction time.

Table 5. Composition of produced precipitate under different time.

Sampling Time/min	Fe (Atom %)	P (Atom %)	Molar Ratio of P/Fe
0	18.2	17.6	0.96
100	19.0	18.3	0.96
320	21.2	21.4	1.00

The solubility of the amorphous form of $FePO_4 \cdot 2H_2O$ is higher than that of the monoclinic form [36]. During the rearrangement of iron phosphate dihydrate, the solution became supersaturated again with the dissolution of amorphous iron phosphate. The iron ions in solution adsorbed on the surface of the agglomerate (were partially neutralized by sulfate ions) or encased in the interior of the agglomerate, leading to a rise in the iron mass fraction (Supporting Information, Table S1). Then iron mass fraction decreased with the nucleation and growth of monoclinic $FePO_4 \cdot 2H_2O$. The rise in mass fraction following transformation could be attributable to the impurity removal due to the decreased content of the S in precipitation.

3.3. Kinetic of Phase Transformation

Raman was used to monitor the transformation of amorphous $FePO_4 \cdot 2H_2O$, as shown in Figure 10. Transformation of amorphous $FePO_4 \cdot 2H_2O$ was followed by the appearance of monoclinic $FePO_4 \cdot 2H_2O$ characteristic peaks and the changes in height of the characteristic peaks: a new peak at 303 cm^{-1} appeared in the Raman spectra and its intensity increased gradually as the monoclinic $FePO_4 \cdot 2H_2O$ content increased during transformation.

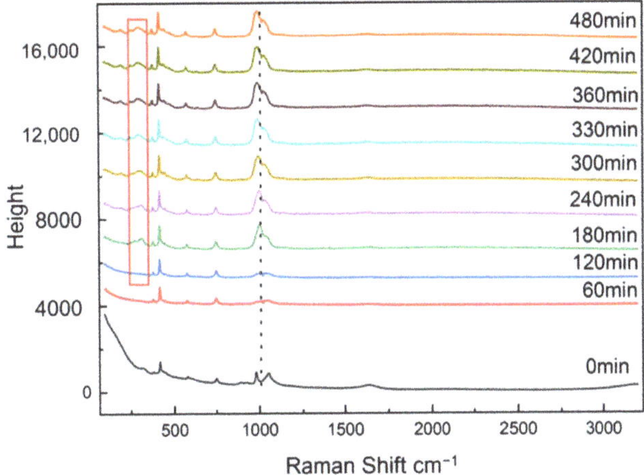

Figure 10. Raman spectra of the precipitates at different reaction times.

The corresponding characteristic peak intensity of the standard solution was measured by Raman spectroscopy (shown in Figure 11) and the standard curve could be obtained according to the relationship between intensity of characteristic peak and the mass fraction of monoclinic $FePO_4 \cdot 2H_2O$, as shown in Figure 12. The transformation process was quantitatively evaluated by the relationship, which can be seen in Figure 13. It shows the changes of polymorphic composition during transformation. The mass fraction of the amorphous $FePO_4 \cdot 2H_2O$ decreased gradually after the induction time of 101 min, which could be attributed to the nucleation of monoclinic $FePO_4 \cdot 2H_2O$ and the transformation of the amorphous $FePO_4 \cdot 2H_2O$. The curves are stable in zone 3 for the completion of transformation.

The Johnson–Mehl–Avrami (JMA) dynamics model was used for describing the phase transformation, as shown in equation 4. It was considered that the solution is well-mixed and the growth rate of crystals is independent of time. It assumed that the nucleation sites were located in the well-mixed reactant bases on solid-state reactions [37,38]. The calculated parameters were shown in the Table 6 and the fitting relevance value of R^2 was over 0.97 indicating that the fitting result was great consistent with experimental results.

$$x(t) = 1 - \exp\left\{-K \times (t - t_{ind})^n\right\} \tag{5}$$

where x is the percentage content of monoclinic $FePO_4 \cdot 2H_2O$ at reaction time t; t_{ind} is the induction period of $FePO_4 \cdot 2H_2O$ transformation; K is the rate constant of transformation; n (called as Avrami exponent) is a constant related to the behaviors of nucleation and growth of $FePO_4 \cdot 2H_2O$; if the value of n is over 1, the nucleation is the key factor for phase transformation or the migration of the chemicals to the nucleation point is dominant.

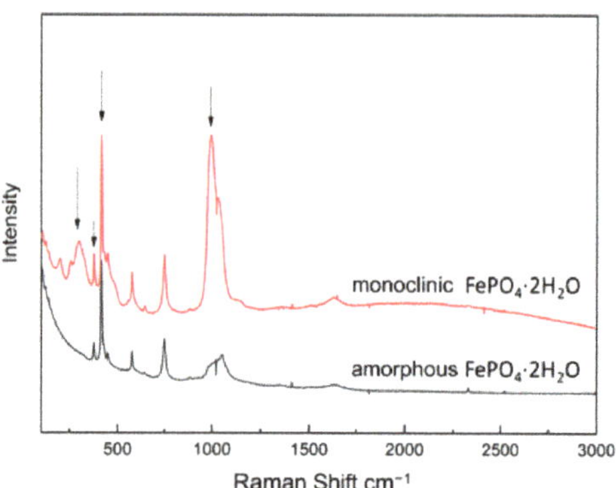

Figure 11. Raman spectra of slurries of monoclinic and amorphous $FePO_4 \cdot 2H_2O$ during the transformation.

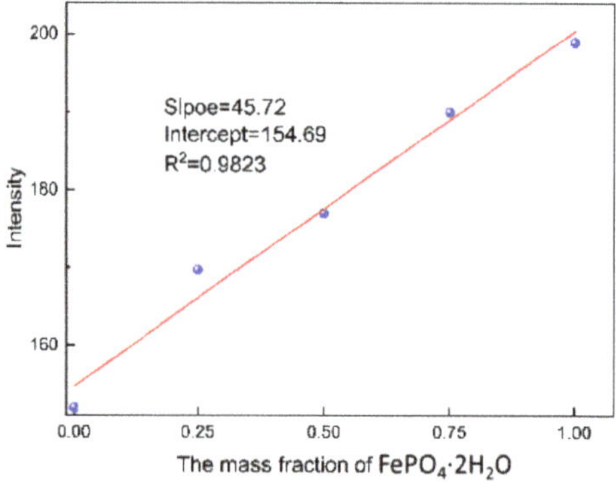

Figure 12. The standard curve of monoclinic $FePO_4 \cdot 2H_2O$.

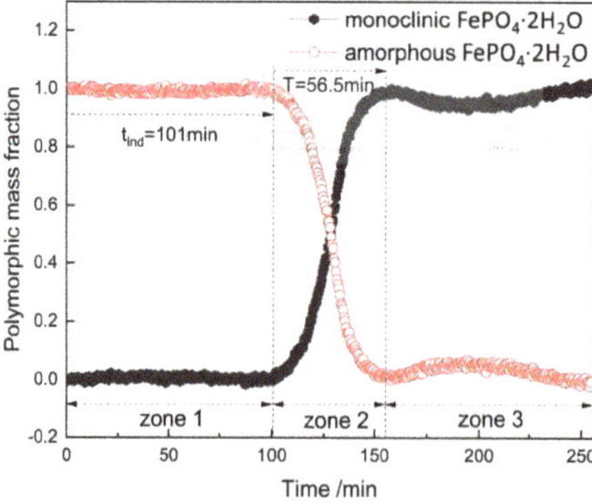

Figure 13. In situ polymorphic mass fraction profiles of monoclinic and amorphous $FePO_4 \cdot 2H_2O$ through the solvent-mediated transformation process.

Table 6. Fitting parameters of JMA model at 90 °C.

Parameters	Value
Temperature/°C	90
t_{ind}/min	101
T/min	56.5
K/h^{-3}	3.4348
n	3.4381
R^2	0.9798

T is the time required for transformation.

According to the spectra presented in Figure 13 and the Equation (5), the relationship of time and crystallinity and the parameters of JMA dynamics model were obtained, which were shown in Figure 14 and Table 6. The time required for transformation was 56.5 min and the induction period of this process was 101 min. Moreover, the concentration of Fe in the solution and the intensity of the corresponding characteristic peak at 303 cm^{-1} changed almost simultaneously: the concentration of Fe in solution decreased accompanying by the increase of the corresponding peak intensity with the time evolution (Supporting Information, Figure S2). It was indicated that the dissolution rate of amorphous $FePO_4 \cdot 2H_2O$ was slowest during phase transformation. Therefore, the dissolution rate of amorphous $FePO_4 \cdot 2H_2O$ is considered to be the control step of the transformation rate.

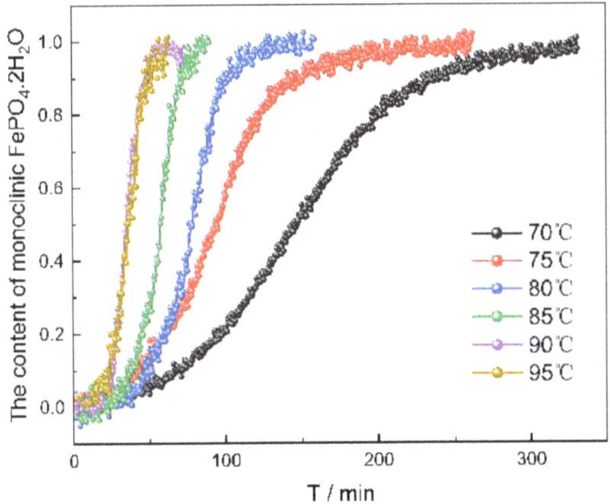

Figure 14. The percentage content of monoclinic $FePO_4 \cdot 2H_2O$ vs. time at different temperatures.

The experiments at different temperatures were carried out in order to evaluate the kinetics of the transformation process. The fitting results at different temperatures evaluated by the JMA model between temperature and the rate constant of transformation are shown in Table 7, and the changes trends of characteristic peak under different temperatures were presented in Figure S2 in Supporting Information. The activation energy of the transformation process was obtained by the Arrhenius equation, which was shown as Equation (6):

$$\ln(K) = -E_a \times (R \times T)^{-1} + \ln(A) \tag{6}$$

where K is the rate constant of transformation; R is the molar gas constant; T is the thermodynamic temperature; E_a is the apparent activation energy and A is the pre-exponential factor which is also called the frequency factor.

Table 7. Fitting parameters of JMA model at different temperature.

Temperature/°C	t_{ind}/min	T/min	K/h^{-3}	n	R^2
70	316	332	0.0742	2.53	0.995
75	259	226	0.2074	2.63	0.996
80	224	114	0.3845	4.34	0.995
85	145	76	0.9724	4.99	0.993
95	75	55	5.1157	3.87	0.983

T is the time required for transformation.

The parameters of reaction calculated by the JMA model are listed in Table 7. With the increase of reaction temperature, the induction period and the time required for transformation decreased, and rate constant of transformation increased. Increasing temperature can enhance molecular movement and reduce the interfacial energy of solid–liquid interface. Therefore, the nucleation rate of monoclinic $FePO_4 \cdot 2H_2O$ was accelerated and correspondingly the induction period and the time required for transformation was shortened. The value of the apparent activation energy E_a in the transformation process can be derived by plotting the logarithm of K against $1/T$ and analyzing the slope and intercept (Figure S3 in Supporting Information). The value of the activation energy was calculated to be 9.619 kJ/mol.

4. Conclusions

The phase transformation from amorphous to monoclinic $FePO_4 \cdot 2H_2O$ was investigated. It was found that Fe and P are predominantly present as $FeHPO_4^+$ and $FeH_2PO_4^{2+}$ and the transformation process of $FePO_4 \cdot 2H_2O$ from amorphous form to monoclinic form was determined by the nucleation rate of the monoclinic form. The corresponding Raman spectroscopy results indicated that the induction period and the time required for transformation reduced as the reaction temperature rose. The kinetic parameters of the $FePO_4 \cdot 2H_2O$ transformation were calculated by the JMA model and the findings demonstrated that the transformation reaction constant increased as the reaction temperature increased. The activation energy of the transformation was 9.619 kJ/mol. These studies were beneficial to controlling and improving properties of $FePO_4$ for meeting the requirements as potential electrode material precursors.

Supplementary Materials: The following supporting information can be downloaded at: https://www.mdpi.com/article/10.3390/cryst12101369/s1, Table S1: The value of different parameter at pH range in equations; Table S2: The concentration of different ions at pH range; Table S3: The concentration of different ions at pH range; Table S4: The concentration of different ions at pH range; Table S5: The content of S and O in precipitation at different time; Figure S1: The mass fraction trend of Fe in solution at different time; Figure S2: Raman spectra of the precipitates at 70 °C (a), 75 °C (b), 80 °C (c), 85 °C (d), 95 °C (e); (f) the change of characteristic peak at 303 cm^{-1} over time under different temperature; Figure S3″ The linear relationship between ln K and (1/T).

Author Contributions: Writing—original draft preparation, H.W.; writing—review and editing, H.W., M.G., Y.N., J.D. and L.Z.; visualization, H.W. and L.Z.; supervision, Q.Y. and L.Z. All authors have read and agreed to the published version of the manuscript.

Funding: This research was funded by Natural Science Foundation of Tianjin, grant number 21JCY-BJC00600.

Institutional Review Board Statement: Not applicable.

Informed Consent Statement: Not applicable.

Data Availability Statement: Not applicable.

Acknowledgments: The authors are grateful for the financial support of the Special project for the transformation of major scientific and technological achievements of Guizhou Province.

Conflicts of Interest: The authors declare no conflict of interest.

References

1. Allen, J.L.; Jow, T.R.; Wolfenstine, J. Analysis of the FePO$_4$ to LiFePO$_4$ Phase Transition. *J. Solid State Electrochem.* **2008**, *12*, 1031–1033. [CrossRef]
2. Wang, Y.; Cao, G. Developments in Nanostructured Cathode Materials for High-Performance Lithium-Ion Batteries. *Adv. Mater.* **2008**, *20*, 2251–2269. [CrossRef]
3. Padhi, A.K.; Nanjundaswamy, K.S.; Goodenough, J.B. Phospho-olivines as Positive-Electrode Materials for Rechargeable Lithium Batteries. *J. Electrochem. Soc.* **1997**, *144*, 1188–1194. [CrossRef]
4. Zhang, H.; Zou, Z.; Zhang, S.; Liu, J.; Zhong, S. A review of the doping modification of LiFePO$_4$ as a cathode material for lithium ion batteries. *Int. J. Electrochem. Sci* **2020**, *15*, 12041–12067. [CrossRef]
5. Zhang, Y.; Huo, Q.; Du, P.; Wang, L.; Zhang, A.; Song, Y.; Lv, Y.; Li, G. Advances in New Cathode Material LiFePO$_4$ for Lithium-Ion Batteries. *Synth. Met.* **2012**, *162*, 1315–1326. [CrossRef]
6. Peng, W.; Jiao, L.; Gao, H.; Qi, Z.; Wang, Q.; Du, H.; Si, Y.; Wang, Y.; Yuan, H. A Novel Sol–Gel Method Based on FePO$_4$·2H$_2$O to Synthesize Submicrometer Structured LiFePO$_4$/C Cathode Material. *J. Power Sources* **2011**, *196*, 2841–2847. [CrossRef]
7. Prosini, P.P.; Lisi, M.; Scaccia, S.; Carewska, M.; Cardellini, F.; Pasquali, M. Synthesis and Characterization of Amorphous Hydrated FePO$_4$ and Its Electrode Performance in Lithium Batteries. *J. Electrochem.* **2002**, *149*, A297–A301. [CrossRef]
8. Zhan, T.T.; Jiang, W.F.; Li, C.; Luo, X.D.; Lin, G.; Li, Y.W.; Xiao, S.H. High Performed Composites of LiFePO$_4$/3DG/C Based on FePO$_4$ by Hydrothermal Method. *Electrochim. Acta* **2017**, *246*, 322–328. [CrossRef]
9. Chen, M.; Du, C.; Song, B.; Xiong, K.; Yin, G.; Zuo, P.; Cheng, X. High-Performance LiFePO$_4$ Cathode Material from FePO$_4$ Microspheres with Carbon Nanotube Networks Embedded for Lithium Ion Batteries. *J. Power Sources* **2013**, *223*, 100–106. [CrossRef]
10. Zhang, T.B.; Lu, Y.C.; Luo, G.S. Iron Phosphate Prepared by Coupling Precipitation and Aging: Morphology, Crystal Structure, and Cr (III) Adsorption. *Cryst. Growth Des.* **2013**, *13*, 1099–1109. [CrossRef]
11. Bouamer, H.; El, M.; Lakhal, M.; Kaichouh, G.; Khalid, O.; Faqir, H.; Hourch, A.; Guessous, A. Growth and Characterization of Electrodeposited Orthorhombic FePO$_4$.2H$_2$O. *J. Mater. Environ. Sci.* **2018**, *9*, 1247–1254.
12. Wang, Z.; Lu, Y. Facile Construction of High-Performance Amorphous FePO$_4$/Carbon Nanomaterials as Cathodes of Lithium-Ion Batteries. *ACS Appl. Mater. Interfaces* **2019**, *11*, 13225–13233. [CrossRef]
13. Guo, J.; Liang, C.; Cao, J.; Jia, S. Synthesis and Electrochemical Performance of Lithium Iron Phosphate/Carbon Composites Based on Controlling the Secondary Morphology of Precursors. *Int. J. Hydrogen Energy* **2020**, *45*, 33016–33027. [CrossRef]
14. Jiang, D.; Zhang, X.; Zhao, T.; Liu, B.; Yang, R.; Zhang, H.; Fan, T.; Wang, F. An Improved Synthesis of Iron Phosphate as a Precursor to Synthesize Lithium Iron Phosphate. *Bull. Mater. Sci.* **2020**, *43*, 50. [CrossRef]
15. Zhang, X.; Zhang, L.; Liu, H.; Cao, B.; Liu, L.; Gong, W. Structure, Morphology, Size and Application of Iron Phosphate. *Rev. Adv. Mater. Sci.* **2020**, *59*, 538–552. [CrossRef]
16. Padhi, A.K.; Nanjundaswamy, K.S.; Masquelier, C.; Okada, S.; Goodenough, J.B. Effect of Structure on the Fe^{3+}/Fe^{2+} Redox Couple in Iron Phosphates. *J. Electrochem. Soc.* **1997**, *144*, 1609–1613. [CrossRef]
17. Song, Y.; Zavalij, P.Y.; Suzuki, M.; Whittingham, M.S. New Iron (III) Phosphate Phases: Crystal Structure and Electrochemical and Magnetic Properties. *Inorg. Chem.* **2002**, *41*, 5778–5786. [CrossRef]
18. Zhang, T.; Lu, Y.; Luo, G. Size Adjustment of Iron Phosphate Nanoparticles by Using Mixed Acids. *Ind. Eng. Chem. Res.* **2013**, *52*, 6962–6968. [CrossRef]
19. Lindsay, W.L.; Moreno, E.C. Phosphate Phase Equilibria in Soils. *Soil Sci. Soc. Am. J.* **1960**, *24*, 177–182. [CrossRef]
20. Chang, S.C.; Jackson, M.L. Solubility Product of Iron Phosphate. *Soil Sci. Soc. Am. J.* **1957**, *21*, 265–269. [CrossRef]
21. Lou, W.; Zhang, Y.; Zhang, Y.; Zheng, S.; Sun, P.; Wang, X.; Qiao, S.; Li, J.; Zhang, Y.; Liu, D.; et al. A Facile Way to Regenerate FePO$_4$·2H$_2$O Precursor from Spent Lithium Iron Phosphate Cathode Powder: Spontaneous Precipitation and Phase Transformation in an Acidic Medium. *J. Alloys Compd.* **2021**, *856*, 158148. [CrossRef]
22. Scholz, F.; Kahlert, H. *Chemical Equilibria in Analytical Chemistry: The Theory of Acid–Base, Complex, Precipitation and Redox Equilibria*; Springer International Publishing: Cham, Switzerland, 2019. [CrossRef]
23. Iuliano, M.; Ciavatta, L.; De Tommaso, G. On the Solubility Constant of Strengite. *Soil Sci. Soc. Am. J.* **2007**, *71*, 1137–1140. [CrossRef]
24. Lindsay, W.L.; Vlek, P.L.G.; Chien, S.H. Phosphate Minerals. In *SSSA Book Series*; Dixon, J.B., Weed, S.B., Eds.; Soil Science Society of America: Madison, WI, USA, 2018; pp. 1089–1130. [CrossRef]
25. Gongyan, W.; Li, L.; Fang, H. Dehydration of FePO$_4$·2H$_2$O for the Synthesis of LiFePO$_4$/C: Effect of Dehydration Temperature. *Int. J. Electrochem. Sci.* **2018**, *13*, 2498–2508. [CrossRef]
26. Zaghib, K.; Julien, C.M. Structure and Electrochemistry of FePO$_4$·2H$_2$O Hydrate. *J. Power Sources* **2005**, *142*, 279–284. [CrossRef]
27. Fan, E.; Li, L.; Zhang, X.; Bian, Y.; Xue, Q.; Wu, J.; Wu, F.; Chen, R. Selective Recovery of Li and Fe from Spent Lithium-Ion Batteries by an Environmentally Friendly Mechanochemical Approach. *ACS Sustain. Chem. Eng.* **2018**, *6*, 11029–11035. [CrossRef]
28. Boonchom, B.; Puttawong, S. Thermodynamics and Kinetics of the Dehydration Reaction of FePO$_4$·2H$_2$O. *Phys. B Condens. Matter.* **2010**, *405*, 2350–2355. [CrossRef]
29. Xia, S.; Li, F.; Chen, F.; Guo, H. Preparation of FePO$_4$ by Liquid-Phase Method and Modification on the Surface of LiNi$_{0.80}$Co$_{0.15}$Al$_{0.05}$O$_2$ Cathode Material. *J. Alloys Compd.* **2018**, *731*, 428–436. [CrossRef]

30. Ma, Y.; Shen, W.; Yao, Y. Preparation of Nanoscale Iron (III) Phosphate by Using Ferro-Phosphorus as Raw Material. *IOP Conf. Ser. Earth Environ. Sci.* **2019**, *252*, 022032. [CrossRef]
31. Tejedor-Tejedor, M.I.; Anderson, M.A. The Protonation of Phosphate on the Surface of Goethite as Studied by CIR-FTIR and Electrophoretic Mobility. *Langmuir* **1990**, *6*, 602–611. [CrossRef]
32. Arai, Y.; Sparks, D.L. ATR–FTIR Spectroscopic Investigation on Phosphate Adsorption Mechanisms at the Ferrihydrite–Water Interface. *J. Colloid Interface Sci.* **2001**, *241*, 317–326. [CrossRef]
33. *Infrared and Raman Spectra of Inorganic and Coordination Compounds*, 1st ed.; John Wiley & Sons, Ltd.: London, UK, 2008. [CrossRef]
34. Khachani, M.; El Hamidi, A.; Kacimi, M.; Halim, M.; Arsalane, S. Kinetic Approach of Multi-Step Thermal Decomposition Processes of Iron(III) Phosphate Dihydrate $FePO_4 \cdot 2H_2O$. *Thermochim. Acta* **2015**, *610*, 29–36. [CrossRef]
35. Chapman, A.C.; Thirlwell, L.E. Spectra of Phosphorus Compounds—I the Infra-Red Spectra of Orthophosphates. *Spectrochim. Acta* **1964**, *20*, 937–947. [CrossRef]
36. Roncal-Herrero, T.; Rodríguez-Blanco, J.D.; Benning, L.G.; Oelkers, E.H. Precipitation of Iron and Aluminum Phosphates Directly from Aqueous Solution as a Function of Temperature from 50 to 200 °C. *Cryst. Growth Des.* **2009**, *9*, 5197–5205. [CrossRef]
37. Özen, M.; Mertens, M.; Snijkers, F.; Cool, P. Hydrothermal Synthesis and Formation Mechanism of Tetragonal Barium Titanate in a Highly Concentrated Alkaline Solution. *Ceram. Int.* **2016**, *42*, 10967–10975. [CrossRef]
38. Málek, J. The Applicability of Johnson-Mehl-Avrami Model in the Thermal Analysis of the Crystallization Kinetics of Glasses. *Thermochim. Acta* **1995**, *267*, 61–73. [CrossRef]

Article

Electroextraction of Ytterbium on the Liquid Lead Cathode in LiCl-KCl Eutectic

Zhuyao Li [1,2], Liandi Zhu [2], Dandan Tang [1,*], Ying Dai [2], Feiqiang He [2], Zhi Gao [2], Cheng Liu [2], Hui Liu [2], Limin Zhou [2], Zhirong Liu [1,2] and Jinbo Ouyang [2,*]

[1] State Key Laboratory of Nuclear Resources and Environment, East China University of Technology, Nanchang 330013, China
[2] School of Chemistry, Biological and Materials Sciences, East China University of Technology, Nanchang 330013, China
* Correspondence: tangdandan@ecut.edu.cn (D.T.); ouyangjinbo@ecut.edu.cn (J.O.)

Abstract: The reduction mechanisms of Yb(III) on W electrodes in molten LiCl-KCl-YbCl$_3$ were explored at 773 K, and the diffusion coefficient of Yb(III) was determined. Then, various electrochemical techniques were employed to investigate the electroreduction of Yb(III) in molten LiCl-KCl on a liquid Pb film and Pb electrode. Electrochemical signals were associated with forming Pb$_3$Yb, PbYb, Pb$_3$Yb$_5$, and PbYb$_2$. The deposition potentials and equilibrium potentials of four Pb-Yb intermetallics were obtained through open-circuit chronopotentiometry. Metallic Yb was extracted by potentiostatic electrolysis (PE) on a liquid Pb electrode, and XRD analyzed the Pb-Yb alloy obtained at different extraction times. The recovered Yb was found in the form of Pb$_3$Yb and PbYb intermetallics. The extraction efficiency of Yb was calculated according to ICP analysis results, and extraction effectivity could attain 94.5% via PE at −1.86 V for 14 h.

Keywords: electrochemical extraction; ytterbium; liquid Pb electrode; Pb-Yb alloy; electrochemical behavior

---

Citation: Li, Z.; Zhu, L.; Tang, D.; Dai, Y.; He, F.; Gao, Z.; Liu, C.; Liu, H.; Zhou, L.; Liu, Z.; et al. Electroextraction of Ytterbium on the Liquid Lead Cathode in LiCl-KCl Eutectic. *Crystals* **2022**, *12*, 1453. https://doi.org/10.3390/cryst12101453

Academic Editor: Stefano Carli

Received: 15 September 2022
Accepted: 11 October 2022
Published: 14 October 2022

Publisher's Note: MDPI stays neutral with regard to jurisdictional claims in published maps and institutional affiliations.

1. Introduction

The increasing demand of human beings for energy and the promotion of global low-carbon energy has increased the number of nuclear power stations in the world [1,2]. However, the management, storage, and disposal of used nuclear fuel (UNF) discharged from nuclear power plants has become a worldwide issue [3,4]. A 1000 MWe nuclear reactor generates about 25–30 tons of UNF annually [5,6]. Based on the International Atomic Energy Agency (IAEA) estimate, around 450 operating nuclear power plants have generated a global inventory of some 270,000 metric tons of UNF [7,8]. Pyroprocessing is one of the most promising technologies for treating UNF, and electrorefining is an essential unit operation in the pyroprocessing technique [9–12].

In electrorefining of UNF in molten LiCl-KCl, the standard potentials of lanthanide elements (Lns) are more negative than those of actinide elements (Ans) [13–15]. U and transuranic elements (TRUs) are deposited onto the cathode, and various lanthanide fission products contained in UNF dissolve in the molten salt. During the electrolysis process, Lns accumulate, which makes difficult the separation of Ans and Lns [16–23]. Therefore, proper electrode materials have a good effect on the separation process of Ans and Lns. Compared with the solid cathode, the liquid metal cathode has been widely studied because of its constant surface area and easy diffusion of deposits to liquid metal. Jiao et al. [24] summarized liquid metals from the aspects of properties, depolarization effects, alloy preparation, etc., such as liquid Bi [25–28], Pb [29–32], Sn [33–36], and Ga [37–40], which provided a reference for molten salt electrorefining.

In addition, lead has a low melting point (327 °C), and through evaluating the vapor pressures of various liquid metals, we discovered that the order of separating Ans and Lns from the liquid metal cathode via distillation is: Cd > Zn > Pb > Bi > Al > Ga > Sn. Liquid

---

Pb appears to be a proper candidate for electrode material in separating and extracting Ans and Lns from molten salt by electrowinning. After electrowinning, a Pb-Ans/Lns alloy can be formed on the liquid Pb electrode. Finally, the cathode deposits can be separated from the liquid Pb cathode by Pb distillation. Therefore, lead was employed as the cathode for ytterbium extraction in this work. In our previous work, liquid lead was used as a cathode due to the underpotential deposition of Dy(III) on the liquid Pb electrode. The reduction potential of Dy on the liquid Pb electrode is -1.35 V (vs. Ag/AgCl), which is more positive than that on the W electrode. Dysprosium was extracted from molten LiCl-KCl-DyCl$_3$, and the extraction efficiency of dysprosium reached 97.2% [41,42]. Ytterbium is a variable valence element, one of the typical rare earth fission elements. Ji et al. [43] studied the electrochemical properties of ytterbium on liquid Zn electrodes in LiCl-KCl melts. Li et al. [44] studied the electrochemical behavior of Yb(III) on Cu electrodes in LiCl-KCl melts. Yan et al. [45] researched ytterbium and zinc co-deposition in LiCl-KCl-ZnCl$_2$-YbCl$_3$ melts on a W electrode. However, there is little information about the extraction of ytterbium from molten chlorides on liquid lead cathodes.

In the present study, different electrochemical techniques were used to explore the reduction mechanisms of Yb(III) on W, liquid Pb film, and liquid Pb electrodes. Moreover, the recovery of Yb from molten salt was executed by PE, and the products were characterized by SEM-EDS and XRD.

2. Experimental

All electrochemical measurements were carried out in an alumina crucible with LiCl-KCl (55.8 wt%) melts in an electric furnace. Yb was introduced into the LiCl-KCl melts as dehydrated YbCl$_3$ powder (analytical grade). The reference electrode consisted of a silver wire (1 mm in diameter) immersed in a 1.5 wt% AgCl solution. A spectral pure graphite rod (6 mm in diameter) was used as a counter electrode. A tungsten wire (1 mm in diameter, 99.95%), a Pb film electrode (Pb coated on a W wire), and liquid lead (99%) were used for the working electrodes. PE and electrochemical measurements were performed (Metrohm, Ltd., Autolab PGSTAT 302N with Nova 1.10 software). The resistance between the reference electrode and the working electrode in molten salt was calculated by electrochemical impedance spectroscopy(EIS). The EIS was performed under the open circuit potential of the LiCl-KCl molten salt system under the condition of 10 mV amplitude over a 0.01-10000 Hz frequency range, as shown in Figure 1. According to Figure 1, the solution resistance(R$_s$) of the molten LiCl-KCl was 0.62 Ω de-termined at 773 K. After the ohmic potential(IR) compensation, cyclic voltammograms (CVs) were obtained. The phase composition of extraction products was characterized using XRD (Philips, Amsterdam, The Netherlands). The metal ion concentration in the melts was analyzed by ICP-AES (USA Thermo Elemental, IRIS Intrepid II XSP). Figure 2 shows the schematic setup of three-electrode electrochemical cells inside the furnace.

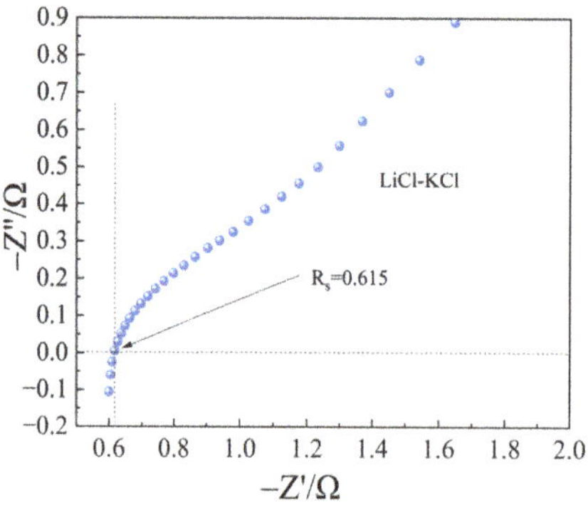

Figure 1. EIS data on the W electrode in molten LiCl-KCl at temperatures of 773 K.

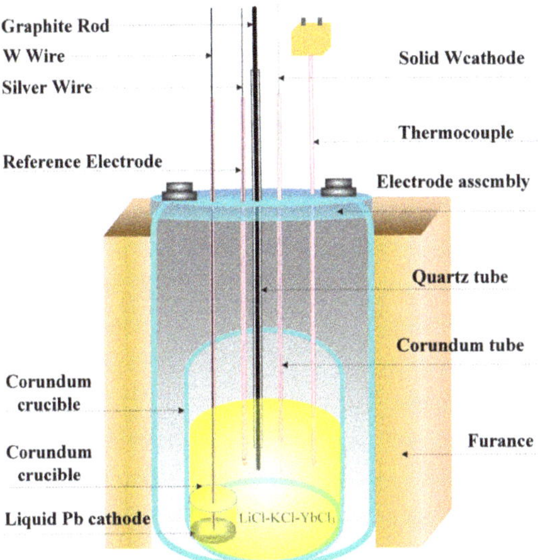

Figure 2. The schematic setup of three-electrode electrochemical cells inside the furnace.

3. Results and Discussion

3.1. Electrode Reaction of Yb(III) on the W Electrode

Figure 3 shows the CVs attained in LiCl-KCl and LiCl-KCl-YbCl$_3$ melts on the W electrode. The redox couples R_A/O_A ($-2.39/-2.28$ V) are considered as the redox of Li(I)/Li(0) on the W electrode. When YbCl$_3$ is added, the newly added electrochemical signal R_B/O_B ($-0.47/-0.32$ V) corresponds to Yb(III)/Yb(II). No electrochemical signal of Yb(II) reduction to Yb was found, which indicates that the reduction potential of Yb(II) is more negative than that of Li (I) [43,44]. In this system, Yb(III) first becomes a low valence ion and is then oxidized to a high valence state. No metal Yb was obtained on the W electrode. They are thought to correspond to the following reaction.

$$\mathrm{Yb(III)} + e^- \rightarrow \mathrm{Yb(II)} \tag{1}$$

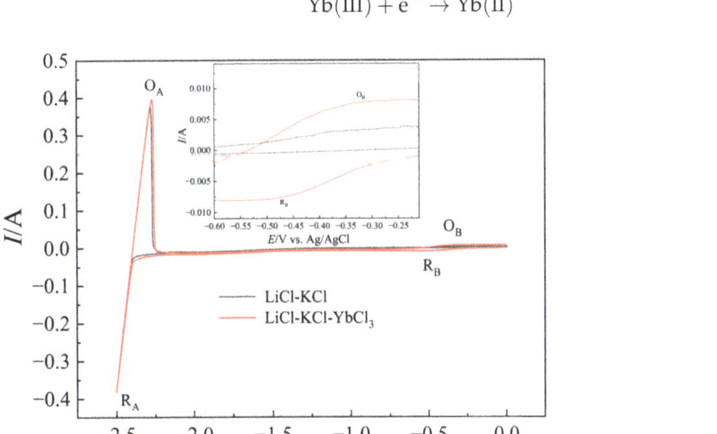

Figure 3. CVs obtained on the W electrode (S = 0.314 cm^2) in the absence and presence of YbCl$_3$ (1.55×10^{-4} mol cm^{-3}) in eutectic LiCl-KCl.

Typical CVs of molten LiCl-KCl-YbCl$_3$ recorded on the W electrodes at various scan rates are shown in Figure 4a. What can be observed is that with an increase in scan rate, the peak potential for R$_B$ was constant (Figure 4b). Therefore, it can be decided that the electrode reduction reaction R$_B$ is a reversible process on scan rate in a range of 0.08–0.28 V s^{-1}. Figure 4b illustrates the linear relationship of I_p and $v^{-1/2}$ of Yb(III) in molten LiCl-KCl-YbCl$_3$, indicating that the redox reaction is controlled by diffusion. In a soluble/soluble system, for the chemical reaction controlled by the mass transfer rate step, there is a relationship between the cathode peak current and the square root of the sweep rate, which can be described by Randles–Sevcik equation as follows [43]:

$$I_P = 0.4463(nF)^{3/2}(RT)^{-1/2}SCD^{1/2}v^{1/2} \tag{2}$$

where S represents the surface area of the working electrode, C_0 represents the bulk concentration of the Yb(III) ion, D corresponds to the diffusion coefficient, I corresponds to the applied current, n corresponds to the number of exchanged electrons, F corresponds to the Faraday constant, and v represents the potential scan rate.

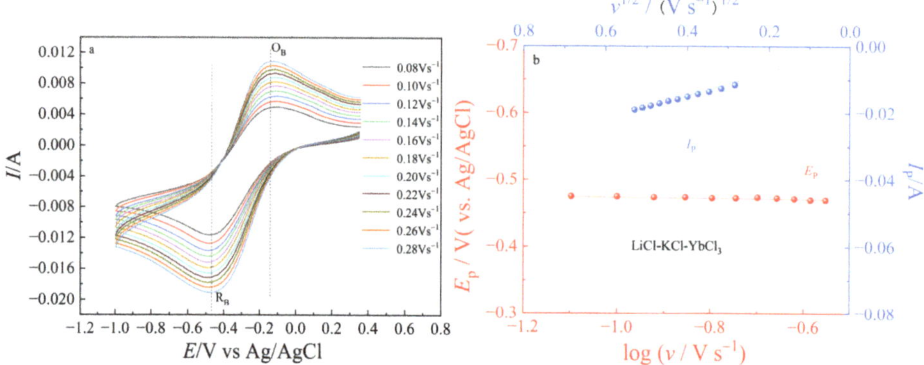

Figure 4. (**a**) CVs obtained on the W electrode at different scan rates in molten LiCl-KCl-YbCl$_3$; (**b**) peak potential of the cathodic R$_B$ vs. log$v$ derived from (**a**); peak current of the cathodic R$_B$ vs. the square root of scan rate derived from (**a**).

The calculated diffusion coefficient of Yb(III) at 773 K was 1.86×10^{-5} cm^2 s^{-1}. The diffusion coefficient of the Yb(III) ion in LiCl KCl molten salt calculated by Smolenski et al. at 848 K is 2.7×10^{-5} cm^2 s^{-1}, which is consistent with our research results [46].

Square wave voltammetry (SWV) was carried out to measure the number of electrons transferred during the reduction process of Yb(III) on the W electrode.

Figure 5a depicts the SWV curves gained in LiCl-KCl-YbCl$_3$ molten salt at different frequencies. One reduction signal, R$_B$ (-0.47 V), correlated with the reduction process of Yb(III) to Yb(II). Furthermore, the shape of R$_B$ was flat and symmetrical, which indicates that the electrode reactions were reversible in a soluble/soluble system. The linear relationship between I_{pc} and $f^{1/2}$ suggests that the reduction of Yb(III) on the W electrode was controlled by diffusion. Therefore, the number of electrons transferred was calculated from $W_{1/2}$ using Equation (3) [42]:

$$W_{1/2} = 3.52\frac{RT}{nF} \tag{3}$$

where $W_{1/2}$ is the half-peak width of the SWV curve, and the calculated number of electrons transferred from the reduction of Yb(III) to Yb(II) is close to 1 (Table 1).

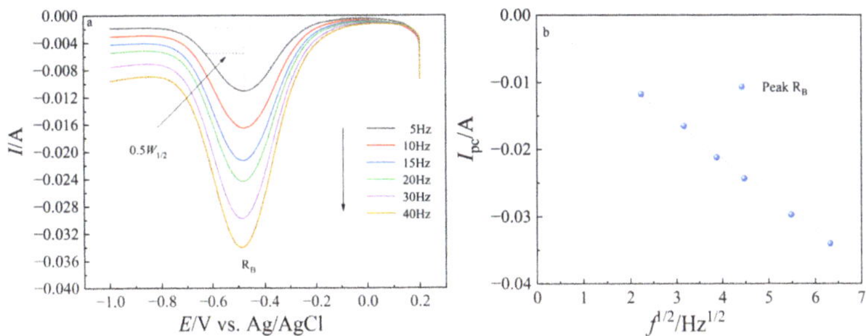

Figure 5. (**a**) SWV curves for the reduction of Yb(III) on the W electrode at 773 K; pulse height: 100 mV; step potential: 5 mV; frequency: 5–40 Hz. (**b**) The linear relationship of I_p versus $f^{1/2}$.

Table 1. Half-peak width and transfer electron number of Yb(III) calculated on a liquid Pb electrode in molten LiCl-KCl-YbCl$_3$ under different frequencies.

f/Hz	$W_{1/2}$/V	n
5	0.259	0.907
10	0.251	0.934
15	0.251	0.934
20	0.243	0.963
30	0.243	0.963
40	0.243	0.963

3.2. Electrochemical Properties of YbCl$_3$ on a Pb Film Electrode

The Pb film was prepared on the surface of the W electrode through PE at -1.0 V for 1 s. Figure 6 shows the CVs on the W electrode before and after adding YbCl$_3$ and PbCl$_2$ to LiCl KCl molten salt. The peaks R$_B$ and R$_C$ in the inset of Figure 6 are related to the reduction of Pb(II) to Pb and Yb(III) to Yb(II), respectively. Based on the binary phase diagram of the Pb-Yb system [47] shown in Figure 7, there are four intermetallics (Pb$_3$Yb, PbYb, Pb$_3$Yb$_5$, and PbYb$_2$) in the study temperature range. Therefore, four reduction peaks (R$_I$, R$_{II}$, R$_{III}$, and R$_{IV}$) in the inset of Figure 6 correspond to four different Pb-Yb intermetallics. Pb(II) is first deposited on the W electrode to form a Pb film electrode. Then, Yb(II) is reduced at the corrected potential on the Pb film electrode due to depolarization to form Pb-Yb intermetallics. The reaction is as follows:

$$\mathrm{Yb(II)} + 2e^- + x\mathrm{Pb} = \mathrm{Pb_x Yb} \tag{4}$$

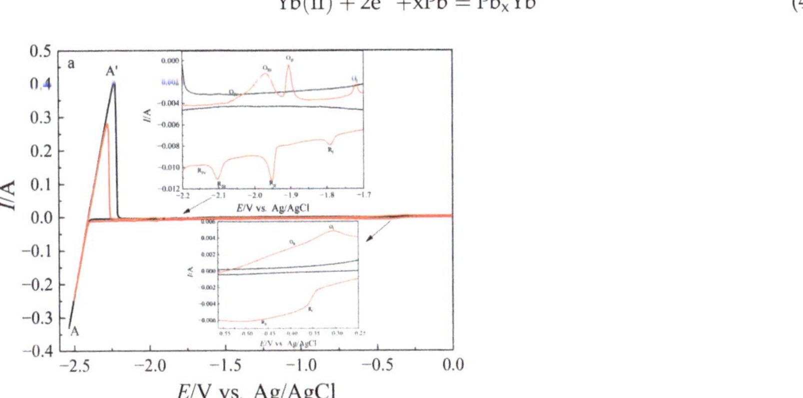

Figure 6. CVs of blank eutectic LiCl-KCl (black line) and molten LiCl-KCl-YbCl$_3$-PbCl$_2$ salts (red line) on the W electrode at 773 K.

reduced to Yb. One signal that appears at −1.71 V corresponds to the formation of a solid Yb(Pb) solution.

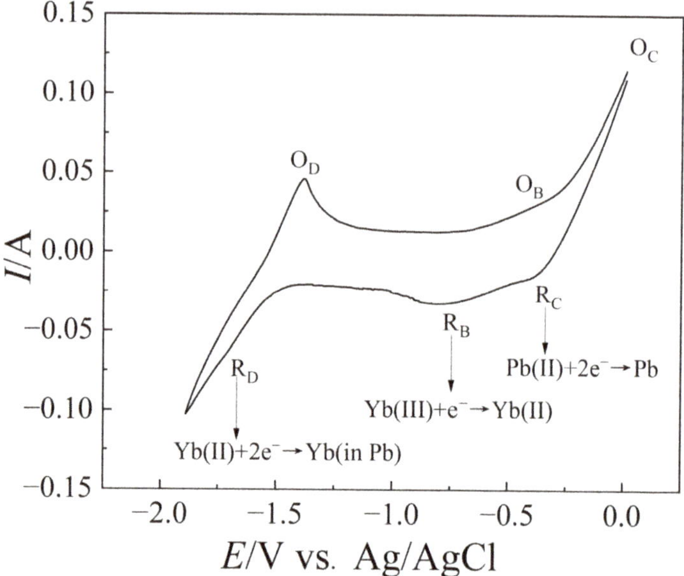

Figure 10. CVs obtained in LiCl-KCl-YbCl$_3$ (1.55×10^{-4} mol cm^{-3}) melt on liquid Pb cathodes (S = 0.949 cm^2) at 773 K.

According to the CV results above, Yb was extracted by PE at −1.86 V on a liquid Pb electrode. Figure 11 shows the alloy's XRD pattern obtained by electrolysis at constant potential −1.86 V for 6, 10, and 14 h, respectively, on the liquid Pb electrode in molten LiCl-KCl-YbCl$_3$ at 773 K. The alloy electrolyzed for 6 h was the Pb$_3$Yb phase, the alloy electrolyzed for 10 h was the Pb$_3$Yb phase, and the alloys electrolyzed for 14 h were Pb$_3$Yb and PbYb phases. This shows that with the increase of electrolysis time, the content of Yb in liquid lead gradually increased, and part of Pb$_3$Yb began to transform into the PbYb phase.

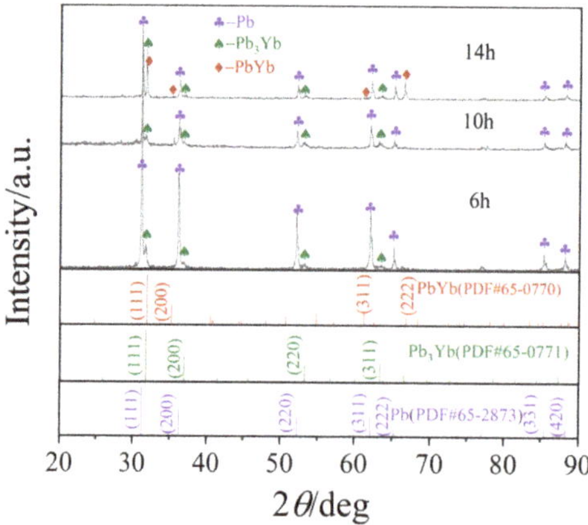

Figure 11. XRD patterns of the cathode products on a liquid Pb electrode in molten LiCl-KCl-YbCl$_3$ salts through PE at −1.86 V for 14 h at 773 K.

After electrolysis for 14 h, the alloy was analyzed by SEM-EDS, and the SEM photograph and element distribution surface scan of the alloy were obtained, as shown in Figure 12. Two zones, dark and bright grey, were on the deposit's surface. From the results of the EDS mapping of the sample (Figure 12b), the element Yb mainly distributed in the dark grey zone. The upper salt was detected by inductively coupled plasma atomic emission spectrometry (ICP-AES). It was calculated that after electrolytic extraction for 14 h, the extraction rate of Yb was 94.5%.

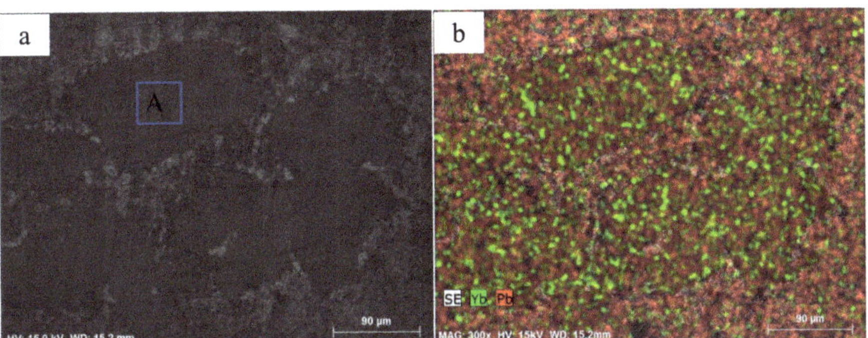

Figure 12. SEM image (**a**) and EDS mapping (**b**) analysis of the deposit obtained by PE at -1.86 V on a liquid Pb electrode in molten LiCl-KCl-YbCl$_3$ salts for 14 h at 773 K.

4. Conclusions

In this work, the electroreduction of Yb(III) ions on the W cathode was explored through CV and SWV. The electroreduction of Yb(III) to Yb(II) was found to be a diffusion-controlled process with one electron exchanged, and the reduction of Yb(III)/Yb(II) was a reversible process. The diffusion coefficient of Yb(III) was 1.86×10^{-5} cm^2 s^{-1}. The electrochemical behavior of Yb(III) on a liquid Pb film cathode was explored by CV, SWV, and OCP. Then, equilibrium potentials and the deposition potentials of four Pb-Yb intermetallics (Pb$_3$Yb, PbYb, Pb$_3$Yb$_5$, and PbYb$_2$) were determined. The reduction of Yb(III) on liquid Pb and liquid Pb film electrodes was proven to be a two-step mechanism.

At last, the electrochemical extraction of Yb(III) on a liquid Pb electrode was studied by CV. Meanwhile, ytterbium was extracted on a liquid Pb electrode from a LiCl-KCl melt by PE at -1.86 V. The extraction product was detected by SEM-EDS and XRD. XRD analysis of the extracted products obtained at different electrolysis times shows that the extracted products gradually changed from a Pb-rich alloy (Pb$_3$Yb) to a Yb-rich alloy (Pb$_3$Yb and PbYb) with the increase in electrolysis extraction time. The extraction efficiency of Yb reached up to 94.5% by PE at -1.86 V for 14 h, which indicates that it is feasible to electrolytically extract Yb from LiCl-KCl melts on a liquid Pb electrode.

Author Contributions: Conceptualization, Investigation, Writing—original draft, Z.L. (Zhuyao Li); Investigation, Software, L.Z. (Liandi Zhu); Methodology, Resources, D.T. Supervision, Y.D.; Formal analysis, Investigation, F.H. and Z.G.; Data curation, C.L. and H.L.; Supervision, L.Z. (Limin Zhou) and Z.L. (Zhirong Liu); Project administration, J.O. All authors have read and agreed to the published version of the manuscript.

Funding: The paper was financially co-supported by the National Natural Science Foundation of China (11905029, 22176032, and 11875105) and the Independent fund of State Key Laboratory of Nuclear Resources and Environment of East China University of Technology (2020Z19). Academic and technical leader training program for major disciplines in Jiangxi Province (20212BCJ23001).

Conflicts of Interest: The authors declare no conflict of interest.

References

1. Karley, D.; Shukla, S.K.; Rao, T.S. Microbiological assessment of spent nuclear fuel pools: An in-perspective review. *J. Environ. Chem. Eng.* **2022**, *10*, 108050. [CrossRef]
2. Grebennikova, T.; Jones, A.N.; Sharrad, C.A. Electrochemical decontamination of irradiated nuclear graphite from corrosion and fission products using molten salt. *Energ. Environ. Sci.* **2021**, *14*, 5501–5512. [CrossRef]
3. Du, Y.; Tang, H.; Zhang, D.; Shao, L.; Li, Y.; Gao, R.; Yang, Z.; Li, B.; Chu, M.; Liao, J. Electro-reduction processes of U_3O_8 to metallic U bulk in LiCl molten salt. *J. Nucl. Mater.* **2021**, *543*, 152627. [CrossRef]
4. Lichtenstein, T.; Nigl, T.P.; Smith, N.D.; Kim, H. Electrochemical deposition of alkaline-earth elements (Sr and Ba) from LiCl-KCl-$SrCl_2$-$BaCl_2$ solution using a liquid bismuth electrode. *Electrochim. Acta* **2018**, *281*, 810–815. [CrossRef]
5. Kurniawan, T.A.; Othman, M.H.D.; Singh, D.; Avtar, R.; Hwang, G.H.; Setiadi, T.; Lo, W.H. Technological solutions for long-term storage of partially used nuclear waste: A critical review. *Ann. Nucl. Energy* **2022**, *166*, 108736. [CrossRef]
6. Zohuri, B. Nuclear Fuel Cycle and Decommissioning. In *Nuclear Reactor Technology Development and Utilization*; Elsevier: Amsterdam, The Netherlands, 2020; pp. 61–120.
7. Taylor, R.; Bodel, W.; Stamford, L.; Butler, G. A Review of Environmental and Economic Implications of Closing the Nuclear Fuel Cycle—Part One: Wastes and Environmental Impacts. *Energies* **2022**, *15*, 1433. [CrossRef]
8. Burek, J.; Nutter, D. Life cycle assessment of grocery, perishable, and general merchandise multi-facility distribution center networks. *Energy Build.* **2018**, *174*, 388–401. [CrossRef]
9. Zhong, Y.K.; Liu, Y.L.; Liu, K.; Wang, L.; Mei, L.; Gibson, J.K.; Chen, J.Z.; Jiang, S.L.; Liu, Y.C.; Yuan, L.Y. In-situ anodic precipitation process for highly efficient separation of aluminum alloys. *Nat. Commun.* **2021**, *12*, 1–6. [CrossRef]
10. Zhang, Y.; Song, J.; Li, X.; Yan, L.; Shi, S.; Jiang, T.; Peng, S. First principles calculation of redox potential for tetravalent actinides in molten LiCl–KCl eutectic based on vertical substitution and relaxation. *Electrochim. Acta* **2019**, *293*, 466–475. [CrossRef]
11. Wang, Y.; Quan, M.; Zhang, S.; Liu, Y.; Wang, Y.; Dai, Y.; Dong, Z.; Cheng, Z.; Zhang, Z.; Liu, Y. Electrochemical extraction of gadolinium on Sn electrode and preparation of Sn-Gd intermetallic compounds in LiCl-KCl melts. *J. Alloy. Compd.* **2022**, *907*, 164220. [CrossRef]
12. Han, W.; Li, Z.; Li, M.; Li, W.; Zhang, M.; Yang, X.; Sun, Y. Reductive extraction of lanthanides (Ce, Sm) and its monitoring in LiCl-KCl/Bi-Li system. *J. Nucl. Mater.* **2019**, *514*, 311–320. [CrossRef]
13. Im, S.; Smith, N.D.; Baldivieso, S.C.; Gesualdi, J.; Liu, Z.K.; Kim, H. Electrochemical recovery of Nd using liquid metals (Bi and Sn) in LiCl-KCl-$NdCl_3$. *Electrochim. Acta* **2022**, *425*, 140655. [CrossRef]
14. Williamson, M.; Willit, J. Pyroprocessing flowsheets for recycling used nuclear fuel. *Nucl. Eng. Technol.* **2011**, *43*, 329–334. [CrossRef]
15. Liu, K.; Chai, Z.F.; Shi, W.Q. Liquid Electrodes for An/Ln Separation in Pyroprocessing. *J. Electrochem. Soc.* **2021**, *168*, 032507. [CrossRef]
16. Solbrig, C.; Westphal, B.; Johnson, T.; Li, S.; Marsden, K.; Goff, K. *Pyroprocessing Progress at Idaho National Laboratory*; Idaho National Lab. (INL): Idaho Falls, ID, USA, 2007.
17. Wang, D.D.; Liu, Y.L.; Yang, D.W.; Zhong, Y.K.; Han, W.; Wang, L.; Chai, Z.F.; Shi, W.Q. Separation of uranium from lanthanides (La, Sm) with sacrificial Li anode in LiCl-KCl eutectic salt. *Sep. Purif. Technol.* **2022**, *292*, 121025. [CrossRef]
18. Han, W.; Li, Z.; Li, M.; Li, W.; Zhang, X.; Yang, X.; Zhang, M.; Sun, Y. Electrochemical extraction of holmium and thermodynamic properties of Ho-Bi alloys in LiCl-KCl eutectic. *J. Electrochem. Soc.* **2017**, *164*, E62. [CrossRef]
19. Han, W.; Li, Z.; Li, M.; Hu, X.; Yang, X.; Zhang, M.; Sun, Y. Electrochemical behavior and extraction of holmium on Cu electrode in LiCl-KCl molten salt. *J. Electrochem. Soc.* **2017**, *164*, D934. [CrossRef]
20. Jiang, S.; Lan, J.; Wang, L.; Liu, Y.; Zhong, Y.; Liu, Y.; Yuan, L.L.Y.; Zheng, L.; Chai, Z.; Shi, W. Competitive Coordination of Chloride and Fluoride Anions Towards Trivalent Lanthanide Cations (La^{3+} and Nd^{3+}) in Molten Salts. *Chem. Eur. J.* **2021**, *27*, 11721–11729. [CrossRef] [PubMed]
21. Jiang, S.; Liu, Y.; Wang, L.; Chai, Z.; Shi, W.Q. The Coordination Chemistry of f-Block Elements in Molten Salts. *Chem. Eur. J.* **2022**, e202201145. [CrossRef] [PubMed]
22. Liu, Y.; Liu, Y.; Wang, L.; Jiang, S.; Zhong, Y.; Wu, Y.; Li, M.; Shi, W. Chemical Species Transformation during the Dissolution Process of U_3O_8 and UO_3 in the LiCl–KCl–AlCl3 Molten Salt. *Inorg. Chem.* **2022**, *61*, 6519–6529. [CrossRef]
23. Liu, Y.C.; Liu, Y.L.; Zhao, Y.; Liu, Z.; Zhou, T.; Zou, Q.; Zeng, X.; Zhong, Y.K.; Li, M.; Sun, Z.X. A simple and effective separation of UO2 and Ln2O3 assisted by NH4Cl in LiCl–KCl eutectic. *J. Nucl. Mater.* **2020**, *532*, 152049. [CrossRef]
24. Jiao, S.Q.; Jiao, H.D.; Song, W.L.; Wang, M.Y.; Tu, J.G. A review on liquid metals as cathodes for molten salt/oxide electrolysis. *Int. J. Min. Met. Mater.* **2020**, *27*, 1588–1598. [CrossRef]
25. Yang, D.W.; Jiang, S.L.; Liu, Y.L.; Geng, J.S.; Li, M.; Wang, L.; Chai, Z.F.; Shi, W.Q. Electrochemical extraction kinetics of Nd on reactive electrodes. *Sep. Purif. Technol.* **2022**, *281*, 119853. [CrossRef]
26. Yin, T.; Liu, Y.; Yang, D.; Yan, Y.; Wang, G.; Chai, Z.; Shi, W. Thermodynamics and kinetics properties of lanthanides (La, Ce, Pr, Nd) on liquid bismuth electrode in LiCl-KCl molten salt. *J. Electrochem. Soc.* **2020**, *167*, 122507. [CrossRef]
27. Yin, T.; Liu, Y.; Jiang, S.; Yan, Y.; Wang, G.; Chai, Z.; Shi, W. Kinetic Properties and Electrochemical Separation of Uranium on Liquid Bismuth Electrode in LiCl–KCl Melt. *J. Electrochem. Soc.* **2021**, *168*, 032503. [CrossRef]
28. Jiang, S.; Liu, K.; Liu, Y.; Yin, T.; Chai, Z.; Shi, W. Electrochemical behavior of Th (IV) on the bismuth electrode in LiCl–KCl eutectic. *J. Nucl. Mater.* **2019**, *523*, 268–275. [CrossRef]

29. Han, W.; Li, W.; Chen, J.; Li, M.; Li, Z.; Dong, Y.; Zhang, M. Electrochemical properties of yttrium on W and Pb electrodes in LiCl–KCl eutectic melts. *RSC Adv.* **2019**, *9*, 26718–26728. [CrossRef] [PubMed]
30. Li, M.; Sun, Z.; Guo, D.; Han, W.; Sun, Y.; Yang, X.; Zhang, M. Electrode reaction of Pr (III) and coreduction of Pr (III) and Pb (II) on W electrode in eutectic LiCl-KCl. *Ionics* **2020**, *26*, 3901–3909. [CrossRef]
31. Han, W.; Wang, W.; Li, M.; Wang, D.; Li, H.; Chen, J.; Sun, Y. Electrochemical separation of La from LiCl-KCl fused salt by forming La-Pb alloys. *Sep. Purif. Technol.* **2021**, *275*, 119188. [CrossRef]
32. Han, W.; Wang, W.; Li, M.; Meng, Y.; Ji, W.; Sun, Y. Electrochemical coreduction of Gd (III) with Pb (II) and recovery of Gd from LiCl-KCl eutectic assisted by Pb metal. *J. Electrochem. Soc.* **2020**, *167*, 142505. [CrossRef]
33. Han, W.; Wang, W.; Li, M.; Wang, J.; Sun, Y.; Yang, X.; Zhang, M. Electrochemical behavior and extraction of zirconium on Sn-coated W electrode in LiCl-KCl melts. *Sep. Purif. Technol.* **2020**, *232*, 115965. [CrossRef]
34. Sun, C.; Xu, Q.; Yang, Y.; Zou, X.; Cheng, H.; Lu, X. Effect of the Li Reduction to Electrodeposition of Nd-Sn Compounds in Liquid Sn Electrode in LiCl-KCl-NdCl$_3$ Melt. *J. Electrochem. Soc.* **2021**, *168*, 102505. [CrossRef]
35. Han, W.; Wang, W.; Zhang, Y.; Wang, Y.; Li, M.; Sun, Y. Electrode reaction of Pr on Sn electrode and its electrochemical recovery from LiCl-KCl molten salt. *Int. J. Energ. Res.* **2021**, *45*, 8577–8592. [CrossRef]
36. Li, M.; Sun, Z.; Han, W.; Sun, Y.; Yang, X.; Wang, W. Electrochemical reaction of Sm (III) on liquid Sn electrode. *J. Electrochem. Soc.* **2020**, *167*, 022502. [CrossRef]
37. Liu, K.; Liu, Y.L.; Chai, Z.F.; Shi, W.Q. Evaluation of the electroextractions of Ce and Nd from LiCl-KCl molten salt using liquid Ga electrode. *J. Electrochem. Soc.* **2017**, *164*, D169. [CrossRef]
38. Yang, D.W.; Jiang, S.L.; Liu, Y.L.; Yin, T.Q.; Li, M.; Wang, L.; Luo, W.; Chai, Z.F.; Shi, W.Q. Electrodeposition Mechanism of La3+ on Al, Ga and Al-Ga Alloy Cathodes in LiCl-KCl Eutectic Salt. *J. Electrochem. Soc.* **2021**, *168*, 062511. [CrossRef]
39. Yang, D.W.; Liu, Y.L.; Yin, T.Q.; Li, M.; Han, W.; Chang, K.-K.; Chai, Z.-F.; Shi, W.Q. Co-reduction behaviors of Ce (III), Al (III) and Ga (III) on a W electrode: An exploration for liquid binary Al-Ga cathode. *Electrochim. Acta* **2019**, *319*, 869–877. [CrossRef]
40. Yang, D.W.; Liu, Y.L.; Yin, T.Q.; Jiang, S.L.; Zhong, Y.K.; Wang, L.; Li, M.; Chai, Z.F.; Shi, W.Q. Application of binary Ga–Al alloy cathode in U separation from Ce: The possibility in pyroprocessing of spent nuclear fuel. *Electrochim. Acta* **2020**, *353*, 136449. [CrossRef]
41. Li, Z.; Liu, Z.; Li, W.; Han, W.; Li, M.; Zhang, M. Electrochemical recovery of dysprosium from LiCl-KCl melt aided by liquid Pb metal. *Sep. Purif. Technol.* **2020**, *250*, 117124. [CrossRef]
42. Li, Z.; Tang, D.; Meng, S.; Gu, L.; Dai, Y.; Liu, Z. Electrolytic separation of Dy from Sm in molten LiCl-KCl using Pb-Bi eutectic alloy cathode. *Sep. Purif. Technol.* **2021**, *276*, 119045. [CrossRef]
43. Wang, P.; Han, W. Electrochemical and thermodynamic properties of ytterbium and formation of Zn–Yb alloy on liquid Zn electrode. *J. Nucl. Mater.* **2019**, *517*, 157–164. [CrossRef]
44. Li, M.; Liu, B.; Ji, N.; Sun, Y.; Han, W.; Jiang, T.; Peng, S.; Yan, Y.; Zhang, M. Electrochemical extracting variable valence ytterbium from LiCl–KCl–YbCl$_3$ melt on Cu electrode. *Electrochim. Acta* **2016**, *193*, 54–62. [CrossRef]
45. Zheng, J.; Yin, T.; Wang, P.; Yan, Y.; Smolenski, V.; Novoselova, A.; Zhang, M.; Ma, F.; Xue, Y. Electrochemical extraction of ytterbium from LiCl–KCl-YbCl$_3$-ZnCl$_2$ melt by forming Zn–Yb alloys. *J. Solid State Electrochem.* **2022**, *26*, 1067–1074. [CrossRef]
46. Smolenski, V.; Novoselova, A.; Osipenko, A.; Caravaca, C.; De Córdoba, G. Electrochemistry of ytterbium (III) in molten alkali metal chlorides. *Electrochim. Acta* **2008**, *54*, 382–387. [CrossRef]
47. Palenzona, A.; Cirafici, S. The Pb-Yb (Lead-Ytterbium) system. *J. Phase Equilib. Diff.* **1991**, *12*, 479–481. [CrossRef]

Article

Silica Nanoparticles-Induced Lysozyme Crystallization: Effects of Particle Sizes

Yuxiao Zhang, Xuntao Jiang, Xia Wu, Xiaoqiang Wang, Fang Huang, Kefei Li, Gaoyang Zheng, Shengzhou Lu, Yanxu Ma, Yuyu Zhou and Xiaoxi Yu *

State Key Laboratory of Heavy Oil Processing, College of Chemistry and Chemical Engineering, China University of Petroleum (East China), Qingdao 266580, China
* Correspondence: yuxiaoxi@upc.edu.cn; Tel.: +86-17606393651

Abstract: This study aimed to explore the effects of nucleate agent sizes on lysozyme crystallization. Silica nanoparticles (SNP) with four different particle sizes of 5 nm, 15 nm, 50 nm, and 100 nm were chosen for investigation. Studies were carried out both microscopically and macroscopically. After adding SNP, the morphological defects of lysozyme crystals decreased, and the number of crystals increases with the size of the SNP. The interaction between SNP and lysozyme was further explored using UV spectroscopy, fluorescence spectroscopy, and Zeta potential. It was found that the interaction between SNP and lysozyme was mainly electrostatic interaction, which increased with the size of SNP. As a result, lysozyme could be attracted to the surface of SNP and aggregated to form the nucleus. Finally, the activity test and circular dichroism showed that SNP had little effect on protein secondary structure.

Keywords: nanoparticle size; lysozyme crystallization; electrostatic interaction

Citation: Zhang, Y.; Jiang, X.; Wu, X.; Wang, X.; Huang, F.; Li, K.; Zheng, G.; Lu, S.; Ma, Y.; Zhou, Y.; et al. Silica Nanoparticles-Induced Lysozyme Crystallization: Effects of Particle Sizes. *Crystals* **2022**, *12*, 1623. https://doi.org/10.3390/cryst12111623

Academic Editors: Jingxiang Yang and Xin Huang

Received: 31 October 2022
Accepted: 9 November 2022
Published: 11 November 2022

Publisher's Note: MDPI stays neutral with regard to jurisdictional claims in published maps and institutional affiliations.

1. Introduction

Single crystal X-ray diffraction technology [1] is a primary technical means to analyze the atomic-level structure of biological macromolecules. The successful application of X-ray [2] crystallography in determining the 3-D structure of proteins relies on the cultivation of high-quality protein crystals [3–5]. However, protein crystallization is more complicated than small molecules due to the complexity and instability of protein molecules. Therefore, conducting in-depth research on the crystallization process of proteins is necessary.

Similar to small molecules, the nucleation process of protein crystallization [6,7] goes in two ways: homogeneous nucleation and heterogeneous nucleation [8]. Kordonskaya et al. [9–11] studied the growth of tetragonal crystals under homogeneous nucleation. It was found that lysozyme forms oligomers prior to crystal formation, which may be intermediates and can serve as growth units in crystal growth. Due to a high potential barrier of oligomer formation, homogeneous nucleation can only occur when the supersaturation in protein solution is high enough, which easily causes precipitates and is limited in low protein concentration systems. In contrast, when foreign substances exist as nucleating agents in the crystallization solution, the nucleation potential barrier will be significantly reduced due to protein adherence to the nucleating agent. As a result, heterogeneous nucleation could occur at relatively low protein concentrations. Therefore, it is beneficial for improving the quality of protein crystals and obtaining better diffraction data.

Nanomaterials refer to materials that have at least one dimension of nanometer size (1–100 nm) in three-dimensional space. Due to their excellent biocompatibility, nanomaterials have attracted much attention in studying the heterogeneous nucleation of proteins. Gold nanoparticles [12], platinum nanoparticles [13] and functionalized carbon nanoparticles [14] were proven to be effective in promoting protein crystallization and increasing crystal quality. Another commonly studied nanomaterial was silica nanoparticles (SNP), which were also confirmed to assist protein crystallization. Compared with other nanomaterials, SNP is cheaper and easier to synthesize. Therefore, SNP has better application potential for the research and industrialization of protein crystallization.

The size of nanoparticles is an important factor affecting its properties. Peukert et al. carried out lysozyme crystallization using silica nanoparticles (SNP) ranging from 10 to 200 nm. The addition of SNP brought about a clear extension in the crystallization window, especially when larger seed particles were used [12]. Interestingly, the work of Delmas investigated the effect of silica particles with sizes from 230 nm to 698 nm and found that the optimal nucleation occurred when particles sizes were 432 nm [15]. Up to now, the mechanism of SNP size on inducing an effect on protein crystallization is still unclear. In addition, the different preparation methods of SNP [16,17] will also lead to differences in their surface properties, affecting the interaction between SNP and proteins. Therefore, it is necessary to deeply explore the effect of silica with different particle sizes on protein crystallization.

In this work, SNP with sizes of 5 nm, 15 nm, 50 nm, and 100 nm were utilized as the nucleating agent, and lysozyme was chosen as the model protein. The influences of SNP sizes on lysozyme crystallization were studied using crystal morphology in the first place. To further reveal the intermolecular interactions between SNP and lysozyme, UV spectroscopy, fluorescence spectroscopy, and Zeta potential were applied. Finally, activity tests and circular dichroism were carried out to determine the effect of SNP on the secondary structure and function of proteins.

2. Materials and Methods

2.1. Materials

Hen egg white lysozyme (HEWL) with a purity of 99% was purchased from Beijing Solarbio Technology of China (Beijing, China) and used directly without further purification. *Micrococcus lysodeikticus* (ATCC No. 4698, Manassas, VA, USA), used for the lysozyme activity assay, was purchased from Sigma-Aldrich. Analytical grade reagents, such as acetic acid, sodium hydrate, sodium chloride, and anhydrous sodium acetate, were purchased from Sinopharm Chemical Reagent Co. Ltd. (Shanghai, China). SNP of different sizes were purchased from Shanghai Jiute Nano Material Technology Co. Ltd. (Shanghai, China). The sizes of SNP were evaluated using transmission electron microscopy (See Supplementary Materials Figure S1). The information of SNP is shown in Table 1. Ultrapure water was used in all experiments.

Table 1. Properties of the SNP used in this investigation.

Particle Size, nm	C_{SiO2}, mg/mL	ρ, g/mL	wt, %	pH
5	10	1.13	20	3–5
15	10	1.21	30	9–10
50	10	1.13	20	9–10
100	10	1.13	20	9–10

2.2. Lysozyme Crystallization

Lysozyme crystallization experiments were carried out in 96-well plates. The concentration of each crystallization solution contained: 15 or 20 mg/mL lysozyme and 3 wt.% sodium chloride. SNP with four different sizes were used as the nucleating agent, respectively, and the concentration was kept at 1 mg/mL. All materials were dissolved in a pH = 4.5, 0.05 M sodium acetate buffer. After gentle mixing, the 96-well plates were sealed with parafilm to prevent evaporation and placed in a refrigerator at T = 4 °C. After every 24 h, the plates were taken out and analyzed using a microscope.

2.3. UV Spectroscopy Experiment

UV spectroscopy was carried out using a lysozyme and SNP solution mixture. The UV spectrophotometer was preheated for 30 min before measuring. At ambient temperature, a

UV cell with a 1 cm light path was selected to measure the UV absorption spectrum of the prepared solution in the 200–800 nm band. Sodium acetate buffer was used for calibration before measuring.

2.4. Fluorescence Spectroscopy Experiment

The samples containing 10 mg/mL lysozyme and SNP with different sizes were incubated for 24 h. Afterward, the solution was added to a 1 cm light path cuvette. The scan mode was set to Emission, and the data mode was selected as Fluorescence [18]. The fluorescence emission spectrum of lysozyme was carried out with an excitation wavelength of 280 nm and a slit of 3 nm.

2.5. Zeta Potentials of Lysozyme and SNP

Zeta potential was measured for both lysozyme and SNP solutions. The concentrations were all 5 mg/mL. A DTS1060 sample cell was used for the measurement. The measurement parameters were the Zeta interface and manual measurement, respectively. The measurements were repeated several times until the data became stable.

2.6. Enzyme Activity Experiment

Lysozyme activity was measured using *Micrococcus lysodeikticus* as the substrate. A cuvette with an optical path length of 1 cm and a volume of 4 mL was used. A 2.5 mL bacterial suspension with an absorbance value (at 450 nm) of about 1.3 was added to the cuvette. Then, 200 µL of lysozyme solution (0.1 mg/mL) was added to the cuvette and mixed immediately. The absorbance values A_1 and A_2 at 450 nm at 1 min and 2 min were recorded at 25 °C. The activity of lysozyme (0.1 mg/mL) and SNP solutions with different particle sizes was also evaluated.

The enzymatic activity E_A was calculated by the following formula [19]:

$$E_A = \frac{\Delta E_{450nm}}{0.001 \times E_W} \tag{1}$$

where ΔE_{450nm} is the change of absorbance per minute at 450 nm, namely $| A_1 - A_2 |$; E_W is the mass of the original enzyme contained in the 0.5 mL detection enzyme solution, mg; 0.001 is a unit in which the absorbance drops; and E_A is the specific activity of the enzyme, with a unit of U/mg.

3. Results and Discussion

3.1. Effect of SNP Sizes on Lysozyme Crystal Morphology

The microscopy images of the lysozyme crystals obtained are shown in Figure 1. Lysozyme crystals were analyzed at different times. Within 24 h, the solution without SNP did not form crystals (Figure 1a). While in crystallization solutions with SNP, tiny crystals formed (Figure 1b–e). When the crystallization time was extended to 48 h and 72 h, the crystals in solutions containing SNP continued to grow (Figure 1g–j,l–o). No crystals could be observed under homogeneous nucleation conditions until 72 h (Figure 1k).

It can be further seen from Figure 1 that the number of crystals changed regularly with the SNP sizes. The numbers of crystals were estimated by image processing and are displayed in Figure 2. The number of crystals increased rapidly within 48 h. The addition of SNP could promote lysozyme nucleation, which agrees with the work of Yamazaki et al. [20]. After 48 h, the increase in the number of crystals slowed down due to the decrease in supersaturation. During the crystallization time investigated, the number of crystals increased with the size of SNP. The size distribution of lysozyme crystals after 48 h and 72 h was estimated and is shown in Figures S2 and S3. It can be found that the particle size of SNP showed no significant effect on the size of lysozyme crystals within 48 h. After 72 h, the crystal size of lysozyme decreased with the increase in SNP size, due to the larger supersaturation consumed. Lysozyme crystallization was further carried out at protein concentrations of 20 mg/mL with SNP of different sizes. A similar phenomenon was observed, as shown in Figure 3.

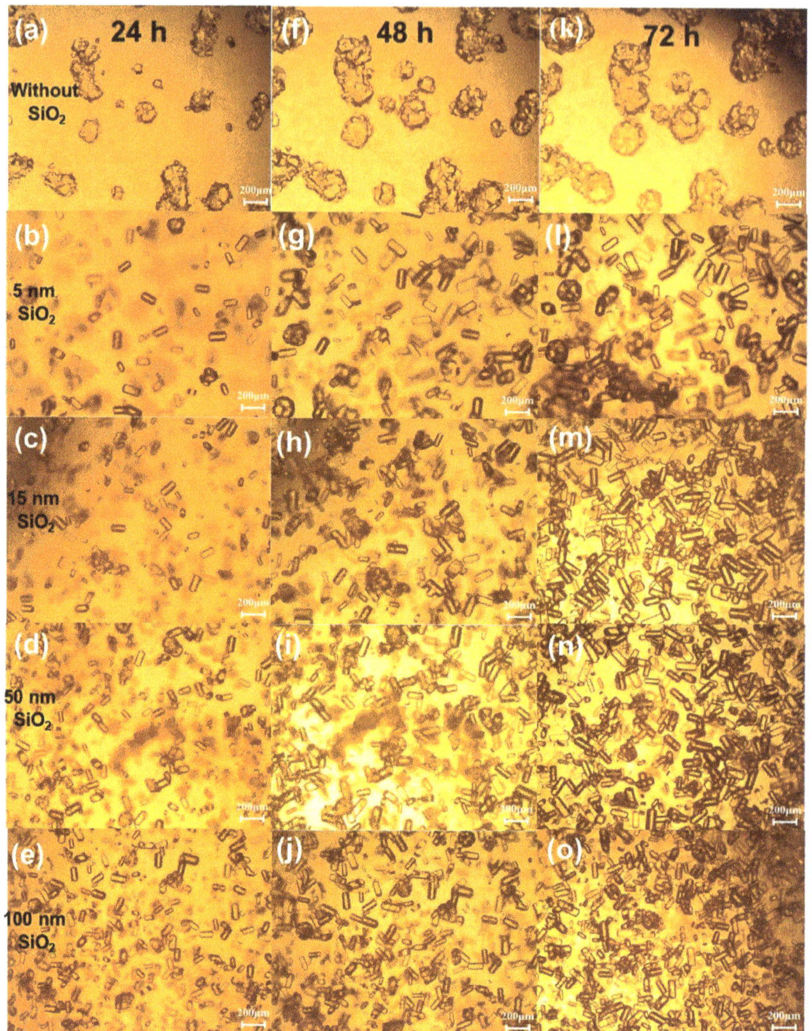

Figure 1. Microscopy photos of crystals obtained with SNP of different sizes. Conditions: 0.05 M, pH 4.5 sodium acetate buffer solution, 3 wt.% sodium chloride, 15 mg/mL lysozyme at 4 °C. Scale bar: 200 μm. (**a**): Crystal micrograph obtained under 24 h without SiO_2; (**b**): Crystal micrograph obtained at 24 h with 1 mg/mL 5 nm SiO_2; (**c**): Crystal micrograph obtained at 24 h with 1 mg/mL 15 nm SiO_2; (**d**): Crystal micrograph obtained at 24 h with 1 mg/mL 50 nm SiO_2; (**e**): Crystal micrograph obtained at 24 h with 1 mg/mL 100 nm SiO_2; (**f**): Crystal micrograph obtained under 48 h without SiO_2; (**g**): Crystal micrograph obtained at 48 h with 1 mg/mL 5 nm SiO_2; (**h**): Crystal micrograph obtained at 48 h with 1 mg/mL 15 nm SiO_2; (**i**): Crystal micrograph obtained at 48 h with 1 mg/mL 50 nm SiO_2; (**j**): Crystal micrograph obtained at 48 h with 1 mg/mL 100 nm SiO_2; (**k**): Crystal micrograph obtained under 72 h without SiO_2; (**l**): Crystal micrograph obtained at 72 h with 1 mg/mL 5 nm SiO_2; (**m**): Crystal micrograph obtained at 72 h with 1 mg/mL 15 nm SiO_2; (**n**): Crystal micrograph obtained at 72 h with 1 mg/mL 50 nm SiO_2; (**o**): Crystal micrograph obtained at 72 h with 1 mg/mL 100 nm SiO_2.

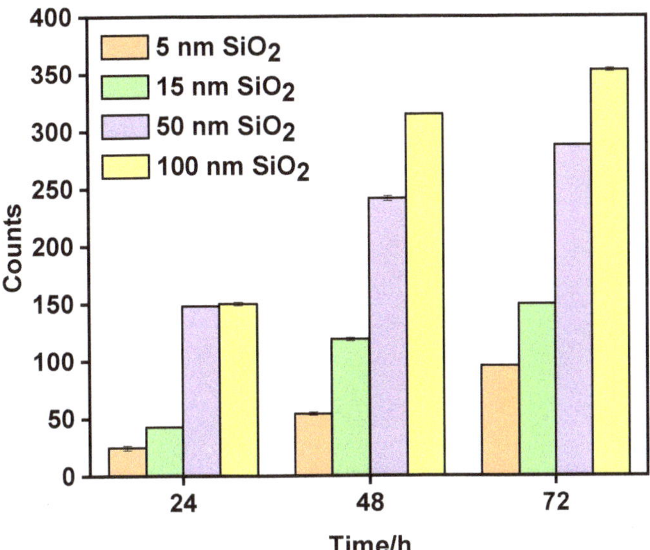

Figure 2. Histogram of the number of crystals with SNP of different sizes as a function of time. Conditions: 0.05 M, pH 4.5 sodium acetate buffer solution, 3 wt.% sodium chloride, 15 mg/mL lysozyme at 4 °C.

In this work, lysozyme crystallization was carried out at 4 °C, in sodium acetate buffer solution near pH = 4.5 with 3 wt.% NaCl. The crystal formed at such conditions is considered to exhibit a tetragonal morphology [21–24]. As stated in other works, no significant effect on the structure of lysozyme crystals was discovered when nanoparticles such as gold particles or carbon quantum dots (GQDs) were used as nucleating agents [12,25]. Therefore, it is speculated here that the addition of nanoparticles may not alter the lysozyme crystal assembly process and that the lysozyme crystals obtained belong to a tetragonal morphology.

As shown in Figures 1 and 3, the crystals obtained with SNP are longer in the 110-face. Forsythe et al. [26] have experimentally demonstrated that the growth of different faces in lysozyme crystals is strongly dependent on the supersaturation of the protein, resulting in changes in the shape of the crystals. The 110-face of the lysozyme crystal grows faster with the increase in supersaturation. As further discovered in Figure S4, some agglomerates could be observed in solutions containing SNP of larger particle sizes and lysozyme. Sun et al. [27] found out that lysozyme could adsorb on SNP and the adsorption was related to the surface curvature of SNP. The adsorption experiments of lysozyme on SNP with different particle sizes proved that the saturated adsorption capacity of lysozyme increased with SNP sizes [28]. Therefore, it can be assumed that the addition of SNP plays a key role in protein aggregation, which improves the local concentration of lysozyme, and is beneficial to the growth of the 110-face.

In addition, as reported in previous work [28], lysozyme crystals could only be observed when the size of SNP reached 100 nm at a supersaturation of 4. Here in this work, lysozyme crystals were obtained with all sizes of SNP used at the same supersaturation, proving that the different properties of SNP could cause the difference in protein crystallization. Therefore, the mechanism of SNP inducing lysozyme crystallization needs to be further explored.

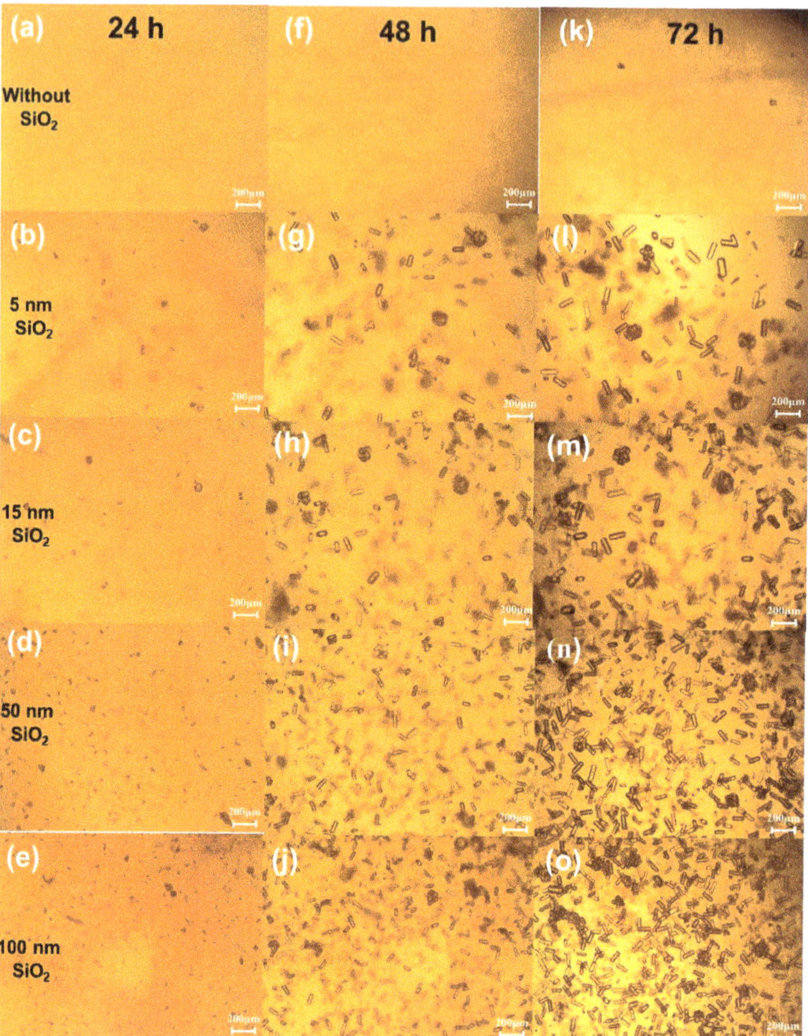

Figure 3. Microscopy photos of crystals obtained with SNP of different sizes. Conditions: 0.05 M, pH 4.5 sodium acetate buffer solution, 3 wt.% sodium chloride, 20 mg/mL lysozyme at 4 °C. Scale bar: 200 μm. (**a**): Crystal micrograph obtained under 24 h without SiO_2; (**b**): Crystal micrograph obtained at 24 h with 1 mg/mL 5 nm SiO_2; (**c**): Crystal micrograph obtained at 24 h with 1 mg/mL 15 nm SiO_2; (**d**): Crystal micrograph obtained at 24 h with 1 mg/mL 50 nm SiO_2; (**e**): Crystal micrograph obtained at 24 h with 1 mg/mL 100 nm SiO_2; (**f**): Crystal micrograph obtained under 48 h without SiO_2; (**g**): Crystal micrograph obtained at 48 h with 1 mg/mL 5 nm SiO_2; (**h**): Crystal micrograph obtained at 48 h with 1 mg/mL 15 nm SiO_2; (**i**): Crystal micrograph obtained at 48 h with 1 mg/mL 50 nm SiO_2; (**j**): Crystal micrograph obtained at 48 h with 1 mg/mL 100 nm SiO_2; (**k**): Crystal micrograph obtained under 72 h without SiO_2; (**l**): Crystal micrograph obtained at 72 h with 1 mg/mL 5 nm SiO_2; (**m**): Crystal micrograph obtained at 72 h with 1 mg/mL 15 nm SiO_2; (**n**): Crystal micrograph obtained at 72 h with 1 mg/mL 50 nm SiO_2; (**o**): Crystal micrograph obtained at 72 h with 1 mg/mL 100 nm SiO_2.

3.2. Zeta Potential of Lysozyme and SNP

Zeta potential measurements of SNP with different particle sizes and lysozyme were further performed. Figure 4a shows that SNP with different particle sizes was negatively charged. With the increase in the SNP size, the absolute value of the potential increased. Figure 4b shows that lysozyme was positively charged. Therefore, there should be an electrostatic interaction between SNP and lysozyme, increasing with the size of the SNP.

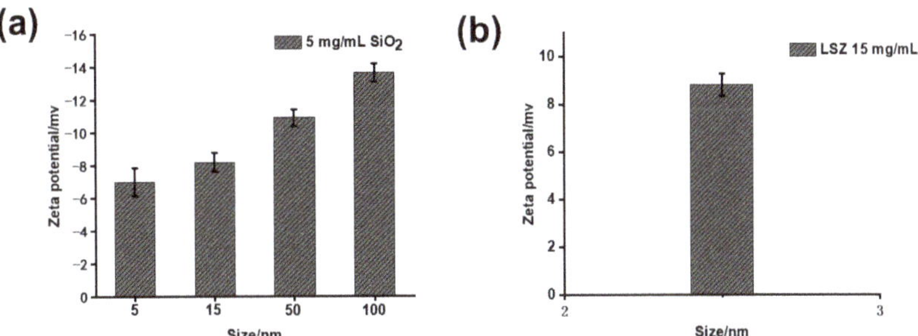

Figure 4. (**a**) Zeta potential of SNP with different particle sizes. (**b**) Zeta potential of lysozyme.

3.3. Effect of SNP Sizes on the UV Spectroscopy of Lysozyme

The UV spectra of lysozyme and SNP with different particle sizes were measured. As shown in Figure 5A, the spectrum exhibited absorbance peaks at 220 nm and 280 nm, respectively, which were caused by the presence of lysozyme. The UV spectrum of SNP was measured and showed no absorbance in the measured wavelength range (Figure S5). For the convenience of observation and discussion, the spectrum from 200 to 350 nm was intercepted (Figure 5B). An apparent increase in absorbance occurred with the addition of SNP at the same lysozyme concentration. According to the Lambert–Beer law, the absorbance value protein is in positive correlation with its concentration. Hence, it is assumed here that the increase in absorbance was mainly caused by an increase in local concentration of lysozyme. Due to the electrostatic interaction between lysozyme and SNP, lysozyme could be attracted and gathered at the surface of SNP, leading to an increase in local lysozyme concentration in solution. With the increase in SNP sizes, the interaction between lysozyme and SNP was enhanced, and the amount of lysozyme absorbed was further increased, resulting in a higher UV intensity.

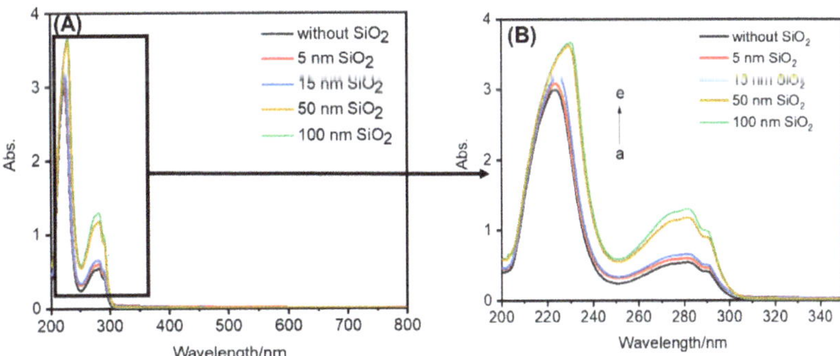

Figure 5. (**A**) UV spectrum from 200 nm to 800 nm. (**B**) UV spectrum from 200 nm to 350 nm. UV spectra of lysozyme solutions with and without SNP. From bottom to top: a, without SNP; b, 0.025 mg/mL 5 nm SNP; c, 0.025 mg/mL 15 nm SNP; d, 0.025 mg/mL 50 nm SNP; e, 0.025 mg/mL 100 nm SNP.

To verify the assumption above, the UV spectrum of pure lysozyme with different concentrations was measured (Figure S6). As shown in Figure S6, the UV absorbance increased with lysozyme concentration. The increasing trend and shape of the spectrum were consistent with Figure 5B. Hence, the presence of SNP can induce lysozyme aggregation and increase the local lysozyme concentration, which is more conducive to lysozyme crystallization.

3.4. Effect of SNP Sizes on the Fluorescence Spectroscopy Experiment of Lysozyme

Fluorescence experiments were carried out to further reveal the interaction between SNP and lysozyme. As shown in Figure 6, the fluorescence intensity of lysozyme increased after adding SNP of different sizes. For the same SNP size, the fluorescence intensity of lysozyme increased with the increase in SNP concentration. When the concentration of SNP was the same, the fluorescence intensity of lysozyme was higher after adding larger SNP.

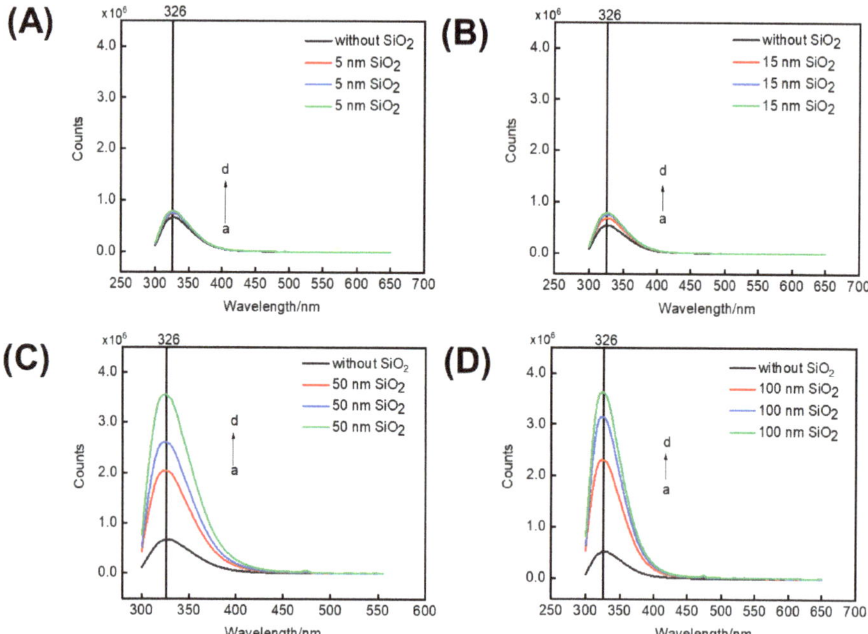

Figure 6. Fluorescence spectra of lysozyme solutions with and without SNP. (**A**) 5 nm SiO_2; (**B**) 15 nm SiO_2; (**C**) 50 nm SiO_2; (**D**) 100 nm SiO_2. From bottom to top: a, no SNP; b, 1 mg/mL; c, 2 mg/mL; d, 3 mg/mL.

There may be two possible reasons for the enhancement in fluorescence intensity [29,30]. Due to the aggregation of lysozyme caused by SNP, the probability of non-radiative relaxation of the excited state was reduced. Meanwhile, the contact probability between oxygen molecules with a quenching effect in water and chromogenic groups was reduced. As a result, the fluorescence intensity was enhanced.

3.5. Effect of SNP on Lysozyme Activity

In order to clarify the influence of the SNP–lysozyme interaction on lysozyme structure and function, the lysozyme activity and secondary structure were further determined. The activity test results are shown in Figure 7, and the results from circular dichroism are displayed in Figure S7. It can be found that the addition of SNP had almost no effect on the secondary structure and function of lysozyme.

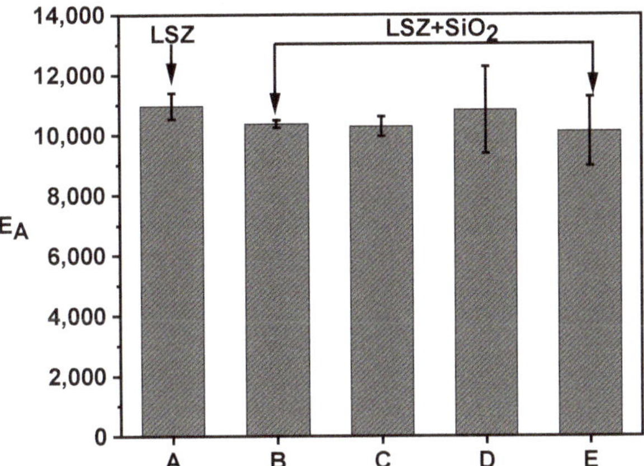

Figure 7. Effect of SNP on lysozyme activity. A: without SNP; B: 0.005 mg/mL 5 nm SiO_2; C: 0.005 mg/mL 15 nm SiO_2; D: 0.005 mg/mL 50 nm SiO_2; E: 0.005 mg/mL 100 nm SiO_2.

3.6. Mechanism of SNP-Induced Lysozyme Crystallization

Based on the results shown above, a mechanism of SNP-induced lysozyme crystallization is proposed and shown in Figure 8.

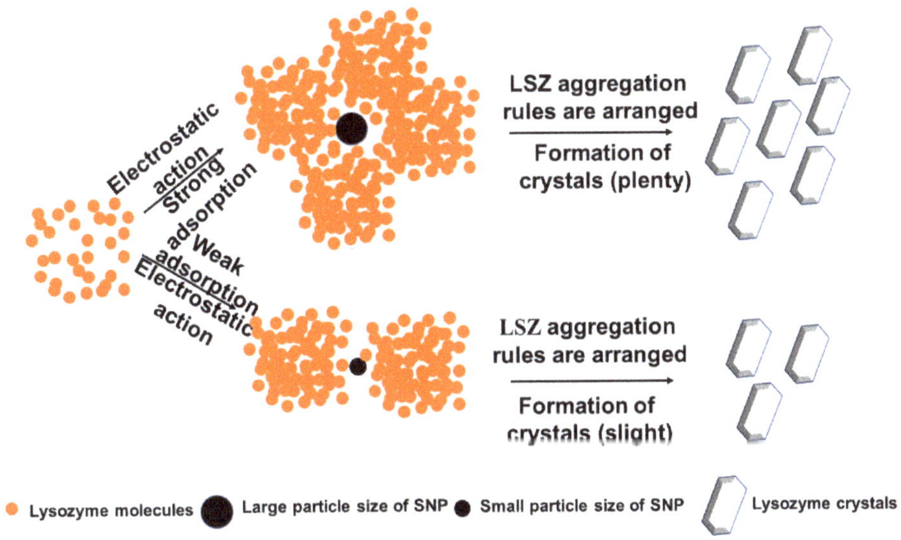

Figure 8. Mechanism of interaction between SNP with different particle sizes and lysozyme. (Note: the figure only shows the mechanism of a single SNP and lysozyme. There are multiple SNP and lysozymes in the specific crystallization process.).

Driven by electrostatic attraction, lysozyme adsorbs and aggregates on the surface of SNP, leading to an increase in lysozyme local concentration. As a result, the nucleation of lysozyme is easier when SNP is present. Moreover, the space occupied by each lysozyme molecule (asymmetric unit) in the crystal is a rectangular block with a size of 28.0 Å × 28.0 Å × 37 Å [31], which is smaller than the SNP sizes used here. It can be assumed that SNP only play the role of gathering lysozyme and do not participate in protein assembly. The initial nucleation conditions of lysozyme will not change by the addition of

SNP. With the increase in SNP size, the adsorption capacity of lysozyme increases, leading to a rise in the number of crystals.

4. Conclusions

The effects of SNP size on lysozyme crystallization were investigated. It was found that the morphological defects of lysozyme crystals were reduced after SNP addition. Further investigations showed that SNP can induce lysozyme aggregation and increase local supersaturation by electrostatic interaction, which increased with SNP size. As a result, lysozyme crystallizes easier with the addition of SNP, especially with SNP of larger size. Furthermore, the addition of SNP does not affect the secondary structure and function of lysozyme.

Supplementary Materials: The following supporting information can be downloaded at: https://www.mdpi.com/article/10.3390/cryst12111623/s1, Figure S1. Transmission electron microscopy images of SNP with different particle sizes. (a): 5 nm SiO_2, (b): 15 nm SiO_2, (c): 50 nm SiO_2, (d): 100 nm SiO_2; Figure S2. Crystal size distribution of 0.05 M, pH 4.5 sodium acetate buffer solution, 3 wt. % sodium chloride solution, 15 mg/mL lysozyme at 4 °C for 48 h. (a) 5 nm SiO_2, (b) 15 nm SiO_2, (c) 50 nm SiO_2, (d) 100 nm SiO_2; Figure S3. Crystal size distribution of 0.05 M, pH 4.5 sodium acetate buffer solution, 3 wt. % sodium chloride solution, 15 mg/mL lysozyme at 4 °C for 72 h. (a) 5 nm SiO_2, (b) 15 nm SiO_2, (c) 50 nm SiO_2, (d) 100 nm SiO_2; Figure S4. (a): Corresponding microscope image after mixing 1 mg/mL 50 nm SiO_2 and 20 mg/mL LSZ. (b): Corresponding microscope image after mixing 1 mg/mL 100 nm SiO_2 and 20 mg/mL LSZ; Figure S5. UV spectra of silica with different particle sizes; Figure S6. UV spectra of lysozyme at different concentrations; Figure S7. CD spectra of lysozyme in the presence and absence of SNP.

Author Contributions: Conceptualization, Y.Z. (Yuxiao Zhang), X.Y., F.H. and X.W. (Xiaoqiang Wang); methodology, Y.Z. (Yuxiao Zhang), X.Y., F.H. and X.W. (Xiaoqiang Wang); investigation, Y.Z. (Yuxiao Zhang), X.J., X.W. (Xia Wu), Y.M., S.L., K.L. and G.Z., Y.Z. (Yuyu Zhou); writing—original draft preparation, Y.Z. (Yuxiao Zhang); writing—review and editing, X.Y., F.H. and X.W. (Xiaoqiang Wang) All authors have read and agreed to the published version of the manuscript.

Funding: The work was supported by the National Natural Science Foundation of China (22178387).

Data Availability Statement: Not applicable.

Conflicts of Interest: The authors declare no conflict of interest.

References

1. Stevens, R.C. High-throughput protein crystallization. *Curr. Opin. Struct. Biol.* **2000**, *10*, 558–563. [CrossRef]
2. Dora, G.C. Crystal structure analysis: A primer. *Crystallogr. Rev.* **2011**, *17*, 157–160. [CrossRef]
3. Chayen, N.E. Turning protein crystallisation from an art into a science. *Curr. Opin. Struct. Biol.* **2004**, *14*, 577–583. [CrossRef] [PubMed]
4. Chayen, N.E.; Saridakis, E. Protein crystallization: From purified protein to diffraction-quality crystal. *Nat. Methods* **2008**, *5*, 147–153. [CrossRef] [PubMed]
5. Basu, S.K.; Govardhan, C.P.; Jung, C.W.; Margolin, A.L. Protein crystals for the delivery of biopharmaceuticals. *Expert Opin. Biol. Ther.* **2004**, *4*, 301–317. [CrossRef]
6. L'vov, P.E.; Umantsev, A.R. Two-Step Mechanism of Macromolecular Nucleation and Crystallization: Field Theory and Simulations. *Cryst. Growth Des.* **2020**, *21*, 366–382. [CrossRef]
7. Sear, R.P. Nucleation: Theory and applications to protein solutions and colloidal suspensions. *J. Phys. Condens. Matter* **2007**, *19*, 033101. [CrossRef]
8. Zhou, R.-B.; Cao, H.-L.; Zhang, C.-Y.; Yin, D.-C. A review on recent advances for nucleants and nucleation in protein crystallization. *CrystEngComm* **2017**, *19*, 1143–1155. [CrossRef]
9. Kordonskaya, Y.V.; Marchenkova, M.A.; Timofeev, V.I.; Dyakova, Y.A.; Pisarevsky, Y.V.; Kovalchuk, M.V. Precipitant ions influence on lysozyme oligomers stability investigated by molecular dynamics simulation at different temperatures. *J. Biomol. Struct. Dyn.* **2021**, *39*, 7223–7230. [CrossRef]
10. Kovalchuk, M.V.; Blagov, A.E.; Dyakova, Y.A.; Gruzinov, A.Y.; Marchenkova, M.A.; Peters, G.S.; Pisarevsky, Y.V.; Timofeev, V.I.; Volkov, V.V. Investigation of the initial crystallization stage in lysozyme solutions by small-angle X-ray scattering. *Cryst. Growth Des.* **2016**, *16*, 1792–1797. [CrossRef]

11. Kovalchuk, M.V.; Boikova, A.S.; Dyakova, Y.A.; Ilina, K.B.; Konarev, P.V.; Kryukova, A.E.; Marchenkova, M.A.; Pisarevsky, Y.V.; Timofeev, V.I. Pre-crystallization phase formation of thermolysin hexamers in solution close to crystallization conditions. *J. Biomol. Struct. Dyn.* **2019**, *37*, 3058–3064. [CrossRef]

12. Ko, S.; Kim, H.Y.; Choi, I.; Choe, J. Gold nanoparticles as nucleation-inducing reagents for protein crystallization. *Cryst. Growth Des.* **2017**, *17*, 497–503. [CrossRef]

13. Takeda, Y.; Mafuné, F. Induction of protein crystallization by platinum nanoparticles. *Chem. Phys. Lett.* **2016**, *647*, 181–184. [CrossRef]

14. Leese, H.S.; Govada, L.; Saridakis, E.; Khurshid, S.; Menzel, R.; Morishita, T.; Clancy, A.J.; White, E.R.; Chayen, N.E.; Shaffer, M.S. Reductively PEGylated carbon nanomaterials and their use to nucleate 3D protein crystals: A comparison of dimensionality. *Chem. Sci.* **2016**, *7*, 2916–2923. [CrossRef] [PubMed]

15. Delmas, T.; Roberts, M.M.; Heng, J.Y.Y. Nucleation and crystallization of lysozyme: Role of substrate surface chemistry and topography. *J. Adhes. Sci. Technol.* **2011**, *25*, 357–366. [CrossRef]

16. Lindberg, R.; Sjöblom, J.; Sundholm, G. Preparation of silica particles utilizing the sol-gel and the emulsion-gel processes. *Colloids Surf. A Physicochem. Eng. Asp.* **1995**, *99*, 79–88. [CrossRef]

17. Zhuravlev, L.T. Surface characterization of amorphous silica—A review of work from the former USSR. *Colloids Surf. A Physicochem. Eng. Asp.* **1993**, *74*, 71–90. [CrossRef]

18. Qian, R.; Ding, L.; Ju, H. Switchable fluorescent imaging of intracellular telomerase activity using telomerase-responsive mesoporous silica nanoparticle. *J. Am. Chem. Soc.* **2013**, *135*, 13282–13285. [CrossRef]

19. Ramanaviciene, A.; Zukiene, V.; Acaite, J.; Ramanavicius, A. Influence of caffeine on lysozyme activity in the blood serum of mice. *Acta Med. Litu.* **2002**, *9*, 241–244.

20. Yamazaki, T.; Kimura, Y.; Vekilov, P.G.; Furukawa, E.; Shirai, M.; Matsumoto, H.; Van Driessche, A.E.S.; Tsukamoto, K. Two types of amorphous protein particles facilitate crystal nucleation. *Proc. Natl. Acad. Sci. USA* **2017**, *114*, 2154–2159. [CrossRef]

21. Bhamidi, V.; Skrzypczak-Jankun, E.; Schall, C.A. Dependence of nucleation kinetics and crystal morphology of a model protein system on ionic strength. *J. Cryst. Growth* **2001**, *232*, 77–85. [CrossRef]

22. Cacioppo, E.; Pusey, M.L. The solubility of the tetragonal form of hen egg white lysozyme from pH 4.0 to 5.4. *J. Cryst. Growth* **1991**, *114*, 286–292. [CrossRef]

23. Tsekova, D.S. Formation and Growth of Tetragonal Lysozyme Crystals at Some Boundary Conditions. *Cryst. Growth Des.* **2009**, *9*, 1312–1317. [CrossRef]

24. Müller, C.; Ulrich, J. A more clear insight of the lysozyme crystal composition. *Cryst. Res. Technol.* **2011**, *46*, 646–650. [CrossRef]

25. Kang, M.; Lee, G.; Jang, K.; Jeong, D.W.; Lee, J.-O.; Kim, H.; Kim, Y.J. Graphene Quantum Dots as Nucleants for Protein Crystallization. *Cryst. Growth Des.* **2022**, *22*, 269–276. [CrossRef]

26. Forsythe, E.L.; Nadarajah, A.; Pusey, M.L. Growth of (101) faces of tetragonal lysozyme crystals: Measured growth-rate trends. *Acta Crystallogr. Sect. D Biol. Crystallogr.* **1999**, *55*, 1005–1011. [CrossRef] [PubMed]

27. Sun, X.; Feng, Z.; Zhang, L.; Hou, T.; Li, Y. The selective interaction between silica nanoparticles and enzymes from molecular dynamics simulations. *PLoS ONE* **2014**, *9*, e107696. [CrossRef] [PubMed]

28. Weichsel, U.; Segets, D.; Janeke, S.; Peukert, W. Enhanced nucleation of lysozyme using inorganic silica seed particles of different Sizes. *Cryst. Growth Des.* **2015**, *15*, 3582–3593. [CrossRef]

29. Luo, Z.; Yuan, X.; Yu, Y.; Zhang, Q.; Leong, D.T.; Lee, J.Y.; Xie, J. From aggregation-induced emission of Au (I)–thiolate complexes to ultrabright Au (0)@ Au (I)–thiolate core–shell nanoclusters. *J. Am. Chem. Soc.* **2012**, *134*, 16662–16670. [CrossRef]

30. Zhu, J.; He, K.; Dai, Z.; Gong, L.; Zhou, T.; Liang, H.; Liu, J. Self-assembly of luminescent gold nanoparticles with sensitive pH-stimulated structure transformation and emission response toward lysosome escape and intracellular imaging. *Anal. Chem.* **2019**, *91*, 8237–8243. [CrossRef]

31. Nadarajah, A.; Li, M.; Pusey, M.L. Growth mechanism of the (110) face of tetragonal lysozyme crystals. *Acta Crystallogr. Sect. D Biol. Crystallogr.* **1997**, *53*, 524–534. [CrossRef] [PubMed]

Article

Enhancing the Water Solubility of 9-Fluorenone Using Cyclodextrin Inclusions: A Green Approach for the Environmental Remediation of OPAHs

Yue Niu [1], Ling Zhou [1,*], Huiqi Wang [1], Jiayu Dai [1], Ying Bao [1,2], Baohong Hou [1,2] and Qiuxiang Yin [1,2,*]

[1] School of Chemical Engineering and Technology, State Key Laboratory of Chemical Engineering, Tianjin University, Tianjin 300072, China; niuyue1994@tju.edu.cn (Y.N.); whq001_1@tju.edu.cn (H.W.); daijiayu@tju.edu.cn (J.D.); yingbao@tju.edu.cn (Y.B.); houbaohong@tju.edu.cn (B.H.)

[2] Haihe Laboratory of Sustainable Chemical Transformations, Tianjin 300072, China

* Correspondence: zhouling@tju.edu.cn (L.Z.); qxyin@tju.edu.cn (Q.Y.)

Abstract: Oxygenated polycyclic aromatic hydrocarbons (OPAHs) are toxic and carcinogenic compounds widely present in the natural environment, posing a serious threat to the environment and human health. However, the removal of OPAHs is mainly hindered by their low water solubility. Cyclodextrins (CDs) are frequently used to form inclusion complexes (ICs) with hydrophobic molecules to improve their solubility. In this study, we investigated the solubility enhancement ability of different CDs on 9-fluorenone, a common OPAH, through phase solubility experiments. We successfully prepared three solid ICs of 9-fluorenone with β-, hydroxypropyl-β-(HP-β-) and sulfobutylether-β-CD (SBE-β-CD) using the cooling crystallization method for the first time and characterized them via powder X-ray diffraction, Fourier transform infrared spectroscopy, scanning electron microscopy, etc. Molecular dynamics simulations were employed to investigate the binding modes and stable configurations of the ICs in the liquid phase and to explore the factors affecting their solubility enhancement ability. The results showed that all the CDs had a solubility enhancement effect on 9-fluorenone, with SBE-β-CD displaying the strongest effect, increasing the solubility of 9-fluorenone by 146 times. HP-β-CD, β-CD, α-CD, and γ-CD followed in decreasing order. Moreover, 9-fluorenone formed a ratio of 1:1 ICs to CDs. In addition, the interaction energy between SBE-β-CD and 9-fluorenone was the lowest among the CDs, which further validated the results of the phase solubility experiments from a theoretical perspective. Overall, this study provides a green method for the removal of 9-fluorenone pollutants in the environment and is expected to be applied to the removal and environmental remediation of other OPAHs.

Keywords: cyclodextrin inclusion complex; OPAHs; 9-fluorenone; solubilization; molecular dynamics simulation

Citation: Niu, Y.; Zhou, L.; Wang, H.; Dai, J.; Bao, Y.; Hou, B.; Yin, Q. Enhancing the Water Solubility of 9-Fluorenone Using Cyclodextrin Inclusions: A Green Approach for the Environmental Remediation of OPAHs. *Crystals* **2023**, *13*, 775. https://doi.org/10.3390/cryst13050775

Academic Editor: Waldemar Maniukiewicz

Received: 17 April 2023
Revised: 28 April 2023
Accepted: 4 May 2023
Published: 6 May 2023

1. Introduction

Polycyclic aromatic hydrocarbons (PAHs) are a class of persistent and toxic organic pollutants that are commonly present in the environment [1]. Oxygenated polycyclic aromatic hydrocarbons (OPAHs) are derivatives of PAHs that contain at least one carbonyl oxygen on the aromatic ring and frequently coexist with PAHs in soils, air, and surface water [2]. OPAHs are primarily released into the environment through the incomplete combustion of solid fuels and the transformation of PAHs [3–5]. OPAHs are known for their high stability and resistance to degradation, and studies have found that OPAHs may exhibit stronger cytotoxicity and developmental toxicity than their parent PAHs [6,7]. OPAHs in the soil can accumulate in agricultural plants and be consumed by humans, while OPAHs in the atmosphere can be adsorbed onto particles and inhaled by humans [8,9]. These pollutants can cause allergies and even lead to cancer [10,11]. Therefore, the efficient removal of PAHs and OPAHs from soil and air is an important environmental issue and a necessary step for sustainable development.

Although OPAHs have better water solubility compared to their parent PAHs, their low water solubility is still a major challenge in removing them from contaminated soils [12,13]. However, recent studies have suggested that cyclodextrins (CDs) and their derivatives hold great promise as agents for improving the solubility of these persistent organic pollutants [14–16]. CDs are cyclic molecules with a hydrophobic interior cavity and a hydrophilic exterior, which allows them to form inclusion complexes (ICs) with hydrophobic molecules, thus enhancing their solubility [17]. Natural CDs such as α-, β-, and γ-CD have been shown to form such complexes with organic molecules, as have modified CD derivatives such as hydroxypropyl-β-(HP-β-) and sulfobutylether-β-CD(SBE-β-CD) [18,19]. These modified derivatives exhibit significantly improved water solubility compared to natural CDs, thus providing a more effective means of increasing the solubility of insoluble molecules [20].

In recent years, numerous studies have demonstrated that the preparation of CD ICs can lead to the effective removal of PAHs from the environment. T. Badr and K. Hanna et al. [14] studied the solubilization and extraction effects of CDs on the organic pollutants naphthalene and phenanthrene in soil and confirmed the formation of PAH ICs. Inga Tijunelyte et al. [16] investigated the ability to encapsulate two PAHs, naphthalene and fluoranthene, in CDs' cavities. Fuat Topuz et al. [21] described a novel concept of PAH removal from aqueous solutions using CD-functionalized mesostructured silica nanoparticles and pristine mesostructured silica nanoparticles. However, there is currently a dearth of research on the use of CD ICs to remove OPAHs from the environment. Meanwhile, the solubilization mechanism of cyclodextrins for pollutants has received relatively little attention in research. Therefore, further investigation is highly necessary. Molecular dynamics (MD) simulations have emerged as a powerful tool for investigating the mechanism of CD ICs' formation, which allows for more intuitive information on how the ICs are bound at the molecular level [22]. Various methods have been employed for the preparation of ICs, including mechanochemical methods, cooling crystallization, ultrasonication, and freeze drying. Cooling crystallization has gained considerable attention due to its low energy consumption and the high purity of the resulting ICs. Recent advances in the field have led to the development of novel techniques such as supercritical anti-solvent and supercritical assisted atomization, which have also been explored for the preparation of ICs [23,24].

9-Fluorenone (Figure 1A), also known as diphenylene ketone, is a common small-molecule OPAH and a frequently used fine chemical raw material [7]. 9-Fluorenone in the environment is usually derived from the combustion of solid fuels or from the oxidation of its parent compound, fluorene [25,26]. Found in the environment, it typically exhibits direct mutagenicity and stronger toxicity and persistence, often being widely detected in soils, groundwater, and the atmosphere [27–29]. Due to its poor solubility and limited bioavailability, it appears to be difficult to effectively remove from the environment [30]. Therefore, the preparation of ICs of 9-fluorenone with CDs can significantly enhance its solubility in water, thereby effectively removing the pollutants from the soil. Additionally, for 9-fluorenone that is already present in the effluent, it can be precipitated out through the formation of solid ICs, which leads to environmental restoration.

The purpose of this study was to investigate the solubilization effects of different CDs on 9-fluorenone. We prepared and characterized 9-fluorenone/CD ICs and elucidated the structural details of the complexes in aqueous solution through molecular dynamics simulations, identifying the optimal host molecule for improving the water solubility of 9-fluorenone. This information provides valuable support for the removal of 9-fluorenone from the environment and also offers a green method for the removal of other OPAHs.

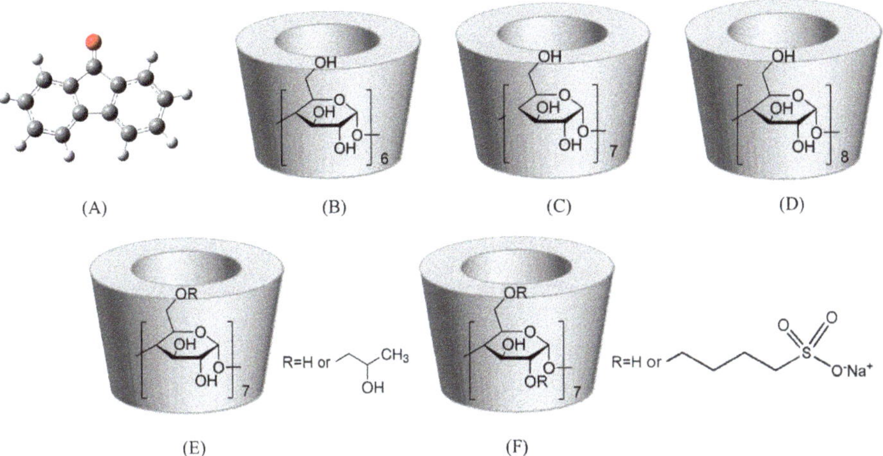

Figure 1. Molecular structure of (**A**) 9-fluorenone, (**B**) α-CD, (**C**) β-CD, (**D**) γ-CD, (**E**) (2-hydroxypropyl)-β-CD, (**F**) sulfobutylether-β-CD.

2. Materials and Methods

2.1. Materials

9-Fluorenone was purchased from Tianjin Heowns Biochem LLC (purity 99%, Tianjin, China). α-CD (purity 99%, average MW 973 g/mol), β-CD (purity > 98%, average MW 1135 g/mol), γ-CD (purity > 98%, average MW 1297 g/mol), (2-hydroxypropyl)-β-CD (purity > 98%, average MW 1542 g/mol), and sulfobutylether-β-CD (purity > 98%, average MW 1451 g/mol) were purchased from Tianjin Hiensi Biochemical Technology Co. (Tianjin, China). Distilled–deionized water was formed in the laboratory using a NANOPURE system from BARNSTEAD (Thermo Scientific Co., Tianjin, China). All chemicals were used as received. Ethanol of analytical grade was from Tianjin Kewei Chemical Technique Co., Ltd. (Tianjin, China).

2.2. Phase Solubility Experiments

2.2.1. Standard Curve Construction

9-Fluorenone was accurately weighed (5 mg) and dissolved in anhydrous ethanol to prepare a 0.1 mg/mL stock solution. Aliquots of the stock solution (2.0, 4.0, 6.0, 8.0, and 10.0 mL) were transferred into 25 mL volumetric flasks and diluted with anhydrous ethanol to the mark, with anhydrous ethanol as the reference blank. The solutions were shaken well and subjected to full-spectrum scanning analysis using a UV–visible spectrophotometer (UV-3600, Shimadzu, Kyoto, Japan). The absorption peak of 9-fluorenone appeared at 298 nm, while β-CD did not show any absorption peak at this wavelength. A standard curve was constructed by plotting the absorbance as the ordinate against the mass concentration (ρ) as the abscissa.

2.2.2. Solubility Curve Construction

Solutions of various CDs with concentrations ranging from 0 to 10 mM (10 mL) were prepared. Excess 9-fluorenone was added to the CD solutions, and the resulting solutions were stirred at 300 rpm and 25 °C for 48 h. After reaching the solid–liquid equilibrium, the supernatants were filtered using a 0.45 μm aqueous syringe filter, diluted to a certain degree, and the absorbance values of the samples at 298 nm were measured using a UV–visible spectrophotometer. The solubility of 9-fluorenone in the different CD concentrations was calculated from the corresponding standard curve, and the solubility of fluorenone was plotted as the ordinate against the CD concentration as the abscissa to construct the solubility curve.

2.2.3. Calculation of Apparent Stability Constants

The binding constants for each CD were calculated using the Higuchi–Connor method [31,32]. According to the Higuchi–Connor formula for the binding constant K:

$$K = \frac{Slope}{S_0 \times (1 - Slope)}$$

where $Slope$ represents the slope of the phase solubility curve and S_0 represents the solubility of the guest molecule in water.

Meanwhile, the solubilization efficiency (S_e) is calculated using the following equation:

$$S_e = \frac{S_{CD}}{S_0}$$

where S_{CD} is the apparent solubility of 9-fluorenone at a fixed concentration of 100 mM CD.

2.3. Preparation of 9-Fluorenone/CD ICs

2.3.1. Solid Samples

For the different CDs, saturated solutions (200mL) were prepared at 50 °C. 9-Fluorenone (1:1 molar ratio of host to guest) was dissolved in 5 mL of ethanol (as a co-solvent) and added to the prepared aqueous solution of CDs. The mixed solution was mechanically stirred at 300 rpm for 6 h while avoiding exposure to light. Afterward, the solution was cooled down to 5 °C at a rate of 8 °C/h, resulting in a pale-yellow precipitate. The precipitate was washed with a small amount of ethanol, filtered, and dried for 12 h in a vacuum-drying oven at 40 °C.

2.3.2. Physical Mixture

CDs and 9-fluorenone (at a ratio of 1:1) were lightly ground manually using a mortar and pestle for 10 min to obtain a homogeneous mixture.

2.4. Characterization

Crystal shapes were observed using an Olympus BX51 polarized light microscope (PLM, Olympus, Tokyo, Japan). Powder X-ray diffraction (PXRD) was performed with a D/MAX-2500 (Rigaku, Tokyo, Japan) using Cu-Kα radiation (0.15405 nm) in the 2θ between 2° and 40° with a scanning rate of 8°/min. The applied voltage and current were 40 kV and 100 mA, respectively. Differential scanning calorimetry (DSC) curves were measured with a DSC 1/500 (Mettler Toledo, Co., Zurich, Switzerland) under the protection of nitrogen, and the temperature range was from 25 to 300 °C at a heating rate of 10 °C/min. Fourier transform infrared spectroscopy (FTIR) spectra were collected via the KBr tablet method using a Nexus Fourier transform infrared spectrometer (FTIR, Thermo Fisher, Waltham, MA, USA), focusing on the wavelength range of 4000 to 400 cm^{-1} at a resolution of 4 cm^{-1} under ambient conditions. The crystal morphology of the raw materials, physical mixture, and ICs were observed using scanning electron microscopy (SEM, TM3000, Hitachi, Tokyo, Japan).

2.5. Molecular Dynamics (MD) Simulation

MD simulations were carried out using GROMACS [33] to investigate the IC conformations in the solution. All molecular topology and structure files were obtained using Automated Topology Builder (ATB) [34]. The structural modifications of HP-β-CD and SBE-β-CD were obtained based on the β-CD framework acquired from ATB and further refined using Guessview6.0. The resulting structures were submitted to ATB for topology and molecular structure file generation via computation. Molecular modeling was performed using the open-source software Packmol [35]. The GROMOS96 force field was employed. The default steps and protocols of the MD were selected to optimize the system with an equilibrium of 100ps, and the production run was carried out for 10ns (NPT ensemble) in triplicate with a step of 2 fs at constant temperature (323K). The energies and guest dispositions were registered every 1000 steps. VMD [36] software was utilized to visualize the simulation results, ensuring that the outcomes were easily interpretable.

3. Results and Discussion

3.1. Phase Solubility Study

A standard curve equation was obtained by fitting the absorbance values of 9-fluorenone with the mass concentration, which is depicted in Figure 2A. The equation was A = 13.6875 m + 0.015, with a high fitting accuracy of $R^2 = 0.99986$. The solubility curves of the various CDs at 25 °C are presented in Figure 2B, and the fitting curves and apparent stability constants K of the ICs were calculated and are listed in Table 1 and Table S1. The results indicated that increasing the concentration of CDs could linearly enhance the solubility of 9-fluorenone, and the enhancing capacity order was SBE-β- > HP-β- > β- > α- > γ-CD. Based on the Higuchi–Connors classification [37], the solubility curves belong to the A_L type, which indicates that the stoichiometric ratio of 9-fluorenone and CDs in the liquid state was 1:1. The highest K value was detected for 9-fluorenone/SBE-β-CD (1528 M^{-1}), followed by 9-fluorenone/HP-β-CD (799 M^{-1}) and 9-fluorenone/β-CD (346 M^{-1}). Notably, γ-CD showed almost no significant solubility-enhancing effect on 9-fluorenone, possibly due to insufficient intermolecular interaction forces between the γ-CD and 9-fluorenone molecules, resulting in weak inclusion effects. The solubility of 9-fluorenone was significantly increased by approximately 17-fold, 37-fold, 82-fold, 146-fold, and 2-fold, respectively, after forming complexes with α-, β-, HP-β-, SBE-β-, and γ-CD, as shown in Table 1. These findings suggest that natural CDs and their derivatives can significantly enhance the solubility of hydrophobic guest molecules, especially SBE-β-CD, which exhibits a more significant improvement in the stability and solubility of 9-fluorenone than natural CDs.

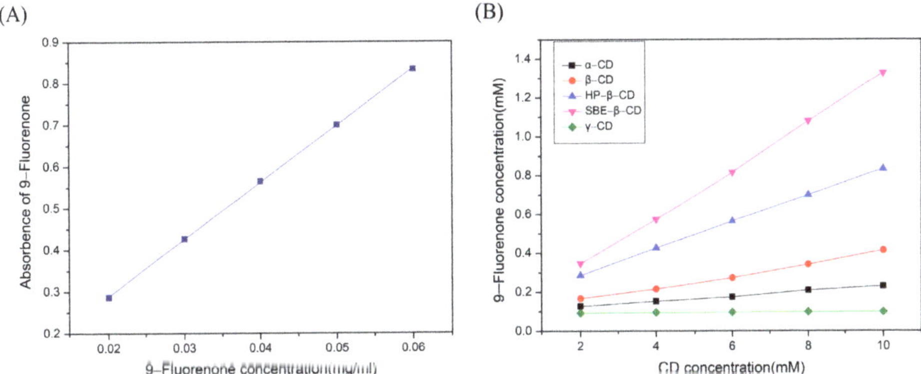

Figure 2. (**A**) Standard UV curve of 9-fluorenone. (**B**) The phase solubility curve of 9-fluorenone/CDs.

Table 1. Apparent stability constants (K, M^{-1}) and solubilization efficiency (S_e) for the 9-fluorenone/CD ICs.

CD Types	Slope	R^2	K(M^{-1})	Se
α-CD	0.0130	0.9935	144.2	17
β-CD	0.03084	0.9911	345.9	37
HP-β-CD	0.0684	0.9989	798.6	82
SBE-β-CD	0.12324	0.9991	1527.9	146
γ-CD	0.000703	0.9018	7.6	2

3.2. Characterization of 9-Fluorenone/CD ICs

Solid ICs of 9-fluorenone with β-CD, HP-β-CD, and SBE-β-CD were prepared using the cooling crystallization method. However, the solids of 9-fluorenone with α-CD and

γ-CD obtained using the same method were determined to be physical mixtures after analysis, which may be due to their weaker binding ability in the liquid state. Figure S1 shows photographs of all the prepared ICs. As 9-fluorenone has a yellow crystalline form, the liquid ICs exhibit varying degrees of yellow coloration depending on their solubility. The solid ICs obtained are also pale-yellow powders.

3.2.1. Powder X-ray Diffraction

The PXRD patterns of 9-fluorenone, the CDs, physical mixtures, and ICs are presented in Figure 3. The diffraction peaks observed in the physical mixtures of all three CDs represent an overlay of the 9-fluorenone and CD spectra. The XRD pattern of the 9-fluorenone/β-CD IC shows new distinct diffraction peaks at $2\theta = 4.9°, 7.1°, 7.5°, 10.0°, 14.0°$, and $21.5°$, which indicate that a new crystalline structure was formed. The disappearance of some of the characteristic peaks of β-CD and 9-fluorenone in the IC pattern suggests that the two compounds underwent a molecular rearrangement. The XRD pattern of 9-fluorenone/HP-β-CD IC shows two broad peaks at $2\theta = 9.5°$ and $18.1°$, which differ from those observed in HP-β-CD and the physical mixtures. The disappearance of the crystalline peak corresponding to 9-fluorenone in the X-ray diffraction pattern indicates the formation of an inclusion complex between 9-fluorenone and HP-β-CD. The complexation process leads to the amorphization of the guest molecule upon encapsulation within the CD cavity [38]. The amorphous state observed in the XRD pattern of 9-fluorenone/SBE-β-CD IC may indicate that 9-fluorenone is either present on the surface of SBE-β-CD or is present in a disordered state within the cavity of SBE-β-CD. Overall, the XRD results confirm the formation of solid ICs and provide evidence of the structural changes that occurred during complexation.

3.2.2. Fourier Transform Infrared Spectroscopy

Figure 4 shows the FTIR spectra of 9-fluorenone, the CDs, their physical mixtures, and the ICs. The FTIR spectrum of 9-fluorenone displays characteristic peaks at 667 cm^{-1}, 725 cm^{-1}, 758 cm^{-1}, 915 cm^{-1}, 1086 cm^{-1}, 1297 cm^{-1}, 1450 cm^{-1}, 1592 cm^{-1}, and 1710 cm^{-1}. The peaks observed at 1450 cm^{-1} and 1592 cm^{-1} correspond to the stretching and bending vibrations of the benzene ring skeleton, respectively. The peak observed at 1710 cm^{-1} is attributed to the stretching vibration of the C=O bond [39]. The peaks observed at 667 cm^{-1} and 725 cm^{-1} are attributed to the bending vibrations of the =C-H bond [40]. A significant and broad absorption band is observed at 3280 cm^{-1} in the spectrum of β-CD, which is attributed to the stretching vibration of the O-H bond of the hydroxyl group. A small characteristic peak is observed at 2925 cm^{-1}, which is attributed to the asymmetric stretching vibration of -CH_2. The spectrum of HP- exhibits a characteristic peak of -CH_3 stretching vibration at 2930 and 2837 cm^{-1}. In the spectrum of SBE-CD, the peak observed at 1153 cm^{-1} is attributed to the stretching vibration of the S=O bond [41]. Distinct differences in characteristic peaks were observed upon comparing the spectra of the ICs and physical mixtures composed of compounds with the same dose ratio, indicating the successful preparation of ICs. In the spectrum of β-IC, the presence of C=O (1710 cm^{-1}) and C=C (1592 cm^{-1}) indicates that the basic skeleton of the 9-fluorenone structure has not changed. A comparison of the infrared spectra of the IC and prednisolone did not reveal any new peaks, indicating the absence of new chemical bonds. The presence of an -OH peak in the region of 3600 to 3100 cm^{-1} in the IC spectrum is due to the abundance of hydroxyl groups in β-CD. Many typical, characteristic peaks of 9-fluorenone are masked by the characteristic peaks of β-CD, indicating that 9-fluorenone is encapsulated within the cavity of β-CD. The -OH absorption band is observed to shift from 3290 to 3261 cm^{-1}, a redshift phenomenon, indicating the formation of new hydrogen bonds between 9-fluorenone and β-CD [42].

Similarly, the -OH absorption band of HP-β-IC shifted from 3347 to 3310 cm^{-1}, and that of SBE-β-IC shifted from 3372 to 3353 cm^{-1}. This also indicates the formation of hydrogen bonds between 9-fluorenone and CDs. The intensity of the peak at 1710 cm^{-1}

in the IC spectrum is also weak compared to that in the spectrum of 9-fluorenone alone, suggesting that the C=O bond in 9-fluorenone is involved in the inclusion complexation with CDs. The formation of hydrogen bonds between CDs and 9-fluorenone molecules can improve the solubility and stability of 9-fluorenone in aqueous solution by reducing the intermolecular attractive forces between 9-fluorenone molecules and enhancing the intermolecular interaction forces between 9-fluorenone and water molecules. Overall, the FTIR spectra provide strong evidence for the successful formation of ICs and the involvement of hydrogen bonding in the inclusion process, confirming the results of the solubility studies.

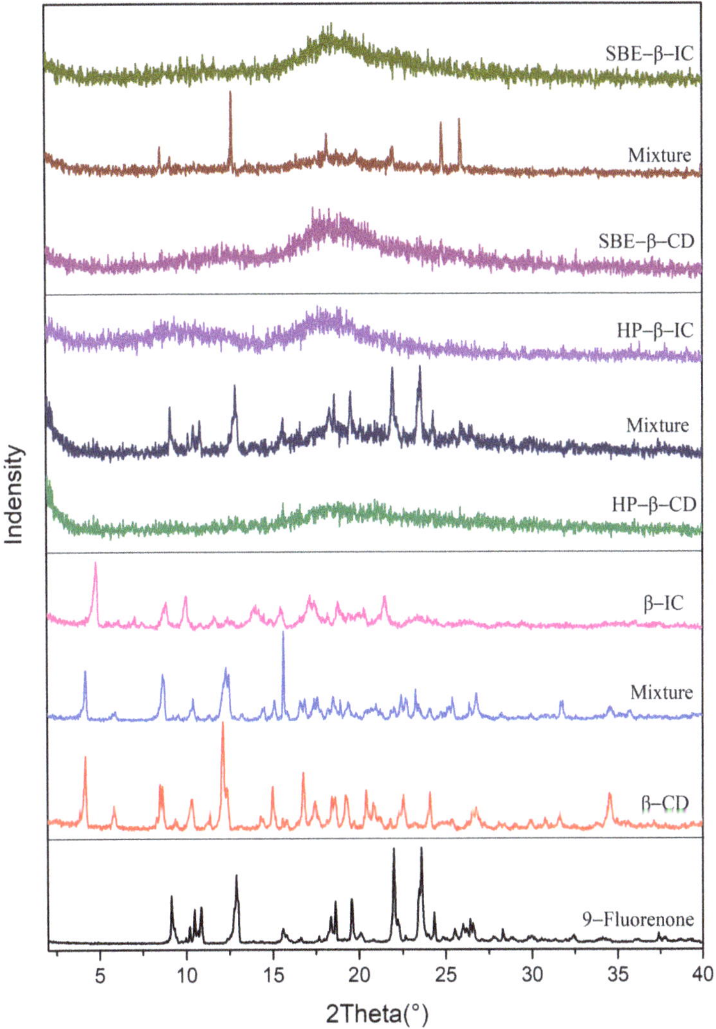

Figure 3. Powder X-ray diffraction pattern of 9-fluorenone, CDs, physical mixtures, and ICs.

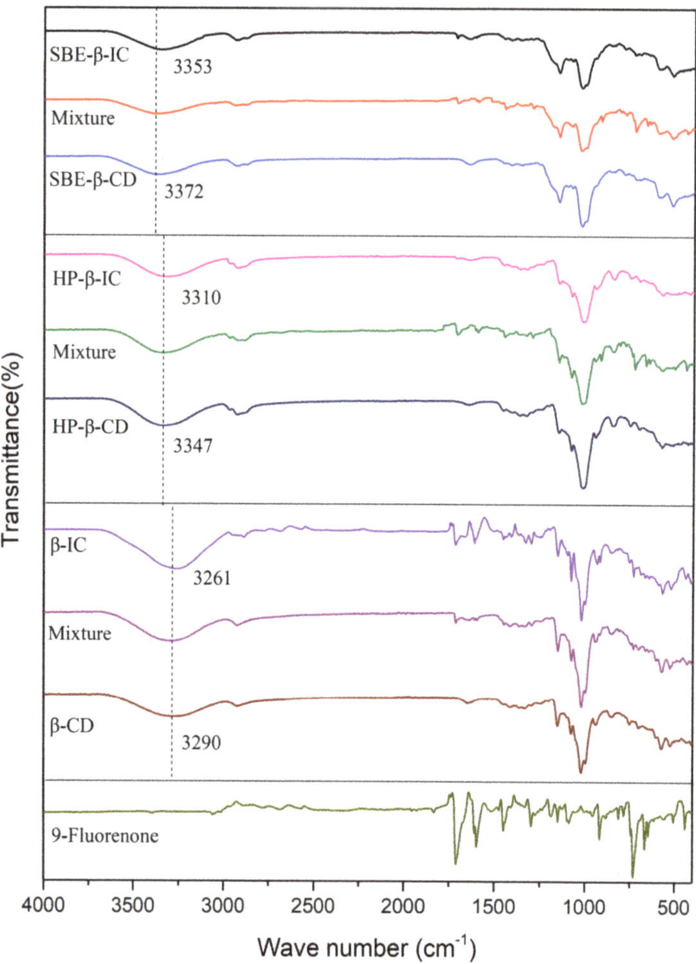

Figure 4. FTIR spectra of 9-fluorenone, CDs, physical mixtures, and ICs.

3.2.3. Differential Scanning Calorimetry

The DSC thermal spectra of 9-fluorenone, CDs, physical mixtures, and ICs are presented in Figure 5 (For detailed thermal analysis data see Table S2). The DSC curve of 9-fluorenone exhibits an endothermic peak at approximately 86 °C due to its melting. The endothermic peak of β-CD corresponds to the release of water from its cavity at approximately 65 °C. In the DSC curve of the β-CD IC sample, the melting peak of 9-fluorenone vanishes due to the formation of the IC, and the endothermic peak represents the evaporation of water in the IC, resulting in an increase in the dehydration temperature to approximately 75 °C [43]. In the DSC curve of HP-β-CD, two endothermic peaks emerge, corresponding to the evaporation of water and the melting temperature of the CD, respectively. In HP-β-CD IC, the dehydration temperature increases to approximately 70 °C, while the melting peak of HP-β-CD disappears, which might indicate that the melting point of the IC rose to above 300 °C. A similar situation occurs in SBE-β-CD IC, except that the dehydration temperature is lower than that of SBE-β-CD. The DSC spectra do not reveal the presence of 9-fluorenone, providing additional evidence that 9-fluorenone was fully incorporated into the IC. These findings demonstrate that the thermodynamic behavior of 9-fluorenone can be altered by combining it with different types of CDs.

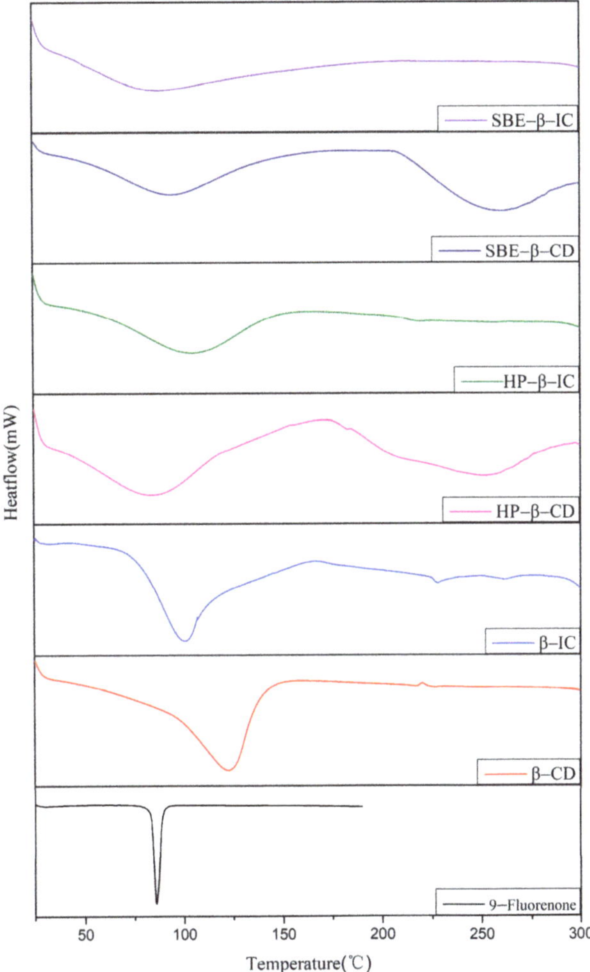

Figure 5. DSC thermogram of 9-fluorenone, CDs, physical mixtures, and ICs.

3.2.4. Scanning Electron Microscopy and Polarized Light Microscopy

The SEM images provide valuable information about the morphology and particle size of the materials. Figure 6A demonstrates that pure 9-fluorenone exists as block-shaped crystals, whereas the morphology of the CDs varies depending on the type of CD. Specifically, β-CD appears as a particulate structure with a relatively small particle size, while HP-β-CD and SBE-β-CD exhibit spherical particulate structures. Both 9-fluorenone and CDs are present alone in the physical mixtures, according to Figure 6B, and they do not combine to produce a new solid form. The SEM images of β-CD IC show that its crystal morphology is one of massive, aggregated particles with a more uniform particle size distribution. The SEM images of HP-β-IC and SBE-β-IC show that they are both massive structures with a large particle size. Additionally, their morphology is completely different from the original, spherical, amorphous structure of CDs. Overall, these SEM images demonstrate the differences in morphology and structure between the ICs and their corresponding raw materials.

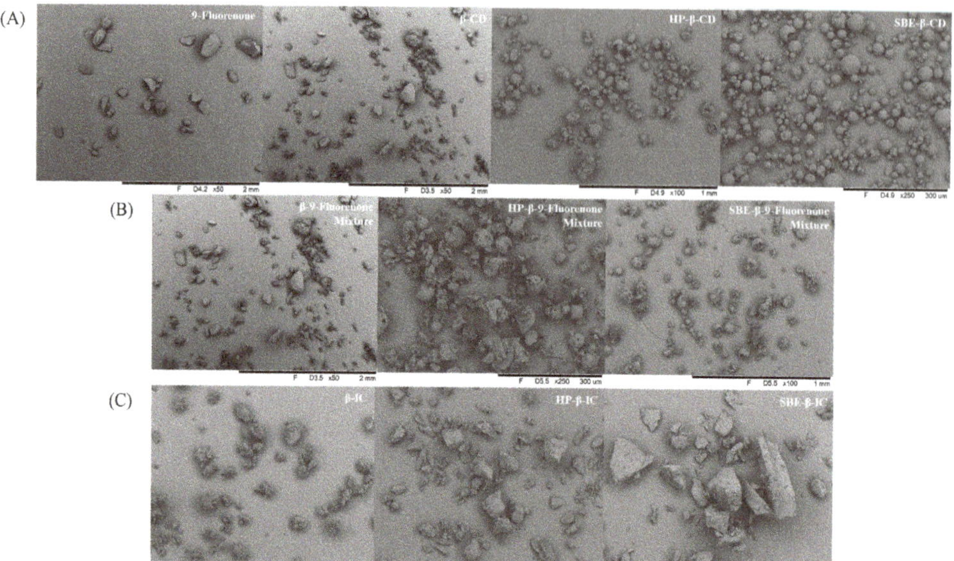

Figure 6. SEM images of (**A**) raw materials, (**B**) physical mixtures, and (**C**) ICs.

3.2.5. Polarized Light Microscopy

The crystal morphology of 9-fluorenone/β-CD IC in the liquid and solid states was observed using optical microscopy (Figure 7). It can clearly be seen that the IC is a rectangular, sheet-like crystal in the liquid state. When solid inclusion occurs, the flake crystals agglomerate together to form a granular product due to the small particle size, which is consistent with the SEM observation.

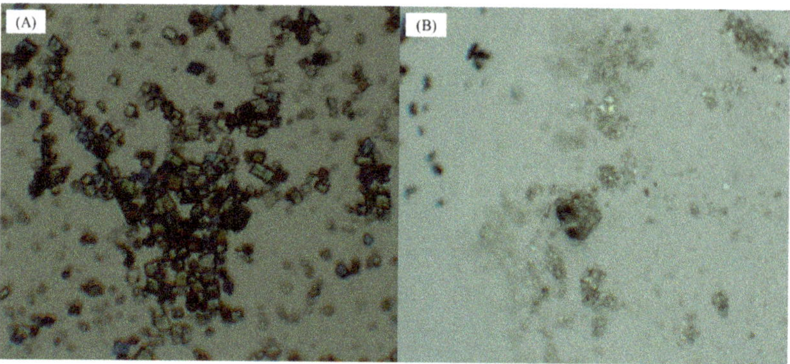

Figure 7. Light micrographs of 9-fluorenone/β-CD IC: (**A**) in the liquid state and (**B**) in the solid state (a drop of cedar oil for better observation).

3.3. Molecular Dynamics Simulations

To investigate the binding mode of 9-fluorenone with the CDs in solution, MD simulations were performed. A water box of 4 × 4 nanometers was established, and the 9-fluorenone and CD molecules were placed inside it. Two initial position settings were chosen for each CD system to study the binding mode of the guest and host molecules (Figure 8). The final structures of different CD systems are presented in Figure 9. During the simulation process, the 9-fluorenone molecule could easily enter the CD cavity of α-, β-, and SBE-β-CD, regardless of the initial position settings. However, for HP-β-CD and γ-CD, 9-fluorenone could only enter the CD cavity from a specific direction. In the case of HP-β-CD, the substitution of the hydroxyl group at the CD's primary hydroxyl end affects the entry of the guest molecule; hence, it can only enter the CD molecule through the secondary hydroxyl end. Meanwhile, γ-CD's cavity size is significantly larger than that of the 9-fluorenone molecule, which prevents it from stably encapsulating the guest molecule (Figure S2). This explains the poorer solubilization effect of γ-CD. These findings provide insights into the binding mechanisms of CDs with guest molecules in solution.

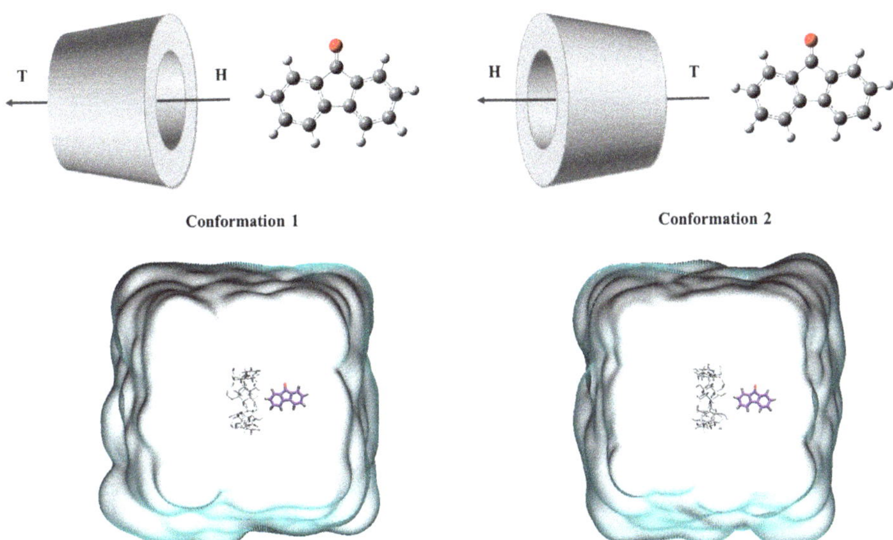

Figure 8. Two conformations of the initial position setting of 9-fluorenone toward the head (H) or tail (T) of different CDs.

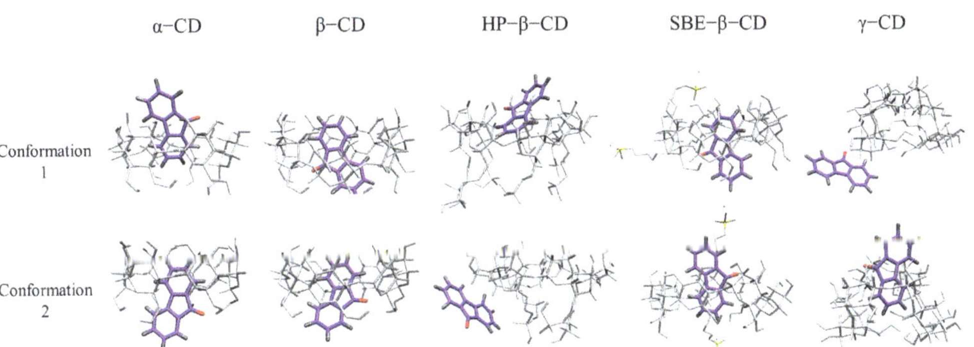

Figure 9. The final structure of 9-fluorenone/CDs based on simulations at two initial positions. (For clearer viewing, the C, O, and H atoms of 9-fluorenone are shown in purple, red, and silver, respectively. The C and H atoms of the CDs are shown in white, the S atom in yellow, and the Na$^+$ in blue).

Through MD simulations, we determined the stable configuration of the 9-fluorenone/CD ICs by comparing the binding energies between the host and guest molecules (Figure 10). The 9-fluorenone molecule was found to be upright inside the cavity of α-, β-, and SBE-β-CD, whereas it was tilted in HP-β-CD. The ICs were stabilized by hydrogen bonds formed between the 9-fluorenone molecule and the O or H atoms inside the CD cavity. Figure 11 illustrates the simulation results of the lower interaction energy for each CD at different initial positions. The root-mean-square deviation (RMSD) of the system was calculated to estimate the equilibration of the system (Figure 11A), and it was observed that each system reached equilibrium before 500 ps of the MD simulation, indicating that the simulation time was sufficient for each system. The number of hydrogen bonds formed between the 9-fluorenone molecule and the CD molecule is illustrated in Figure 11B. During the binding process with α-, β-, and SBE-β-CDs, the 9-fluorenone molecule formed one or two hydrogen bonds, whereas with HP-β-CD, it formed three hydrogen bonds at certain times, which may be attributed to the orientation of the 9-fluorenone molecule inside the cavity. This is consistent with the red shift of the hydroxyl peaks in the FTIR spectrum. It was found that hydrogen bonds were not the absolute factor affecting the water solubility of the ICs but were important for the stable existence of the structure of the ICs. The interaction energy between the host and guest molecules at different initial positions is shown in Figure S3. Van der Waals interactions contribute far more to the total interaction of the IC system than electrostatic interactions [44]. The collective interaction energies between the host and guest molecules is presented in Figure 11C. Among the four CDs, the interaction energy between SBE-β-CD and the 9-fluorenone molecule is the lowest, with an average value of approximately 120 kJ/mol. This indicates that the complex structure of SBE-β-CD and 9-fluorenone is the most stable among the four CDs, which also explains why it has the strongest solubilization ability. Finally, the solvent-accessible surface area (SASA) of the 9-fluorenone molecule was analyzed during the simulation. The SASA of 9-fluorenone could be seen to decrease and then remain stable, indicating that the guest molecule was encapsulated, thereby reducing its area of contact with the solvent. After reaching the equilibrium state, the SASA of 9-fluorenone was almost identical after its inclusion in the different CDs, indicating that each type of CD can stably contain 9-fluorenone molecules within its cavity to an almost equal extent. In the MD experiments, we did not search for signs of the unstable formation of α-CD with 9-fluorenone in the liquid state; hence, the reason for its inability to form a solid IC may be related to the mechanism of supramolecular self-assembly during crystallization, and further studies are needed.

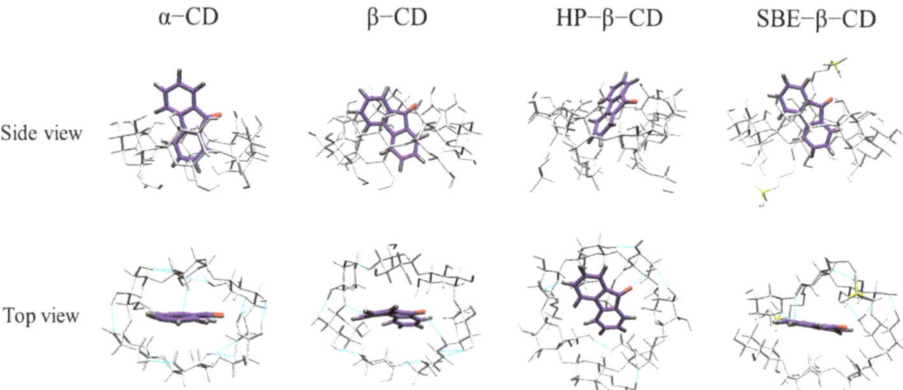

Figure 10. Stable conformation and hydrogen bonding (blue dotted line) of 9-fluorenone with CDs.

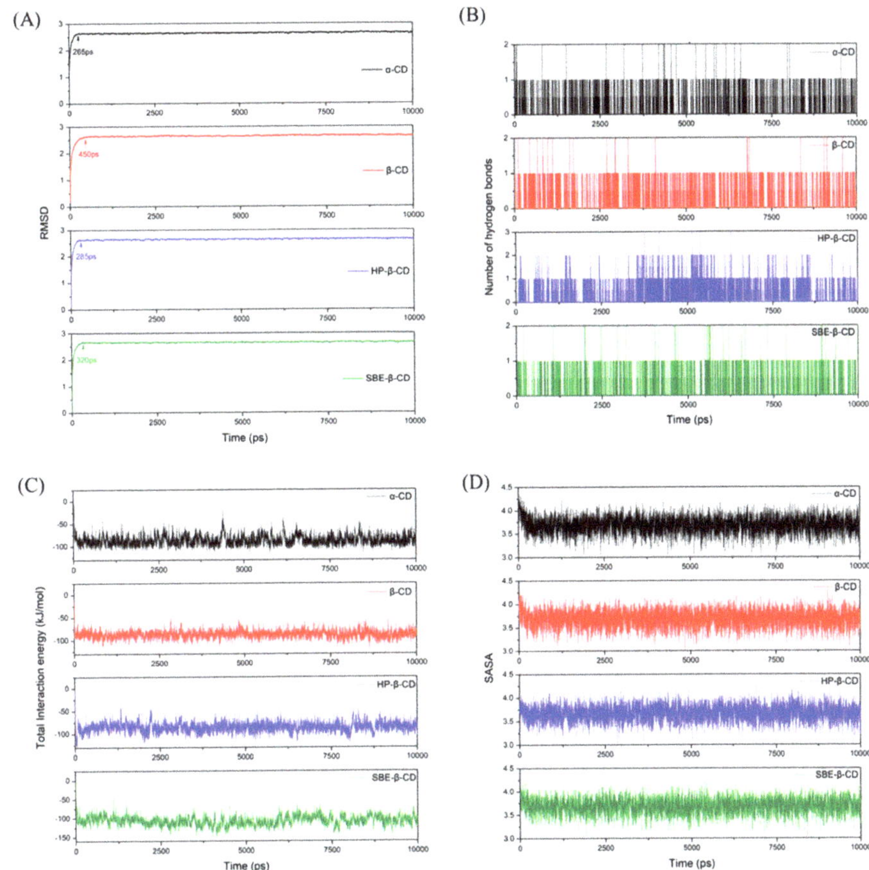

Figure 11. MD results: (**A**) RMSD of the system with the simulation time (ps); (**B**) the number of hydrogen bonds during the simulation; (**C**) total interaction energy between 9–fluorenone and CDs; (**D**) solvent-accessible and surface area (SASA) values of 9-fluorenone with the simulation time (ps).

4. Conclusions

In this study, we utilized phase solubility analysis to investigate the solubilization effects of different CDs on 9-fluorenone. Our results revealed significant differences in the solubilization efficiency of the different CDs towards 9-fluorenone. Specifically, SBE-β-CD exhibited the highest solubilization ability, with an apparent stability constant of 1527 M^{-1}, followed by HP-β-CD, β-CD, and α-CD, while γ-CD showed the weakest solubilization efficacy. Additionally, it was confirmed that ICs were formed between 9-fluorenone and the CDs in a 1:1 ratio in the liquid phase. We successfully prepared three solid ICs of 9-fluorenone using the cooling crystallization method and characterized them using various techniques, including PXRD, FTIR, DSC, and SEM. MD simulation was employed to investigate the binding modes between 9-fluorenone and the CDs in the liquid phase. Our results revealed that the binding between SBE-β-CD and 9-fluorenone was the most stable, which was consistent with the findings of the phase solubility experiments. Hydrogen bonding and van der Waals interactions were found to play crucial roles in stabilizing the ICs. Overall, our study provides theoretical and experimental support for the design and preparation of ICs of OPAHs in the environment and offers a green and environmentally friendly method for their removal from soil. Furthermore, the successful production of solid ICs provides guidance for the removal of these pollutants from water.

Supplementary Materials: The following are available online at https://www.mdpi.com/article/10.3390/cryst13050775/s1, Figure S1: Photographs of the prepared 9-fluorenone/CD ICs in (A) the liquid state and (B) the solid state; Figure S2: Binding process of 9-fluorenone with γ-CD; Figure S3: The interaction energy of host and guest molecules of different conformations; Table S1: Data and fitted equations for 9-fluorenone at different CD concentrations; Table S2: Thermal analysis data.

Author Contributions: Conceptualization, Y.N.; methodology, Y.N.; validation, L.Z., H.W. and J.D.; formal analysis, Y.N.; data curation, H.W. and L.Z.; writing—original draft preparation, Y.N.; writing—review and editing, Y.N., L.Z. and Q.Y. visualization, B.H. and Y.B. All authors have read and agreed to the published version of the manuscript.

Funding: The authors are grateful for the financial support of the Tianjin Municipal Natural Science Foundation (No. 21JCYBJC00600).

Data Availability Statement: Not applicable.

Conflicts of Interest: The authors declare no conflict of interest.

References

1. Zhang, X.Y.; Yu, T.; Li, X.; Yao, J.Q.; Liu, W.G.; Chang, S.L.; Chen, Y.G. The fate and enhanced removal of polycyclic aromatic hydrocarbons in wastewater and sludge treatment system: A review. *Crit. Rev. Environ. Sci. Technol.* **2019**, *49*, 1425–1475. [CrossRef]
2. Qiao, M.; Qi, W.; Liu, H.; Qu, J. Oxygenated polycyclic aromatic hydrocarbons in the surface water environment: Occurrence, ecotoxicity, and sources. *Environ. Int.* **2022**, *163*, 107232. [CrossRef]
3. Pulleyblank, C.; Cipullo, S.; Campo, P.; Kelleher, B.; Coulon, F. Analytical progress and challenges for the detection of oxygenated polycyclic aromatic hydrocarbon transformation products in aqueous and soil environmental matrices: A review. *Crit. Rev. Environ. Sci. Technol.* **2019**, *49*, 357–409. [CrossRef]
4. Krzyszczak, A.; Czech, B. Occurrence and toxicity of polycyclic aromatic hydrocarbons derivatives in environmental matrices. *Sci. Total Environ.* **2021**, *788*, 147738. [CrossRef]
5. Yu, H.T. Environmental carcinogenic polycyclic aromatic hydrocarbons: Photochemistry and phototoxicity. *J. Environ. Sci. Health Part C-Environ. Carcinog. Ecotoxicol. Rev.* **2002**, *20*, 149–183. [CrossRef]
6. Clerge, A.; Le Goff, J.; Lopez, C.; Ledauphin, J.; Delepee, R. Oxy-PAHs: Occurrence in the environment and potential genotoxic/mutagenic risk assessment for human health. *Crit. Rev. Toxicol.* **2019**, *49*, 302–328. [CrossRef]
7. Ma, X.; Wu, S. Oxygenated polycyclic aromatic hydrocarbons in food: Toxicity, occurrence and potential sources. *Crit. Rev. Food Sci. Nutr.* **2022**, 1–22. [CrossRef] [PubMed]
8. Cao, W.; Yuan, J.; Geng, S.; Zou, J.; Dou, J.; Fan, F. Oxygenated and Nitrated Polycyclic Aromatic Hydrocarbons: Sources, Quantification, Incidence, Toxicity, and Fate in Soil—A Review Study. *Processes* **2022**, *11*, 52. [CrossRef]
9. Bandowe, B.A.M.; Lueso, M.G.; Wilcke, W. Oxygenated polycyclic aromatic hydrocarbons and azaarenes in urban soils: A comparison of a tropical city (Bangkok) with two temperate cities (Bratislava and Gothenburg). *Chemosphere* **2014**, *107*, 407–414. [CrossRef] [PubMed]
10. Fu, P.P.; Xia, Q.; Sun, X.; Yu, H. Phototoxicity and environmental transformation of polycyclic aromatic hydrocarbons (PAHs)-light-induced reactive oxygen species, lipid peroxidation, and DNA damage. *J. Environ. Sci. Health C Environ. Carcinog. Ecotoxicol. Rev.* **2012**, *30*, 1–41. [CrossRef] [PubMed]
11. Albinet, A.; Leoz-Garziandia, E.; Budzinski, H.; Villenave, E. Polycyclic aromatic hydrocarbons (PAHs), nitrated PAHs and oxygenated PAHs in ambient air of the Marseilles area (South of France): Concentrations and sources. *Sci. Total Environ.* **2007**, *384*, 280–292. [CrossRef]
12. Lammel, G.; Kitanovski, Z.; Kukucka, P.; Novak, J.; Arangio, A.M.; Codling, G.P.; Filippi, A.; Hovorka, J.; Kuta, J.; Leoni, C.; et al. Oxygenated and Nitrated Polycyclic Aromatic Hydrocarbons in Ambient Air-Levels, Phase Partitioning, Mass Size Distributions, and Inhalation Bioaccessibility. *Environ. Sci. Technol.* **2020**, *54*, 2615–2625. [CrossRef] [PubMed]
13. Ji, B.; Liu, Y.; Wu, Y.; Gao, S.; Zeng, X.; Yu, Z. Occurrence, Source and Potential Ecological Risk of Parent and Oxygenated Polycyclic Aromatic Hydrocarbons in Sediments of Yangtze River Estuary and Adjacent East China Sea. *Ecol. Environ. Sci.* **2022**, *31*, 1400–1408.
14. Badr, T.; Hanna, K.; de Brauer, C. Enhanced solubilization and removal of naphthalene and phenanthrene by cyclodextrins from two contaminated soils. *J. Hazard. Mater.* **2004**, *112*, 215–223. [CrossRef]
15. Fai, P.B.; Grant, A.; Reid, B.J. Compatibility of hydroxypropyl-beta-cyclodextrin with algal toxicity bioassays. *Environ. Pollut.* **2009**, *157*, 135–140. [CrossRef]
16. Tijunelyte, I.; Dupont, N.; Milosevic, I.; Barbey, C.; Rinnert, E.; Lidgi-Guigui, N.; Guenin, E.; de la Chapelle, M.L. Investigation of aromatic hydrocarbon inclusion into cyclodextrins by Raman spectroscopy and thermal analysis. *Environ. Sci. Pollut. Res. Int.* **2017**, *24*, 27077–27089. [CrossRef]
17. Crini, G. Review: A history of cyclodextrins. *Chem. Rev.* **2014**, *114*, 10940–10975. [CrossRef]

18. Harada, A.; Takashima, Y.; Yamaguchi, H. Cyclodextrin-based supramolecular polymers. *Chem. Soc. Rev.* **2009**, *38*, 875–882. [CrossRef] [PubMed]
19. Cid-Samamed, A.; Rakmai, J.; Mejuto, J.C.; Simal-Gandara, J.; Astray, G. Cyclodextrins inclusion complex: Preparation methods, analytical techniques and food industry applications. *Food Chem.* **2022**, *384*, 132467. [CrossRef]
20. Ferreira, L.; Campos, J.; Veiga, F.; Cardoso, C.; Paiva-Santos, A.C. Cyclodextrin-based delivery systems in parenteral formulations: A critical update review. *Eur. J. Pharm. Biopharm.* **2022**, *178*, 35–52. [CrossRef]
21. Topuz, F.; Uyar, T. Cyclodextrin-functionalized mesostructured silica nanoparticles for removal of polycyclic aromatic hydrocarbons. *J. Colloid Interface Sci.* **2017**, *497*, 233–241. [CrossRef]
22. Mazurek, A.H.; Szeleszczuk, L.; Gubica, T. Application of Molecular Dynamics Simulations in the Analysis of Cyclodextrin Complexes. *Int. J. Mol. Sci.* **2021**, *22*, 9422. [CrossRef] [PubMed]
23. Wu, H.T.; Chuang, Y.H.; Lin, H.C.; Hu, T.C.; Tu, Y.J.; Chien, L.J. Immediate Release Formulation of Inhaled Beclomethasone Dipropionate-Hydroxypropyl-Beta-Cyclodextrin Composite Particles Produced Using Supercritical Assisted Atomization. *Polymers* **2022**, *14*, 2114. [CrossRef] [PubMed]
24. Franco, P.; De Marco, I. Preparation of non-steroidal anti-inflammatory drug/β-cyclodextrin inclusion complexes by supercritical antisolvent process. *J. CO2 Util.* **2021**, *44*, 101397. [CrossRef]
25. Gbeddy, G.; Goonetilleke, A.; Ayoko, G.A.; Egodawatta, P. Transformation and degradation of polycyclic aromatic hydrocarbons (PAHs) in urban road surfaces: Influential factors, implications and recommendations. *Environ. Pollut.* **2020**, *257*, 113510. [CrossRef] [PubMed]
26. Ding, Z.Z.; Yi, Y.Y.; Zhang, Q.Z.; Zhuang, T. Theoretical investigation on atmospheric oxidation of fluorene initiated by OH radical. *Sci. Total Environ.* **2019**, *669*, 920–929. [CrossRef]
27. Bandowe, B.A.M.; Sobocka, J.; Wilcke, W. Oxygen-containing polycyclic aromatic hydrocarbons (OPAHs) in urban soils of Bratislava, Slovakia: Patterns, relation to PAHs and vertical distribution. *Environ. Pollut.* **2011**, *159*, 539–549. [CrossRef]
28. Yadav, I.C.; Devi, N.L.; Singhd, V.K.; Li, J.; Zhang, G. Concentrations, sources and health risk of nitrated- and oxygenated-polycyclic aromatic hydrocarbon in urban indoor air and dust from four cities of Nepal. *Sci. Total Environ.* **2018**, *643*, 1013–1023. [CrossRef]
29. Shen, G.F.; Tao, S.; Wang, W.; Yang, Y.F.; Ding, J.N.; Xue, M.A.; Min, Y.J.; Zhu, C.; Shen, H.Z.; Li, W.; et al. Emission of Oxygenated Polycyclic Aromatic Hydrocarbons from Indoor Solid Fuel Combustion. *Environ. Sci. Technol.* **2011**, *45*, 3459–3465. [CrossRef]
30. Qiao, M.; Qi, W.X.; Liu, H.J.; Qu, J.H. Occurrence, behavior and removal of typical substituted and parent polycyclic aromatic hydrocarbons in a biological wastewater treatment plant. *Water Res.* **2014**, *52*, 11–19. [CrossRef]
31. Connors, K.A. Population Characteristics of Cyclodextrin Complex Stabilities in Aqueous Solution. *J. Pharm. Sci.* **1995**, *84*, 843. [CrossRef] [PubMed]
32. Connors, K.A. The Stability of Cyclodextrin Complexes in Solution. *Chem. Rev.* **1997**, *97*, 1325–1357. [CrossRef] [PubMed]
33. Van der Spoel, D.; Lindahl, E.; Hess, B.; Groenhof, G.; Mark, A.E.; Berendsen, H.J.C. GROMACS: Fast, flexible, and free. *J. Comput. Chem.* **2005**, *26*, 1701–1718. [CrossRef]
34. Stroet, M.; Caron, B.; Visscher, K.M.; Geerke, D.P.; Malde, A.K.; Mark, A.E. Automated Topology Builder Version 3.0: Prediction of Solvation Free Enthalpies in Water and Hexane. *J. Chem. Theory Comput.* **2018**, *14*, 5834–5845. [CrossRef] [PubMed]
35. Martinez, L.; Andrade, R.; Birgin, E.G.; Martinez, J.M. PACKMOL: A Package for Building Initial Configurations for Molecular Dynamics Simulations. *J. Comput. Chem.* **2009**, *30*, 2157–2164. [CrossRef] [PubMed]
36. Humphrey, W.; Dalke, A.; Schulten, K. VMD: Visual molecular dynamics. *J. Mol. Graph.* **1996**, *14*, 33–38. [CrossRef]
37. Connors, K.; Higuchi, T. Phase solubility techniques. *Adv. Anal. Chem. Instrum.* **1965**, *4*, 117–212.
38. Narayanan, G.; Boy, R.; Gupta, B.S.; Tonelli, A.E. Analytical techniques for characterizing cyclodextrins and their inclusion complexes with large and small molecular weight guest molecules. *Polym. Test.* **2017**, *62*, 402–439. [CrossRef]
39. Celebioglu, A.; Wang, N.; Kilic, M.E.; Durgun, E.; Uyar, T. Orally Fast Disintegrating Cyclodextrin/Prednisolone Inclusion-Complex Nanofibrous Webs for Potential Steroid Medications. *Mol. Pharm.* **2021**, *18*, 4486–4500. [CrossRef]
40. Park, K.H.; Choi, J.M.; Cho, E.; Jung, S. Enhanced Solubilization of Fluoranthene by Hydroxypropyl beta-Cyclodextrin Oligomer for Bioremediation. *Polymers* **2018**, *10*, 111. [CrossRef]
41. Huang, T.; Zhao, Q.; Su, Y.; Ouyang, D. Investigation of molecular aggregation mechanism of glipizide/cyclodextrin complexation by combined experimental and molecular modeling approaches. *Asian J. Pharm. Sci.* **2019**, *14*, 609–620. [CrossRef] [PubMed]
42. Ma, Y.; Niu, Y.; Yang, H.; Dai, J.; Lin, J.; Wang, H.; Wu, S.; Yin, Q.; Zhou, L.; Gong, J. Prediction and design of cyclodextrin inclusion complexes formation via machine learning-based strategies. *Chem. Eng. Sci.* **2022**, *261*, 117946. [CrossRef]
43. Roik, N.; Belyakova, L. IR Spectroscopy, X-ray diffraction and thermal analysis studies of solid 'b-cyclodextrin-para-aminobenzoic acid' inclusion complex. *Phys. Chem. Solid State* **2011**, *12*, 168–173.
44. Mahalapbutr, P.; Wonganan, P.; Charoenwongpaiboon, T.; Prousoontorn, M.; Chavasiri, W.; Rungrotmongkol, T. Enhanced Solubility and Anticancer Potential of Mansonone G By beta-Cyclodextrin-Based Host-Guest Complexation: A Computational and Experimental Study. *Biomolecules* **2019**, *9*, 545. [CrossRef] [PubMed]

Article

The Influence of Different Types of SiO$_2$ Precursors and Ag Addition on the Structure of Selected Titania-Silica Gels

Anna Adamczyk

Material Science and Ceramics Department, AGH University of Science and Technology, Al. Mickiewicza 30, 30-059 Kraków, Poland; aadamcz@agh.edu.pl

Abstract: In this paper, samples of titania-silica system were obtained by the sol-gel method as bulk materials, using titanium propoxide Ti(C$_3$H$_7$O)$_4$ to introduce titania and two precursors of SiO$_2$: TEOS tetraethoxysilane Si(OC$_2$H$_5$)$_4$ and DDS dimethyldiethoxysilane (CH$_3$)$_2$(C$_2$H$_5$O)$_2$Si. To enhance antibacterial properties, Ag was added to gels of selected compositions. The main aim of the performed studies was to find the correlations between the structural changes and the applied precursor of silica. Simultaneously, the influence of different compositions of gels and the addition of Ag on the specimens' structure was investigated. To study the structure, two complementary methods, FTIR (Fourier Transform InfraRed) spectroscopy and X-ray diffraction, were applied. The analysis of the FTIR spectra and the XRD patterns made it possible to confirm the amorphous state of all dried gels and establish the presence of TiO$_2$ polymorphs: anatase and rutile in all annealed samples. The addition of Ag atoms into the gels caused the crystallization of cristobalite phase in addition to titania polymorphs. The presence of crystalline Ag phase in the annealed gels allowed the calculation dimensions of Ag crystallites based on the Scherrer equation. The use of DDS as a silica precursor led to easier and faster crystallization of different TiO$_2$ phases in the annealed samples and parallel increases in the depolymerization of silica lattice.

Keywords: TiO$_2$-SiO$_2$ system; sol-gel method; silica precursors; FTIR spectroscopy; X-ray diffraction

1. Introduction

Titania-silica materials are widely applied in many fields because of their unique properties and usually simple preparation. They exhibit high self-cleaning and deodorising abilities [1–5], and possess good mechanical properties as well as high thermal stability [6,7]. Materials of TiO$_2$-SiO$_2$ systems can be applied in the form of monoliths or as coatings. As TiO$_2$-SiO$_2$ coatings, they can be anticorrosive and/or antibacterial layers [8]. More importantly, such thin films can be used in the field of human health protection because of their bioactivity, biocompatibility and antibacterial as well as anticorrosive behaviours [9–11].

Studies of the structure of TiO$_2$-SiO$_2$ samples have shown that they mainly consist of TiO$_2$ molecules dispersed in a silica matrix. When titania concentration is low, one can observe TiO$_2$ microcrystals that grow and thus cause the depolymerization of silica lattice during annealing [3,12]. These small titania crystals can act as crystallisation nuclei and decrease the temperature of crystallization in all phases of a sample.

One of the most often used methods of titania-silica material synthesis is the sol-gel method, which obtains highly homogenous samples of precisely defined compositions. Meanwhile, the structure of TiO$_2$-SiO$_2$ materials is very sensitive to the annealing procedure and the type of titania and silica precursors applied. In particular, the silica precursor plays an important role in porosity and influences the temperature of anatase-rutile transformation, as well as the rate of anatase crystallization [13,14]. Anatase and rutile are two of the most common titania polymorphs. They both crystallize in the tetragonal crystallographic system. There is also another form of TiO$_2$, called brucite. Among these three polymorphs, rutile is the most thermodynamically stable. The anatase-rutile and brucite-rutile transformations are one-directional and irreversible.

Citation: Adamczyk, A. The Influence of Different Types of SiO$_2$ Precursors and Ag Addition on the Structure of Selected Titania-Silica Gels. *Crystals* **2023**, *13*, 811. https://doi.org/10.3390/cryst13050811

Academic Editors: Jingxiang Yang and Xin Huang

Received: 28 February 2023
Revised: 3 May 2023
Accepted: 5 May 2023
Published: 13 May 2023

Two SiO_2 precursors were selected for the syntheses described in this work. TEOS (tetraethoxysilane $Si(OC_2H_5)_4$) seems to be the most often applied component in the sol-gel method. The rate of hydrolysis and polycondensation of sols obtained using this silane depends on the pH of the solution and the catalysts involved. The quantity of water added during the synthesis also determines the structure of the polymer obtained [15–17]. DDS (dimethyldiethoxysilane $(CH_3)_2(C_2H_5O)_2Si$) belongs to the group of siloxanes. The addition of DDS to sols containing TEOS can increase the time of gelation and the degree of hydrolysis and the polycondensation [17–19]. Materials containing TEOS and DDS possess hydrophobic and anti-reflective properties.

To enhance antibacterial properties of titania–silica gels, Ag was introduced into their structure. The antibacterial properties of silver have been well known for many years and have described in numerous articles [20–22]. Silver nanoparticles can penetrate the cell membranes of bacteria, destroying them or restricting their proliferation. Ag particles are characterised by high dispersion in the silica matrix, but can also act as the nuclei of crystallization [21].

The main aim of this work is to study the influence of different silica precursors on the structure of titania-silica gels of selected compositions. Additionally, the effect of Ag additives on the structure of selected gels is investigated.

2. Materials and Methods

As already mentioned, all materials in this study were synthesized by the sol-gel method. Gels of two compositions were planned to be obtained. In these samples, TiO_2:SiO_2 = 1:2 and 1:3 molar ratios were selected (Table 1). As the TiO_2 precursor, commonly applied titanium propoxide $Ti(C_3H_7O)_4$ (Aldrich 98%) was used, whereas, as SiO_2 precursors—TEOS: tetraethoxysilane $Si(OC_2H_5)_4$ (Aldrich 98%) and DDS: dimethyldiethoxysilane $(CH_3)_2(C_2H_5O)_2Si$ (Aldrich 97%) were applied. To study the influence of silver addition on the structure of the synthesized gels, Ag was introduced using silver nitrate $AgNO_3$ (Chempur 99.9%) into the gel of TiO_2:SiO_2 = 1:3 ratio composition. In the samples where SiO_2 was introduced simultaneously using two precursors, the constant TEOS:DDS = 1 molar ratio was maintained [18]. The content of Ag in the prepared sols corresponded to an Ag:Si = 1:24 molar ratio value.

Table 1. Compositions of gels obtained with and without Ag addition.

Sample	TiO_2 Content [% mol]	SiO_2 Content [% mol]	TEOS:DDS Ratio [mol]	Ag Addition Ag:Si Ratio [mol]
TiO_2:SiO_2 = 1:2	33.3	66.7	-	-
TiO_2:SiO_2 = 1:3	25	75	-	-
TiO_2:SiO_2 = 1:2	33.3	66.7	1:1	-
TiO_2:SiO_2 = 1:3	25	75	1:1	-
TiO_2:SiO_2 = 1:3 + Ag	25	75	-	1:24
TiO_2:SiO_2 = 1:3 + Ag	25	75	1:1	1:24

As the first step, two one-component sols, the first containing silica and the second containing titania, were prepared. Then, they were mixed to obtain the planned TiO_2:SiO_2 = 1:2 and 1:3 molar ratios.

To synthesize a pure titania sol (2.5 wt. %), two separate solutions were prepared as the first step. The first contained 26.3 mL of ethanol (99.8%) and 4.4 mL of titanium propoxide. The second solution consisted of 2.9 mL of a redistilled water, 26.3 mL of ethanol (99.8%) and a low amount (0.9 mL) of an acetic acid (30% Aldrich). Each solution was homogenized and then the second solution was added dropwise to the one containing TiO_2 precursor. The final solution was then homogenized for 30 min.

There were two procedures to obtain a silica sol, one only involved TEOS, and the second one used both TEOS and DDS. To obtain 5% silica sol, two separate solutions were again prepared. In the first, 48 mL of ethanol (99.8%) and 19 mL of TEOS were thoroughly homogenized. In parallel, the suspension of 6 mL of redistilled water, 48 mL of ethanol (99.8%) and 0.17 mL of HCL (30% Fluka) as a catalyst was obtained. After 10 min of homogenization, this last suspension was added dropwise to the solution of TEOS and homogenized for another 2 h.

To prepare SiO_2 sol synthesised using both TEOS and DDS silica precursors, two solutions were prepared. The first solution contained 60 mL of ethanol (99.8%), 4.6 mL of TEOS and 3.5 mL of DDS, which were thoroughly homogenized for 20 min. The second prepared solution contained 33.3 mL of ethanol (99.8%) as a solvent, 0.9 mL of redistilled water and the addition of 0.3 mL of ammonia (30% Fluka) as a catalyst. After 20 min of homogenization, this last solution was very slowly added to the initial solution. The final silica sol was then stirred for another two and half of hours.

The addition of Ag atoms into the titania-silica structure was realized during the synthesis of sols. The proper amount of $AgNO_3$ was weighed in the appropriate amount to attain the assumed Ag:Si = 1:24 ratio and the amount of silica in the prepared sol with TiO_2:SiO_2 = 1:3 (molar ratio) composition. This amount of $AgNO_3$ (in the case of typical synthesis with TEOS and TEOS/DDS, 0.08409 g or 0.02242 g, respectively) was dissolved in 14 mL of ethanol (99.8%) and added dropwise to the final TiO_2-SiO_2 sol.

As previously mentioned, both sols, titania and silica (in two versions) were mixed at a proper molar ratio to synthesize gels of the selected compositions. After one month of drying in air, all samples were ground in an agate mortar and annealed at 1200 °C for one hour, in air.

The main goal of this work was to study the dependence of the structure of the synthesized materials on their composition and the applied SiO_2 precursor. Thus, two complementary methods: IR spectroscopy and X-ray diffraction were applied. X-ray diffraction is suitable for crystalline samples, whereas FTIR (Fourier Transform InfraRed) spectroscopy allows the study of not only crystalline samples, but also amorphous or partially amorphous ones. The results obtained by both methods provide complete information on the far and near order in the crystalline lattice of materials.

FTIR spectra were collected on a Bruker 70 V IR spectrometer (Bruker, Billerica, MA, USA) at a resolution of 4 cm^{-1}; measurements were run for the samples prepared as KBr pellets. All graphs were prepared using OPUS software bought together using the spectrometer.

All X-ray diffraction patterns were measured with an X'Pert diffractometer (Panalytical, Almelo, The Netherlands) using CuK_α radiation and the standard Brag-Brentano configuration. All analyses and graphs were prepared using HighScore Plus software bought together with the equipment. The calculations of silver crystallite sizes were conducted based on the Scherrer equation and the Scherrer Calculator in HighScore Plus software:

$$D_{hkl} = \frac{k\lambda}{\beta cos\theta} \tag{1}$$

where:

β—the full width at half maximum, $\beta = \beta_{obs} - \beta_{stand}$, [rad] or [°],

λ—the length of radiation applied (CuK_α), $\lambda = 0.15406$ [nm],

θ—the peak position 2 θ [°],

k—Scherrer constant, $k = 0.9$,

D_{hkl}—the average crystallite size [nm] (the dimension perpendicular to the plane, which gave the reflection at the 2 θ position).

β in Scherrer Equation (1) means the corrected full width at half maximum (FWHM) of the selected diffraction peak and is determined by $\beta_{obs} - \beta_{stand}$, where β_{obs} is related to the sample peak and β_{stand} is the FWHM of the standard (corundum α-Al_2O_3).

SEM measurements were run using the desktop scanning electron microscope Phenom XL (manufactured by Thermo Fisher Scientific, Waltham, MA, USA) and software Phenom Prosuite for collecting the EDS spectra. During SEM imaging, as well as during EDS measurements, an accelerating voltage of 15 kV was applied.

3. Results and Discussion

3.1. FTIR Spectroscopy Studies

The FTIR spectra of all gels are presented in Figures 1–6. In each figure, IR spectra of a dried gel and a respective gel annealed at 1200 °C are compared.

In all IR spectra, one can distinguish three main groups of bands, located in the ranges of 940–1220 cm^{-1}, 840–580 cm^{-1} and at around 470 cm^{-1} of what is visible, especially in the spectra of the annealed gels. In the FTIR spectra of dried gels, it is possible to distinguish additional bands at 1350–1540 cm^{-1}, which can be assigned to the bending vibrations of C-H connections, bands assigned to the vibrations of OH^{-} groups at about 3500 cm^{-1} and bands at around 1650–1700 cm^{-1} ascribed to the vibrations in water molecules. All of these groups of bands disappeared after the annealing of gels.

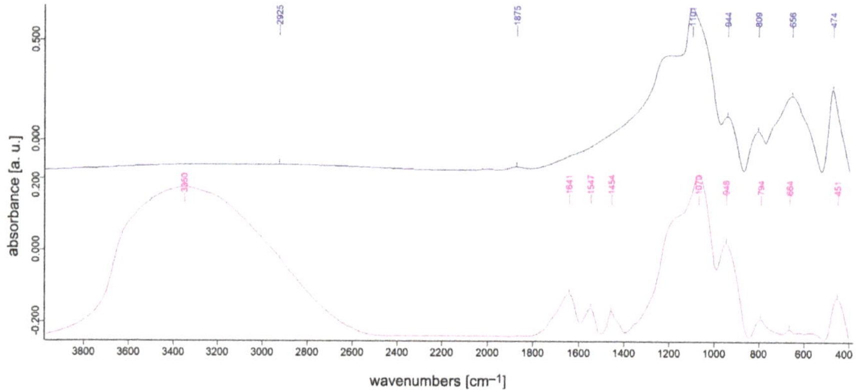

Figure 1. FTIR spectra of TiO$_2$:SiO$_2$ = 1:2 gel synthesized using TEOS, dried (bottom) and annealed at 1200 °C (top).

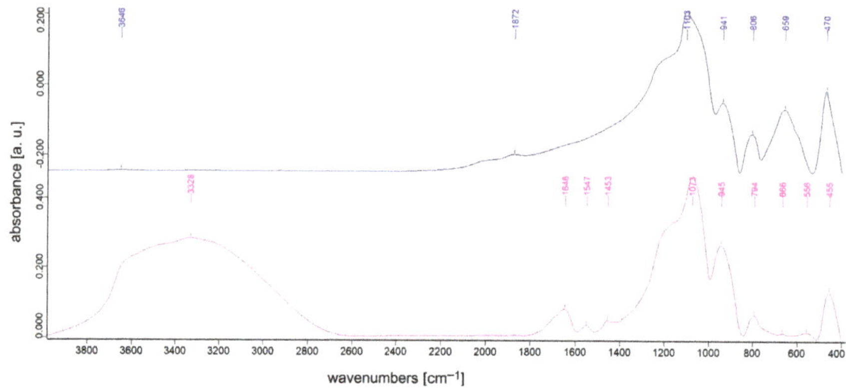

Figure 2. FTIR spectra of TiO$_2$:SiO$_2$ = 1:3 gel synthesized using TEOS, dried (bottom) and annealed at 1200 °C (top).

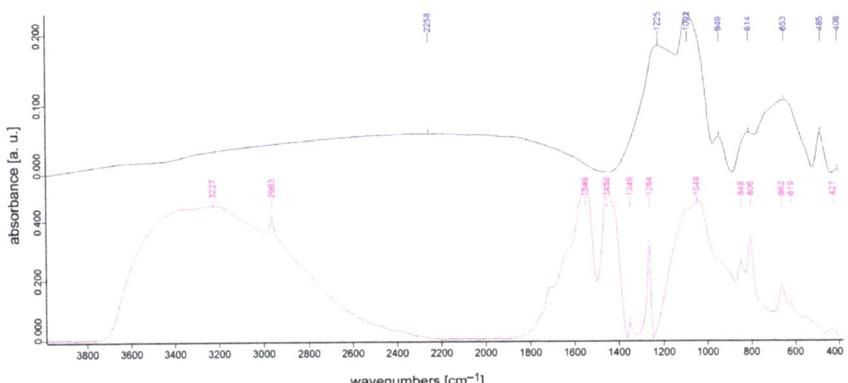

Figure 3. FTIR spectra of TiO$_2$:SiO$_2$ = 1:2 gel synthesized using TEOS and DDS, dried (bottom) and annealed at 1200 °C (top).

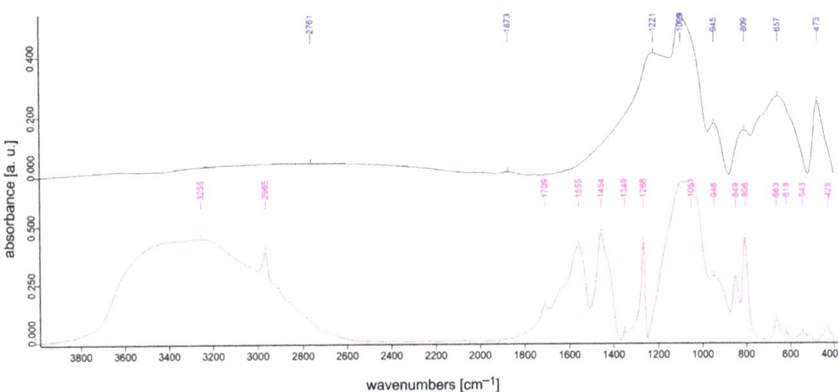

Figure 4. FTIR spectra of TiO$_2$:SiO$_2$ = 1:3 gel synthesized using TEOS and DDS, dried (bottom) and annealed at 1200 °C (top).

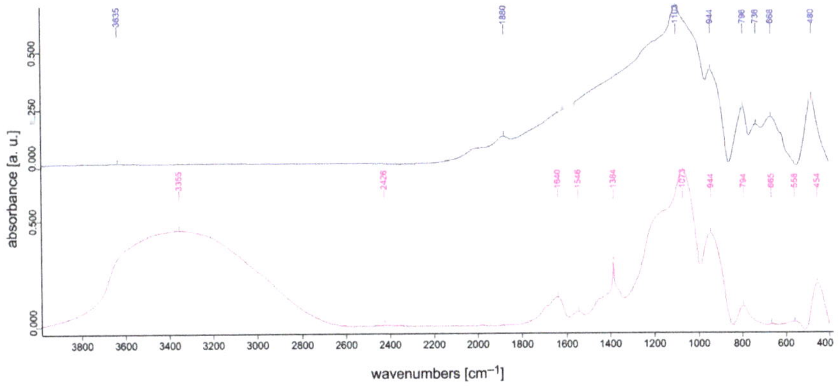

Figure 5. FTIR spectra of TiO$_2$:SiO$_2$ = 1:3 gel synthesized using TEOS with Ag addition, dried (bottom) and annealed at 1200 °C (top).

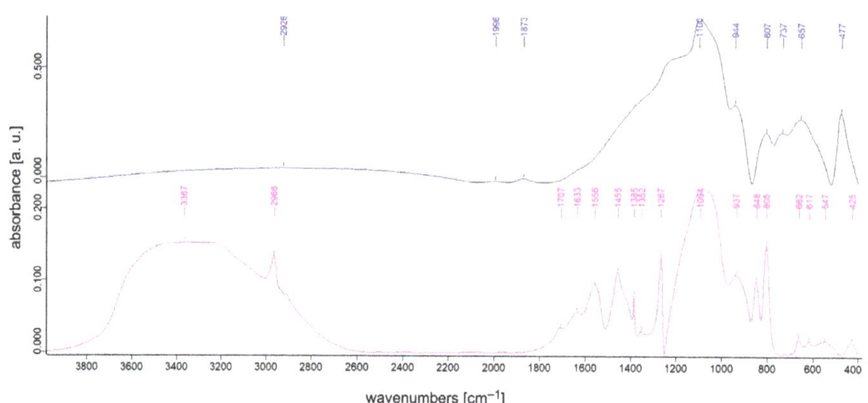

Figure 6. FTIR spectra of TiO_2:SiO_2 = 1:3 gel synthesized using TEOS and DDS with Ag addition, dried (bottom) and annealed at 1200 °C (top).

The intensive band observed at around 470 cm^{-1} in all FTIR spectra can be assigned to bending vibrations of Si-O-Si bridges, while the bands at 790–800 cm^{-1} are probably due to the symmetric stretching vibrations of Si-O bonds. The most intensive band in all spectra, at around 1020–1100 cm^{-1}, originates from the asymmetric stretching vibration of the Si-O bond [22]. There are other bands that can be assigned to the vibrations of Si-O bonds, at around 930 cm^{-1} and at around 1220–1260 cm^{-1}. These bands can be connected with the vibrations of broken Si-O$^-$ bridges and double Si=O bonds, respectively [23]. The band at 1260 cm^{-1} can be also assigned to C-H bond vibration in the Si-CH$_3$ group.

There is one more intensive band located at around 650 cm^{-1}m, which can only be observed in the spectra of the annealed gels, as well as of those synthesized with TEOS and with TEOS and DDS together. This band is probably due to stretching vibrations of the Ti-O bond, and its assignment is confirmed by the crystallization of TiO_2 polymorphs (according the XRD results described in the next section) in the studied samples.

In the spectra of gels synthesized with the use of TEOS and DDS (Figures 3 and 4), one can observe the increase in intensity of the bands at 1250 cm^{-1} and at 650 cm^{-1}, even in the sample with a higher silica content. This change in the mentioned band intensity points to the depolymerization of silica lattice and the parallel increase in Ti-O bonds concentration in the structure of the annealed gels, which can be connected with the easier and faster crystallization of TiO_2 phases in these samples (also confirmed by the XRD results).

The incorporation of Ag atoms into the titania-silica network does not cause distinct changes in the shapes of bands in the IR spectrum of the gel synthesized with the use of TEOS (Figure 5) and TEOS together with DDS (Figure 6), as compared to the IR spectra of samples of the same composition but without Ag addition (Figures 2 and 4, respectively). There are only two small differences observed during the comparison: intensities of the bands at 1250 cm^{-1} and at around 650 cm^{-1} do not increase (Figures 5 and 6), which indicates that, in these samples, the polymerization degree of the silica network does not decrease. The second difference is connected to the presence of a small band at 615 cm^{-1} in the spectrum of the annealed gel synthesized with TEOS (Figure 5). This band occurs in the range of pseudolattice bands in the spectrum and is not observed in the IR spectrum of the annealed gel synthesized with TEOS and DDS. The band at 615 cm^{-1} can be connected to the presence of Ag$^+$ cations in the structure of the studied sample [3].

3.2. X-ray Diffraction Studies

All XRD patterns were collected for the dried samples, as well as for those annealed at 1200 °C. The results are presented in a way similar to the FTIR spectra. Each figure contains two diffraction patterns of the dried and the annealed gel. Results of the XRD studies of all samples are presented in Figures 7–12.

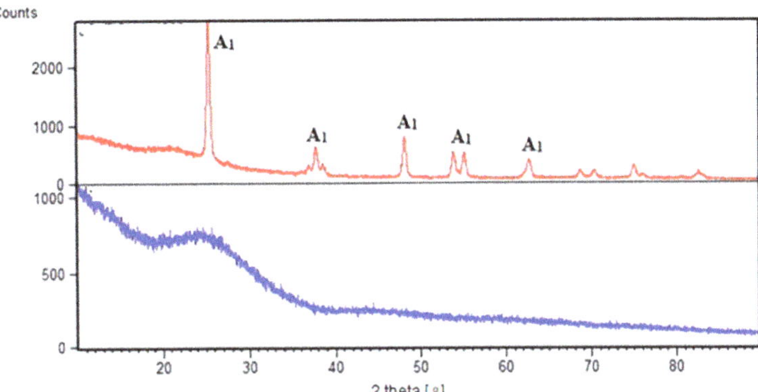

Figure 7. XRD diffraction pattern of TiO_2:SiO_2 = 1:2 gel synthesized using TEOS, dried (bottom) and annealed at 1200 °C (top): A_1—TiO_2 anatase.

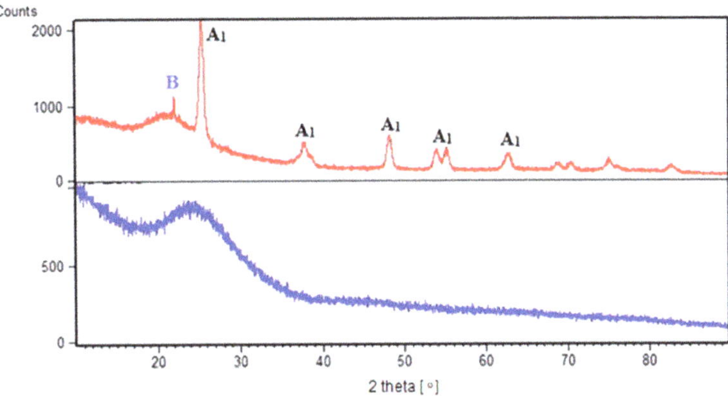

Figure 8. XRD diffraction pattern of TiO_2:SiO_2 = 1:3 gel synthesized using TEOS, dried (bottom) and annealed at 1200 °C (top): A_1—TiO_2 anatase; B—SiO_2 cristobalite.

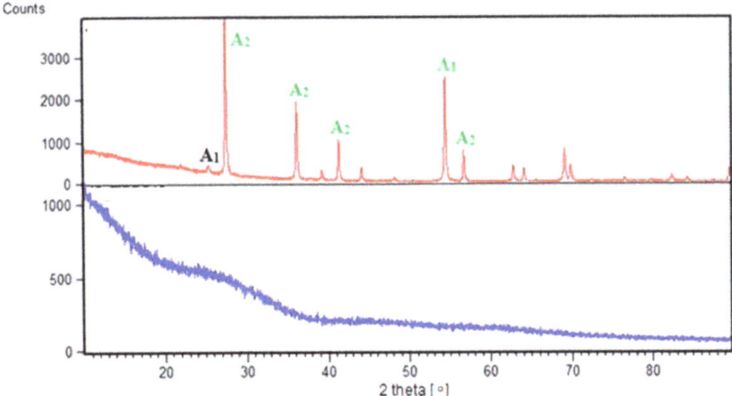

Figure 9. XRD diffraction pattern of TiO_2:SiO_2 = 1:2 gel synthesized using TEOS and DDS, dried (bottom) and annealed at 1200 °C (top): A_1—TiO_2 anatase; A_2—TiO_2 rutile.

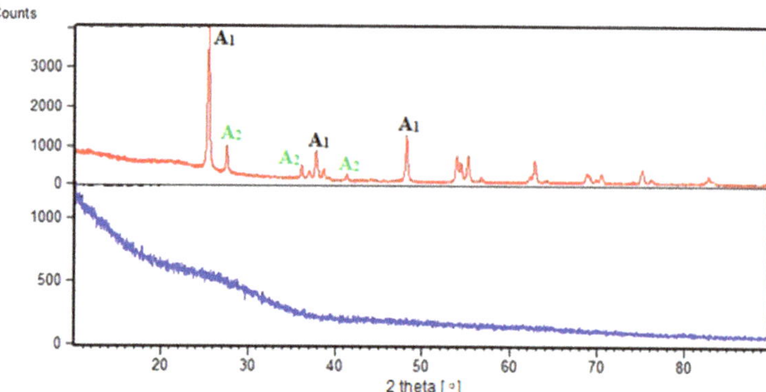

Figure 10. XRD diffraction pattern of TiO_2:SiO_2 = 1:3 gel synthesized using TEOS and DDS, dried (bottom) and annealed at 1200 °C (top): A_1—TiO_2 anatase; A_2—TiO_2 rutile.

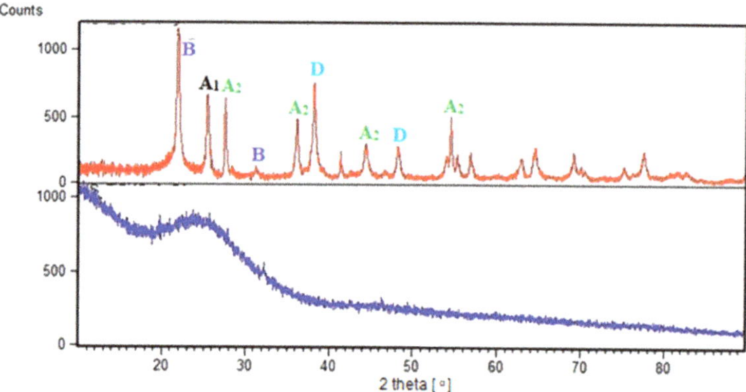

Figure 11. XRD diffraction pattern of TiO_2:SiO_2 = 1:3 gel synthesized using TEOS with Ag addition, dried (bottom) and annealed at 1200 °C (top): A_1—TiO_2 anatase; A_2—TiO_2 rutile; B—SiO_2 cristobalite; D—Ag.

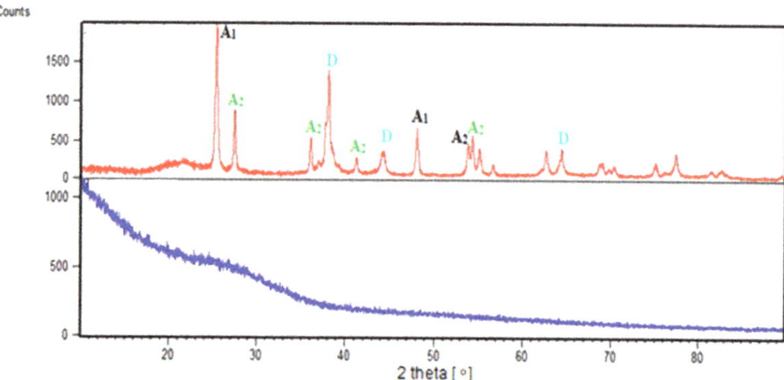

Figure 12. XRD diffraction pattern of TiO_2:SiO_2 = 1:3 gel synthesized using TEOS and DDS with Ag addition, dried (bottom) and annealed at 1200 °C (top): A_1—TiO_2 anatase; A_2—TiO_2 rutile; D—Ag.

All diffraction patterns of the dried gels confirm the amorphous state of the obtained samples. In contrast, all gels annealed up to 1200 °C are crystalline or still contain negligible amounts of amorphous phase, such as, for example, in the case of the sample with higher silica content obtained with the use of TEOS (Figure 8). Based on the analysed diffraction patterns, one can observe that the main phases present in the annealed titania-silica gels without Ag addition are titania polymorphs: anatase and rutile. When Ag atoms were introduced into the structure of the studied materials, the analysis of the XRD patterns of the annealed gels pointed to the presence of cristobalite and the crystalline Ag phase in addition to the anatase and the rutile existing in almost all annealed samples (Figures 11 and 12).

The crystallization of metallic Ag phase enabled the calculation of the dimensions of silver crystallites using Scherrer Equation (1) and the Scherrer Calculator in the HighScore Plus software. It was found that Ag crystallites in the annealed gel of $TiO_2:SiO_2$ = 1:3 composition synthesized with the use of TEOS, with the average dimensions of 9.0–12.0 nm (Table 2), and so they were within the upper limit of nanoparticle sizes (90–100 nm) [24]. When the gel was synthesized using TEOS and DDS as silica precursors, Ag crystallites were bigger than 9.0 nm and reached 13.0 nm. It is worth remembering that the sizes of crystallites are also closely related to the temperature of annealing. Therefore, annealing at a temperature lower than 1200 °C should also reduce the sizes of Ag crystallites.

Table 2. Sizes of silver crystallites in the gels containing Ag.

Sample	Peak Position 2θ [°]	β_{obs} (FWHM) [°]	β_{stand} [°]	D_{hkl} [nm]
$TiO_2:SiO_2$ = 1:3 + Ag synthesized with TEOS	38.169	0.740	0.057	12.3
	44.358	0.925	0.061	9.9
	64.489	0.855	0.078	12.1
$TiO_2:SiO_2$ = 1:3 + Ag synthesized with TEOS:DDS = 1	38.076	0.725	0.057	12.6
	44.215	0.844	0.061	11.0
	64.395	0.803	0.078	12.9

To study the potential influence of Ag addition on the sizes of TiO_2 and SiO_2 crystallites, the dimensions of crystallites were calculated based on the FWHM of the selected, most intensive peak of anatase, rutile and cristobalite phases. Comparing the results (Tables 2–4) one can notice that there is no distinct influence of silver addition on the titania and silica crystallites sizes. More visible is the relation between the crystallite sizes and the application of DDS as the silica precursor. When DDS is applied, the crystallites of anatase crystallites, as well as rutile and silver ones, are bigger than in samples synthesized only using TEOS.

Table 3. Sizes of TiO_2 (anatase) crystallites in the gels annealed at 1200 °C.

Sample	Peak Position 2θ [°]	β_{obs} (FWHM) [°]	β_{stand} [°]	D_{hkl} [nm]
$TiO_2:SiO_2$ = 1:3 synthesized with TEOS	25.401	0.801	0.054	10.9
$TiO_2:SiO_2$ = 1:3 + Ag synthesized with TEOS	25.425	0.525	0.054	17.3
$TiO_2:SiO_2$ = 1:3 synthesized with TEOS:DDS = 1	25.383	0.378	0.054	25.1
$TiO_2:SiO_2$ = 1:3 + Ag synthesized with TEOS:DDS = 1	25.350	0.456	0.054	20.3

Table 4. Sizes of TiO$_2$ (rutile) crystallites in the gels annealed at 1200 °C.

Sample	Peak Position $2\,\theta$ [°]	β_{obs} (FWHM) [°]	β_{stand} [°]	D_{hkl} [nm]
TiO$_2$:SiO$_2$ = 1:3 synthesized with TEOS	-	-	-	-
TiO$_2$:SiO$_2$ = 1:3 + Ag synthesized with TEOS	27.554	0.293	0.058	34.4
TiO$_2$:SiO$_2$ = 1:3 synthesized with TEOS:DDS = 1	27.504	0.276	0.058	37.0
TiO$_2$:SiO$_2$ = 1:3 + Ag synthesized with TEOS:DDS = 1	27.463	0.339	0.066	28.8

3.3. SEM Studies

SEM images and EDS spectra of the gels of TiO$_2$:SiO$_2$ = 1:3 composition, without and with the addition of Ag, are presented in Figures 13 and 14, respectively. The EDS spectrum of the gel obtained with the addition of Ag (Figure 14b) confirms the presence of Ag atoms in its structure. Intensities of the lines characteristic of Ti and Si in the EDS spectra indicate similar Ti:Si ratios in both gels, which agrees with the planned composition of both samples. The surface morphology of both gels suggests their homogeneity and a lack of the distinguished areas of different chemical compositions. This result is also confirmed by FTIR spectra analysis, which suggests a good dispersion of Ti^{+4} cations and TiO$_2$ phases in the silica matrix. Moreover, this conclusion is also confirmed by the calculated sizes of titania, silica and silver crystallites (Tables 2–5). All phases they are similar and do not exceed 37 nm. Therefore, objects observed on the surface of the samples in SEM images (Figures 13a and 14a) with sizes of a few thousand nanometres appear to be small grains of gels rather than crystallites of single TiO$_2$, SiO$_2$ and Ag phases.

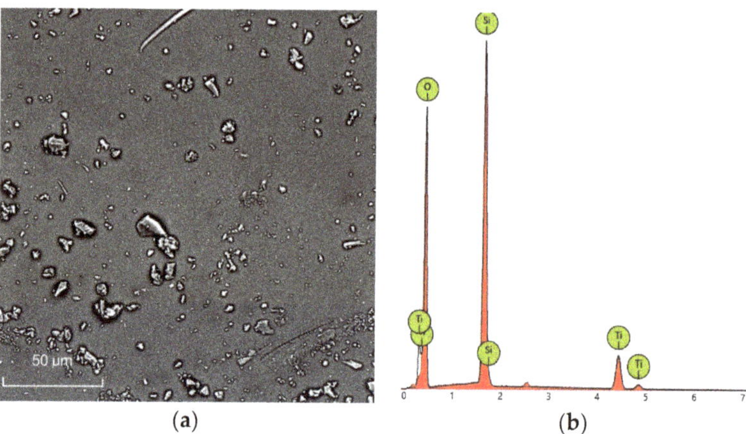

(a) (b)

Figure 13. SEM image of (a) TiO$_2$:SiO$_2$ = 1:3 gel ($\times$1500), synthesized using TEOS and (b) EDS spectrum of this sample.

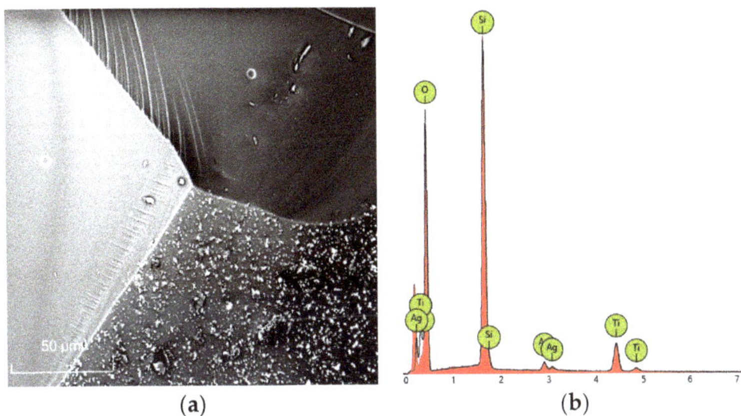

Figure 14. SEM image of (**a**) TiO_2:SiO_2 = 1:3 gel with Ag addition ($\times$1500), synthesized using TEOS and (**b**) EDS spectrum of this sample.

Table 5. Sizes of SiO_2 (crystobalite) crystallites in the gels annealed at 1200 °C.

Sample	Peak Position $2\,\theta$ [°]	β_{obs} (FWHM) [°]	β_{stand} [°]	D_{hkl} [nm]
TiO_2:SiO_2 = 1:3 synthesized with TEOS	22.059	0.328	0.054	29.5
TiO_2:SiO_2 = 1:3 + Ag synthesized with TEOS	21.877	0.655	0.054	13.5
TiO_2:SiO_2 = 1:3 synthesized with TEOS:DDS = 1	-	-	-	-
TiO_2:SiO_2 = 1:3 + Ag synthesized with TEOS:DDS = 1	-	-	-	-

4. Conclusions

Samples of TiO_2-SiO_2 system were synthesized by the sol-gel method using two different silica precursors: TEOS (tetraethoxysilane $Si(OC_2H_5)_4$) and DDS (dimethyldiethoxysilane $(CH_3)_2(C_2H_5O)_2Si$). During the synthesis of the samples, Ag was added to the sols applying $AgNO_3$ as an Ag precursor. All samples were first dried and then annealed at 1200 °C in air. Analysis of FTIR spectra and XRD patterns confirmed the amorphous state of all dried gels and to establish the presence of different TiO_2 polymorphs anatase and rutile, in almost all of the annealed samples. Rutile was not only observed in the annealed gels synthesized using TEOS and without Ag addition. The use of DDS as the second silica precursor resulted in a faster crystallization of gels during heat treatment and caused a parallel increase in the depolymerization of silica lattice, which was confirmed by XRD studies and the increased intensity of bands at 1250 cm^{-1} and at 650 cm^{-1} in the FTIR spectra of selected gels. The incorporation of Ag atoms into the structure of gels caused the crystallization of cristobalite phase together with titania polymorphs. The presence of Ag as crystalline phase enabled the average sizes of Ag crystallites to be calculated based on the Scherrer equation. The sizes of titania and silica crystallites were also calculated. Based on the results obtained, one can draw the conclusion that there is no distinct influence of Ag addition on the sizes of anatase, rutile and crystobalite crystallites, whereas the use of DDS as the second silica precursor leads to larger crystallites of the mentioned phases. SEM studies of selected gels containing Ag, as well as the ones prepared without Ag addition, confirmed their homogeneity and the planned composition of the samples.

Funding: This work was supported by "Excellence Initiative—Research University" of AGH University of Krakow, grant IDUB no 4154.

Data Availability Statement: Data is contained within the article.

Conflicts of Interest: The author declares no conflict of interest.

References

1. Zhang, M.E.L.; Zhang, R.; Liu, Z. The effect of SiO_2 on TiO_2-SiO_2 composite film for self-cleaning application. *Surf. Interfaces* **2019**, *16*, 194–198. [CrossRef]
2. Vishwas, M.; Narasimha Rao, K.; Arjuna Gowda, K.V.; Chakradhar, R.P.S. Optical, electrical and dielectric properties of TiO_2–SiO_2 films prepared by a cost effective sol–gel process. *Spectrochim. Acta Part A Mol. Biomol. Spectrosc.* **2011**, *83*, 614–617. [CrossRef] [PubMed]
3. Adamczyk, A.; Rokita, M. The structural studies of Ag containing TiO_2-SiO_2 gels and thin films deposited on steel. *J. Mol. Str.* **2016**, *1114*, 171–180. [CrossRef]
4. Houmard, M.; Riassetto, D.; Roussel, F.; Bourgeois, A.; Berthomé, G.; Joud, J.C.; Langlet, M. Morphology and natural wettability properties of sol-gel derived TiO_2-SiO_2 composite thin films. *Appl. Surf. Sci.* **2007**, *254*, 1405–1414. [CrossRef]
5. Eldural, B.; Bolukbasi, U.; Karakas, G. Photocatalytic antibacterial activity of TiO_2–SiO_2 thin films. *J. Photochem. Photobiol. A Chem.* **2014**, *283*, 29–37. [CrossRef]
6. Khosravi, H.S.; Veerapandiyan, V.K.; Vallant, R.; Reichmann, K. Effect of processing conditions on the structural properties and corrosion behavior of TiO_2–SiO_2 multilayer coatings derived via the sol-gel method. *Ceram. Intern.* **2020**, *46*, 17741–17751. [CrossRef]
7. Krishna, V.; Padmapreetha, R.; Chandrasekhar, S.B.; Murugan, K.; Johnson, R. Oxidation resistant TiO_2–SiO_2 coatings on mild steel by sol–gel. *Surf. Coat. Technol.* **2019**, *378*, 125031. [CrossRef]
8. Sun, B.; Sun, S.Q.; Li, T.; Zhang, W.Q. Preparation and antibacterial activities of Ag-doped SiO_2–TiO_2 composite films by liquid phase deposition (LPD) method. *J. Mater. Sci.* **2007**, *42*, 10085–10089. [CrossRef]
9. Kawashita, M.; Tsuneyama, S.; Miyaji, F.; Kokubo, T.; Kozuka, H.; Yamamoto, K. Antibacterial silver-containing silica glass prepared by sol–gel method. *Biomaterials* **2000**, *213*, 93–398. [CrossRef]
10. Schierholz, J.M.; Lucas, L.J.; Rump, A.; Pulverer, G. Efficacy of silver-coated medical devices. *J. Hosp. Infect.* **1998**, *40*, 257–262. [CrossRef]
11. Abd Aziz, R.; Sopyan, I. Synthesis of TiO_2-SiO_2 powder and thin film photocatalysts by sol-gel method. *Indian J. Chem.* **2009**, *48A*, 951–957.
12. Zhai, J.; Zhang, L.; Yao, X. Efects of composition and temperature on gel-formed TiO_2/SiO_2 films. *J. Non-Cryst. Solids* **1999**, *260*, 160–163. [CrossRef]
13. Cheng, X.M.; NIE, B.M.; Kumar, S. Preparation and bioactivity of SiO_2 functional films on titanium by PACVD. *Trans. Nonferrous Met. Soc. China* **2008**, *18*, 627–630. [CrossRef]
14. Fatimah, I.; Prakoso, N.I.; Sahroni, I.; Musawwa, M.M.; Sim, Y.L.; Kooli, F.; Muraza, O. Physicochemical characteristics and photocatalytic performance of TiO_2/SiO_2 catalyst synthesized using biogenic silica from bamboo leaves. *Heliyon* **2019**, *5*, e02766. [CrossRef] [PubMed]
15. Bulla, D.A.P.; Morimoto, N.I. Deposition of thick TEOS PECVD silicon oxide layers for integrated optical waveguide applications. *Thin Solid Film.* **1998**, *334*, 60–64. [CrossRef]
16. Choi, D.G.; Yang, S.M. Effect of two-step sol-gel reaction on the mesoporous silica structure. *J. Colloid. Interface Sci.* **2003**, *26*, 127–132. [CrossRef]
17. Wang, F.; Liu, J.; Luo, Z.; Zhang, Q.; Wang, P.; Liang, X.; Li, C.; Chen, J. Effects of dimethyldiethoxysilane addition on the sol–gel process of tetraethylorthosilicate. *J. Non-Cryst. Solids* **2007**, *353*, 321–326. [CrossRef]
18. Zhang, Y.; Wu, D.; Sun, Y. Sol-gel of methyl modified optical silica coatings and gels from DDS and TEOS. *J. Sol-Gel Sci. Technol.* **2005**, *33*, 19–24. [CrossRef]
19. Wang, F.; Liu, J.; Yang, H.; Luo, Z.; Lv, W.; Li, C.; Qing, S. Spherical particles from tetraorthosilicate (TEOS) sol–gel process with dimethyldiethoxysilane (DDS) and diphenyldiethoxysilane (DPDS) addition. *J. Non-Cryst. Solids* **2008**, *354*, 5047–5052. [CrossRef]
20. Peter, A.; Mihaly-Cozmuta, L.; Mihaly-Cozmuta, A.; Nicula, C.; Cadar, C.; Jastrzębska, A.; Kurtycz, P.; Olszyna, A.; Vulpoi, A.; Danciu, V.; et al. Silver functionalized titania-silica xerogels: Preparation, morphostructural and photocatalytic properties, kinetic modeling. *J. Alloys Compd.* **2015**, *648*, 890–902. [CrossRef]
21. Massa, M.A.; Covarrubias, C.; Bittner, M.; Fuentevilla, I.A.; Capetillo, P.; Von Marttens, A.; Carvajal, J.C. Synthesis of new antibacterial composite coating for titanium based on highly ordered nanoporous silica and silver nanoparticles. *Mater. Sci. Eng.* **2014**, *C45*, 146–153. [CrossRef] [PubMed]
22. Handke, M.; Mozgawa, W.; Nocuń, M. Specific features of the IR spectra of silicate glasses. *J. Mol. Str.* **1994**, *325*, 129–136. [CrossRef]

23. Adamczyk, A. The influence of ZrO_2 precursor type on the structure of ZrO_2-TiO_2-SiO_2 gels and selected thin films. *J. Mol. Str.* **2018**, *1171*, 706–771. [CrossRef]
24. Langauer-Lewowicka, H.; Pawlas, K. Nanocząstki, nanotechnologia—Potencjalne zagrożenia środowiskowe i zawodowe. *Med. Sr.-Environ. Med.* **2014**, *17*, 7–14.

MDPI
St. Alban-Anlage 66
4052 Basel
Switzerland
www.mdpi.com

Crystals Editorial Office
E-mail: crystals@mdpi.com
www.mdpi.com/journal/crystals